8B

Maths Links

Teacher's Guide

Geoff Fowler
Sue Muggeridge
Kim Frazer
Sheelagh Raw

OXFORD
UNIVERSITY PRESS

OXFORD
UNIVERSITY PRESS

Great Clarendon Street, Oxford OX2 6DP

Oxford University Press is a department of the University of Oxford.
It furthers the University's objective of excellence in research, scholarship, and
education by publishing worldwide in

Oxford   New York

Auckland   Cape Town   Dar es Salaam   Hong Kong   Karachi
Kuala Lumpur   Madrid   Melbourne   Mexico City   Nairobi
New Delhi   Shanghai   Taipei   Toronto

With offices in

Argentina   Austria   Brazil   Chile   Czech Republic   France   Greece
Guatemala   Hungary   Italy   Japan   Poland   Portugal   Singapore
South Korea   Switzerland   Thailand   Turkey   Ukraine   Vietnam

© Oxford University Press

British Library Cataloguing in Publication Data

Data available

ISBN: 978-0-19-915296-4
10 9 8 7 6 5 4 3 2 1

Printed in Great Britain by Ashford Colour Press Ltd.

Paper used in the production of this book is a natural, recyclable product made
from wood grown in sustainable forests. The manufacturing process conforms to
the environmental regulations to the country of origin.

Acknowledgments
The editors would like to thank Pete Crawford and Charlie Bond
for their excellent work on this book.

About this book

This Teacher's Guide conforms to the renewed Framework for Teaching Mathematics, for first teaching in 2008. The book is designed to accompany MathsLinks Students' Book 8B, which targets Year 8 students who have achieved level 4 at KS2 – the aim is to consolidate level 5 and extend to level 6 as Year 8 progresses.

The series authors are experienced teachers and maths advisers who have an excellent understanding of the KS3 Framework and programme of study, and so are well qualified to help you successfully deliver the Year 8 curriculum in your classroom.

This book is organised by chapter, which have been structured into lesson plans to enable you to deliver the learning objectives in each of the five strands of progression:

- Mathematical processes and applications
- Number
- Algebra
- Geometry
- Statistics

Topics are clearly levelled, enabling you to assess students' progress both collectively and individually. Misconceptions are highlighted and intervention strategies are suggested throughout to anticipate and remove barriers to learning.

𝍢 For those following a condensed KS3 programme, a Fast-track route through the material is provided, along with a suggested scheme of work for Years 1-2.

In addition, full answers are given to the accompanying student book and homework book exercises.

The CD-ROM attached to this book contains all of the lesson plans in customisable Word format, to enable you to create your own lessons to suit your students' needs; also included on the CD are the answers in pdf format, suitable for whole-class display.

PLTS	Personal, learning and thinking skills	MPA	Mathematical processes and applications	I	Introduction
IE	Independent enquirers	1.1	Representing	CS	Case study
CT	Creative thinkers	1.2	Analysing – use mathematical reasoning	UAM	Using and applying mathematics
RL	Reflective learners	1.3	Analysing – use appropriate mathematical procedures	SSM	Shape space measure
TW	Team workers			HD	Handling data
SM	Self-managers	1.4	Interpreting and evaluating	NNS	Numbers and the number system
EP	Effective participators	1.5	Communicating and reflecting		

Contents

1 Integers and decimals — Number

Objectives

- Order decimals **4**
- Use multiples, factors, common factors, highest common factors, lowest common multiples and primes **5**
- Find the prime factor decomposition of a number **5**
- Use squares, square roots and cubes **5**
- Use index notation for small positive integer powers **5**
- Add, subtract, multiply and divide integers **5**

Introduction

This chapter aims to develop skills in identifying and calculating with different types of number including negative numbers, factors, multiples, prime numbers, squares and cubes. The topic extends to finding the prime factor decomposition of a number and the lowest common multiple and highest common factor of two numbers.

Recognising and being able to deal with different types of number is an essential skill in many aspects of daily life, whether dealing with money, measurements, temperature or Sudoku puzzles. The student book discusses tiling a floor but painting a house, laying a carpet and choosing furniture to fit into a room all require the skills and techniques covered in this chapter (CT).

Fast-track
1d

Level

MPA

1.1	1a, c, e, f
1.2	1c, d, f
1.3	1a, b, c, d, e, f
1.4	1a, c, d, f
1.5	1e

PLTS

IE	1c, d, e
CT	1I, b, c, e, f
RL	1a, b, d, e
TW	1b, d
SM	1d, f
EP	1a, b, c, d, e

Extra Resources

1 Start of chapter presentation
1a Animation: Positive and negative numbers
1a Consolidation sheet
1b Starter: Addition with negatives matching
1b Worked solution: Q4k
1b Consolidation sheet
1e Worked solution: Q5a
1e Consolidation sheet

Assessment: chapter 1

- Order decimals (L4)
- Understand negative numbers as positions on a number line (L5)
- Add and subtract integers (L5)

Useful resources
Vertical number line
Mini white boards

Starter – Guess my number

Choose a number between -15 and +15.

Invite students to guess the number. After each guess say whether the next guess should be higher or lower.

Can be extended by using numbers with one decimal place.

Teaching notes

Use students' prior knowledge of negative numbers, place value and ordering to get them to explain how they solve the various types of problems. The number line provides a visual aide to reasoning and common misconceptions should be discussed. Ask how they remember how to use the '<' sign: 'the small end points to the smaller number', 'the crocodile eats the bigger number', *etc*. Point out that -2 < -1 'less than' carries the same information as -1 > -2 'greater than' and both are right. However when ordering several numbers they should keep the signs aligned: -3 < -2 < -1 rather than -2 > -3 < -1.

Ask students to sum three temperature changes +3º + 4º + (-5)º = 2º. Illustrate this on the number line: go to +3, the first number, and move up +4 units then move down +(-5) = –5 units. What would you get if you took away that last temp. change +2 − (-5) = 2 + 5 = 7. Illustrate this as start at +2, the first number, and go up 5 units: − (-5) = +5. Identify the rules:

- adding a negative number is the same as subtracting a positive number
- subtracting a negative number is the same as adding a positive number (1.3)

Bank balances provide a similar illustration.

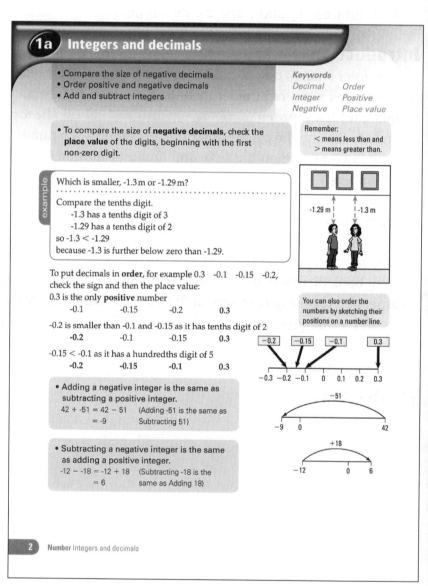

1a Integers and decimals

- Compare the size of negative decimals
- Order positive and negative decimals
- Add and subtract integers

Keywords
Decimal Order
Integer Positive
Negative Place value

- To compare the size of **negative decimals**, check the **place value** of the digits, beginning with the first non-zero digit.

Remember:
< means less than and
> means greater than.

example

Which is smaller, -1.3 m or -1.29 m?
....

Compare the tenths digit.
 -1.3 has a tenths digit of 3
 -1.29 has a tenths digit of 2
so -1.3 < -1.29
because -1.3 is further below zero than -1.29.

-1.29 m   -1.3 m

To put decimals in **order**, for example 0.3 -0.1 -0.15 -0.2, check the sign and then the place value:
0.3 is the only **positive** number
 -0.1   -0.15   -0.2   0.3
-0.2 is smaller than -0.1 and -0.15 as it has tenths digit of 2
 -0.2   -0.1   -0.15   0.3
-0.15 < -0.1 as it has a hundredths digit of 5
 -0.2   -0.15   -0.1   0.3

You can also order the numbers by sketching their positions on a number line.

-0.2   -0.15   -0.1   0.3

-0.3 -0.2 -0.1 0 0.1 0.2 0.3

- Adding a negative integer is the same as subtracting a positive integer.
 42 + -51 = 42 − 51   (Adding -51 is the same as
       = -9   Subtracting 51)

-51
-9   0   42

- Subtracting a negative integer is the same as adding a positive integer.
 -12 − -18 = -12 + 18   (Subtracting -18 is the
       = 6   same as Adding 18)

+18
-12   0   6

2 Number Integers and decimals

Plenary

Students should be challenged, in pairs, to explain their workings to question **5** and appreciate that there were a number of different ways to get to the answer. For example, in part **a**, who added £13 and £8.50 and then wrote the answer as owed £21.50? Who wrote -£13 − £8.50 = -£21.50? (EP, RL)

Simplification

Where students find addition and subtraction of negative numbers difficult use a vertical number line. Using a number line allows you to show that when you have a negative you go down that number of steps.

Encourage students to use a number line, to help them think through what they are doing. All questions (1.3).

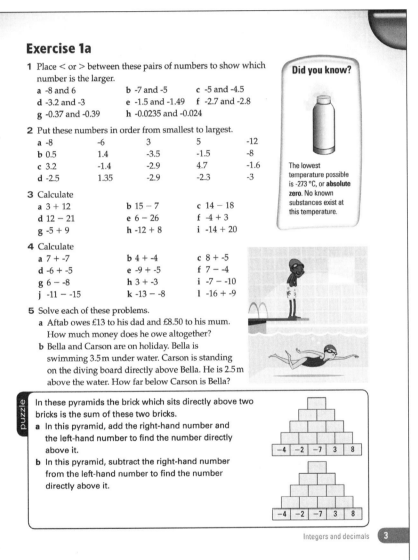

Exercise 1a

1 Place < or > between these pairs of numbers to show which number is the larger.

  a -8 and 6   b -7 and -5   c -5 and -4.5
  d -3.2 and -3   e -1.5 and -1.49   f -2.7 and -2.8
  g -0.37 and -0.39   h -0.0235 and -0.024

2 Put these numbers in order from smallest to largest.

  a -8   -6   3   5   -12
  b 0.5   1.4   -3.5   -1.5   -8
  c 3.2   -1.4   -2.9   4.7   -1.6
  d -2.5   1.35   -2.9   -2.3   -3

3 Calculate

  a 3 + 12   b 15 − 7   c 14 − 18
  d 12 − 21   e 6 − 26   f -4 + 3
  g -5 + 9   h -12 + 8   i -14 + 20

4 Calculate

  a 7 + -7   b 4 + -4   c 8 + -5
  d -6 + -5   e -9 + -5   f 7 − -4
  g 6 − -8   h 3 + -3   i -7 − -10
  j -11 − -15   k -13 − -8   l -16 + -9

5 Solve each of these problems.

  a Aftab owes £13 to his dad and £8.50 to his mum. How much money does he owe altogether?
  b Bella and Carson are on holiday. Bella is swimming 3.5 m under water. Carson is standing on the diving board directly above Bella. He is 2.5 m above the water. How far below Carson is Bella?

puzzle

In these pyramids the brick which sits directly above two bricks is the sum of these two bricks.

  a In this pyramid, add the right-hand number and the left-hand number to find the number directly above it.

| -4 | -2 | -7 | 3 | 8 |

  b In this pyramid, subtract the right-hand number from the left-hand number to find the number directly above it.

| -4 | -2 | -7 | 3 | 8 |

Did you know?

The lowest temperature possible is -273 °C, or **absolute zero**. No known substances exist at this temperature.

Integers and decimals 3

Extension

Challenge students to make up their own addition and subtraction number pyramids together with answers. Swap puzzles with a partner. Together with a partner can they invent a puzzle where values are not just given in the bottom row?

Exercise 1a commentary

Question 1 – Similar to the example. Check that students understand the significance of place value: a larger decimal does not have to be longer 0.2 > 0.196.

Question 2 – Remember that the 'largest' negative number is always the smallest.

Question 3 – Emphasise that when a positive number is taken away from a negative number the answer will be a more negative number. Using a number line helps to emphasise this point.

Question 4 – Remind students to consolidate two signs together into one before they complete the question.

Question 5 – Students must write down the equivalent arithmetic calculation not just an answer (1.1).

Puzzle – Number pyramids are common but some students may need further explanation of how to proceed. In part **b**, the order is important, -4 − -2 ≠ -2 − -4, it may help to write out the calculations separately (1.4).

Assessment criteria – NNS, level 5: order negative numbers in context. Calculating, level 5: solve simple problems involving ordering, adding subtracting negative numbers in context.

Links

Absolute zero is the temperature at which (classically) all molecules and atoms stop moving. The idea was first put forward by the brilliant mathematical physicist and engineer William Thomson, later Lord Kelvin, in the 19th century. There is more information about Lord Kelvin at http://en.wikipedia.org/wiki/William_ Thomson,_1st_Baron_Kelvin

- Multiply and divide (negative) integers   (L5)   ***Useful resources***

Starter – Tables bingo

Ask students to draw a 3 × 3 grid and enter numbers from the 6, 7 and 8 times tables.

Give questions, for example, 7 × 9. The winner is the first student to cross out all their numbers.

The game can be differentiated by the choice of tables.

Teaching notes

To motivate and gain a feel for the signs of the various products consider calculating the area of a rectangle 8 × 6 whose sides are shortened to 5 × 4. (A very quick reminder about areas might be needed.)

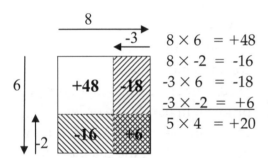

$$8 \times 6 = +48$$
$$8 \times \text{-}2 = \text{-}16$$
$$\text{-}3 \times 6 = \text{-}18$$
$$\underline{\text{-}3 \times \text{-}2 = +6}$$
$$5 \times 4 = +20$$

Its area is given as the starting area, 8 × 6 = +48, take away a narrow horizontal strip, 8 × -2 = -16, take away a narrow vertical strip, -3 × 6 = -18, but this double counts so we must add back in the overlap area, -3 × -2 = +6, which totals 5 × 4 = 20.

This can be used to motivate the multiplication grid in the text, whose three step use can be explained with some examples. Use a number of very simple examples to test understanding of multiplication and then try division.

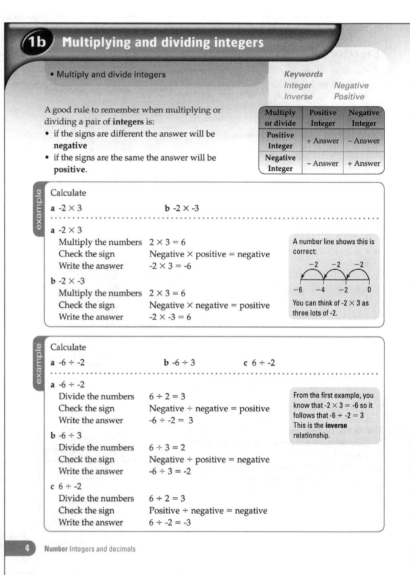

1b Multiplying and dividing integers

- Multiply and divide integers

Keywords
Integer   Negative
Inverse   Positive

A good rule to remember when multiplying or dividing a pair of **integers** is:

- if the signs are different the answer will be **negative**
- if the signs are the same the answer will be **positive**.

Multiply or divide	Positive Integer	Negative Integer
Positive Integer	+ Answer	– Answer
Negative Integer	– Answer	+ Answer

example

Calculate

a -2 × 3   **b** -2 × -3

a -2 × 3
   Multiply the numbers   2 × 3 = 6
   Check the sign   Negative × positive = negative
   Write the answer   -2 × 3 = -6

b -2 × -3
   Multiply the numbers   2 × 3 = 6
   Check the sign   Negative × negative = positive
   Write the answer   -2 × -3 = 6

A number line shows this is correct:

You can think of -2 × 3 as three lots of -2.

example

Calculate

a -6 ÷ -2   **b** -6 ÷ 3   **c** 6 ÷ -2

a -6 ÷ -2
   Divide the numbers   6 ÷ 2 = 3
   Check the sign   Negative ÷ negative = positive
   Write the answer   -6 ÷ -2 = 3

b -6 ÷ 3
   Divide the numbers   6 ÷ 3 = 2
   Check the sign   Negative ÷ positive = negative
   Write the answer   -6 ÷ 3 = -2

c 6 ÷ -2
   Divide the numbers   6 ÷ 2 = 3
   Check the sign   Positive ÷ negative = negative
   Write the answer   6 ÷ -2 = -3

From the first example, you know that -2 × 3 = -6 so it follows that -6 ÷ -2 = 3 This is the **inverse** relationship.

4   **Number** Integers and decimals

Plenary

Put up a multiplication grid, as in the **puzzle**. Go around the class asking students to tell you how to solve it. First they must say what is the corresponding calculation, for example, -4 × -2 = ? or 12 × ? = -3 which can be rewritten 12 ÷ -3 = ?, and then explain how to evaluate this. If any students have produced their own you could use one of these (EP) (1.3).

Simplification

Supply short practice questions without negative signs and then take the next step and follow the three stage solution shown in the examples, that is, multiply(divide) the numbers, check the sign, write the answer.

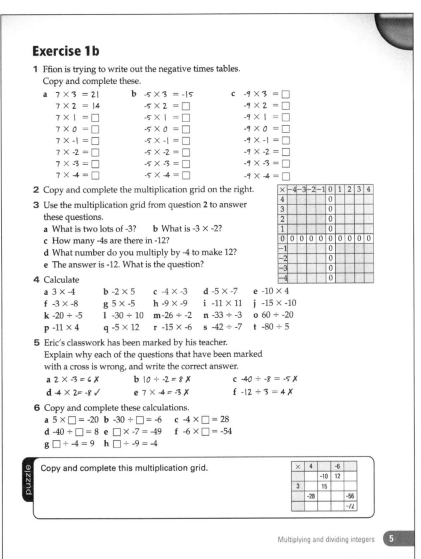

Exercise 1b

1 Ffion is trying to write out the negative times tables.
Copy and complete these.

a		b		c	
7 × 3 = 21		-5 × 3 = -15		-9 × 3 = ☐	
7 × 2 = 14		-5 × 2 = ☐		-9 × 2 = ☐	
7 × 1 = ☐		-5 × 1 = ☐		-9 × 1 = ☐	
7 × 0 = ☐		-5 × 0 = ☐		-9 × 0 = ☐	
7 × -1 = ☐		-5 × -1 = ☐		-9 × -1 = ☐	
7 × -2 = ☐		-5 × -2 = ☐		-9 × -2 = ☐	
7 × -3 = ☐		-5 × -3 = ☐		-9 × -3 = ☐	
7 × -4 = ☐		-5 × -4 = ☐		-9 × -4 = ☐	

2 Copy and complete the multiplication grid on the right.

3 Use the multiplication grid from question **2** to answer these questions.
  a What is two lots of -3?   **b** What is -3 × -2?
  c How many -4s are there in -12?
  d What number do you multiply by -4 to make 12?
  e The answer is -12. What is the question?

×	-4	-3	-2	-1	0	1	2	3	4
4					0				
3					0				
2					0				
1					0				
0	0	0	0	0	0	0	0	0	0
-1					0				
-2					0				
-3					0				
-4					0				

4 Calculate
  a 3 × -4   **b** -2 × 5   **c** -4 × -3   **d** -5 × -7   **e** -10 × 4
  f -3 × -8   **g** 5 × -5   **h** -9 × -9   **i** -11 × 11   **j** -15 × -10
  k -20 ÷ -5   **l** -30 ÷ 10   **m** -26 ÷ -2   **n** -33 ÷ -3   **o** 60 ÷ -20
  p -11 × 4   **q** -5 × 12   **r** -15 × -6   **s** -42 ÷ -7   **t** -80 ÷ 5

5 Eric's classwork has been marked by his teacher.
Explain why each of the questions that have been marked with a cross is wrong, and write the correct answer.
  a 2 × -3 = 6 ✗   **b** 10 ÷ -2 = 8 ✗   **c** -40 ÷ -8 = -5 ✗
  d -4 × 2 = -8 ✓   **e** 7 × -4 = -3 ✗   **f** -12 ÷ 3 = 4 ✗

6 Copy and complete these calculations.
  a 5 × ☐ = -20   **b** -30 ÷ ☐ = -6   **c** -4 × ☐ = 28
  d -40 ÷ ☐ = 8   **e** ☐ × -7 = -49   **f** -6 × ☐ = -54
  g ☐ ÷ -4 = 9   **h** ☐ ÷ -9 = -4

puzzle

Copy and complete this multiplication grid.

×	4		-6
		-10	12
3		15	
	-28		-56
			-12

Extension

Ask students to make up their own grid question involving multiplication of positive and negative numbers as in the **puzzle**, together with an answer. Then, exchange grids with someone else in the class (CT).

Exercise 1b commentary

All questions (1.3)

Questions 1 and **2** – Ask students to explain the patterns that occur in their answers.

Question 3 – The grid developed in question **2** should be used to find the answers. Require students to write out both the calculation and their answer, for example, 2 × -3 = -6.

Question 4 – Parts **a–j** are similar to the first example, parts **k–t** the second example. Emphasise the need to follow the three steps.

Question 5 – Insist that students write down their explanations: it may be necessary to suggest some possible responses. It may also help to go around the class and ask students to read out their answers (RL).

Question 6 – Suggest proceeding in two steps, first, find the number, second, decide its sign.

Puzzle – To get going it may help to write out the calculations associated with individual squares, for example, ? × -6 = 12, *etc.*

Links

The Ancient Babylonians used a number system based on 60. The large number of multiplication facts (60 × 60) made multiplication difficult so the Babylonians developed multiplication tables. The tables were written in cuneiform script on clay tablets and then baked. There is a picture of a Babylonian multiplication tablet for the 35 times table at http://it.stlawu.edu/~dmelvill/mesomath/tablets/36Times.html (TW).

- Use simple tests of divisibility (L5)
- Recognise and use multiples and factors (L5)

Useful resources

Dictionaries (L5)

••

Starter – The answer is -12

Ask students to write down questions where the answer is -12.
Score 1 point for an addition question, 2 points for a subtraction
question, 3 for a multiplication or division question.

Teaching notes

Finding multiples is straightforward,
if tedious, whilst finding factors
requires students to be systematic;
students should be encouraged to
write factors in pairs.

Students may already know some
divisibility tests (for 2, 5, 10, …) and
should be encouraged to explain
these themselves. Others will need
clarifying or introducing with
examples.

- 2, 5 and 10: only the last digit is
  looked at since $10 = 2 \times 5 = 1 \times 10$.
- 3: the test can be applied
  repeatedly: 854 622 sum of digits
  27, sum of digits $9 = 3 \times 3$ is a
  multiple of three (854 622 = 9 ×
  284 874).
- 4: only the last two digits need to
  be tested as $100 = 4 \times 25$.
- 7: must still be checked even if
  there is no simple test.
- 8: only the last three digits need
  to be tested as $1000 = 8 \times 125$.
- 9: this test can also be applied
  repeatedly: 854 622 sum of digits
  27, sum of digits 9 is a multiple
  of nine (854 622 = 9 × 94 958).
- 11: this must be explained: 5291
  alternating sum $= 5 - 2 + 9 - 1 =$
  $(5 + 9) - (2 + 1) = 14 - 3 = 11$ is
  a multiple of 11 (5291 = 11 × 481).

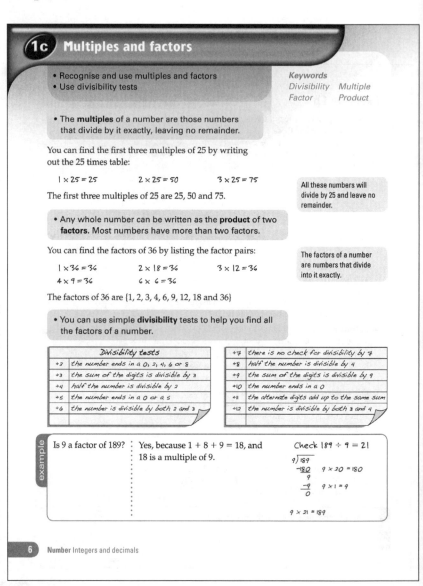

Plenary

Pose a series of missing digit questions, as in question **5**, and
challenge students to find the possible values for the digit to
make the number a multiple.

Simplification

For students who find this work difficult there is a need to find factors for smaller numbers, such as 12 or 16, and to practice divisibility testing for 2, 3, 4, 5, 6 and 8 to develop confidence.

All questions (1.3)

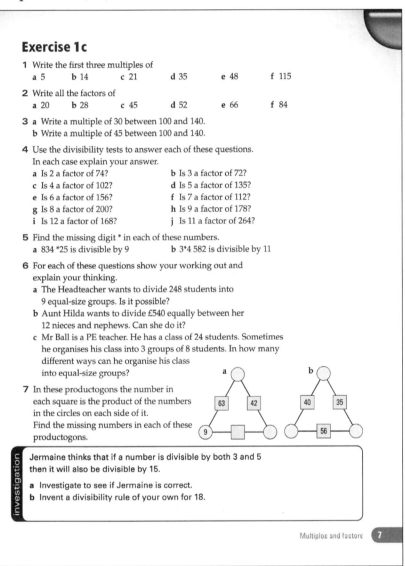

Exercise 1c

1 Write the first three multiples of
   a 5   b 14   c 21   d 35   e 48   f 115

2 Write all the factors of
   a 20   b 28   c 45   d 52   e 66   f 84

3 a Write a multiple of 30 between 100 and 140.
   b Write a multiple of 45 between 100 and 140.

4 Use the divisibility tests to answer each of these questions.
   In each case explain your answer.
   a Is 2 a factor of 74?   b Is 3 a factor of 72?
   c Is 4 a factor of 102?   d Is 5 a factor of 135?
   e Is 6 a factor of 156?   f Is 7 a factor of 112?
   g Is 8 a factor of 200?   h Is 9 a factor of 178?
   i Is 12 a factor of 168?   j Is 11 a factor of 264?

5 Find the missing digit * in each of these numbers.
   a 834 *25 is divisible by 9   b 3*4 582 is divisible by 11

6 For each of these questions show your working out and
   explain your thinking.
   a The Headteacher wants to divide 248 students into
     9 equal-size groups. Is it possible?
   b Aunt Hilda wants to divide £540 equally between her
     12 nieces and nephews. Can she do it?
   c Mr Ball is a PE teacher. He has a class of 24 students. Sometimes
     he organises his class into 3 groups of 8 students. In how many
     different ways can he organise his class
     into equal-size groups?

7 In these productogons the number in
   each square is the product of the numbers
   in the circles on each side of it.
   Find the missing numbers in each of these
   productogons.

investigation

Jermaine thinks that if a number is divisible by both 3 and 5
then it will also be divisible by 15.

a Investigate to see if Jermaine is correct.
b Invent a divisibility rule of your own for 18.

Multiples and factors   7

Extension

Ask students to make their own productogon as in question 7, together with a solution, and ask others to solve it (CT).

Exercise 1c commentary

Question 1 – Students may find it easier to keep adding the first term rather than do multiplications.

Question 2 – Encourage students to be systematic in order to get all factors. It will help to write them as a list in pairs, one column increasing, the other decreasing in size. They should stop, when the columns 'swap over'.

Question 3 – Follows on from question 1.

Question 4 – This could be done as a whole class activity (1.2, 1.3).

Question 5 – Follows on from question 4h and j.

Question 6 – In part c, hint that factor pairs are the key to this question. Remind students that three groups of eight is different from eight groups of three (EP) (1.1).

Question 7 – Part a can be done by looking at the equations associated with each side, for example, $9 \times ? = 63$. Part b will require students to find common factors.

Investigation – In part b, there is more than one answer. Ask why divisible by 2 and 3 is insufficient (IE) (1.2, 1.3).

Assessment Criteria – NNS, level 4: recognize and describe number relationships including multiple, factor and square.

Links

Bring in some dictionaries for the class to use. The word *divisibility* has five i's, the word *indivisibilities* has seven. Ask the class to find other words with at least four i's. Some examples include *infinitesimal* (4), *impossibilities* (5), *invisibility* (5) and *indistinguishability* (6). The dictionary will probably not include *supercalifragilisticexpialidocious* (7)!

- Recognise and use primes (L5)
- Find the prime factor decomposition of a number (L5)

Useful resources

Mini whiteboards

Starter – Make 1 to 15

Throw 3 dice and ask students to make one number between 1 and 15 inclusive using all 3 scores and any operation(s). Throw again; students make another number between 1 and 15. Repeat 13 further times. The winner is the student who makes the most different numbers.

Teaching notes

To make links with students' prior knowledge, initiate a paired discussion and then take feedback to agree a whole class definition of what is a prime number. Consolidate understanding by inviting students to account for why 1 is not defined to be a prime number and why 2 is the only even prime number.

When developing discussion of factors, choose a number and use mini whiteboards to invite students to contribute the factors of that number and ensure all are involved. Pose the question why should factors be used in pairs and tease out that it is helpful to make sure no factors are missed and to know when the list is exhausted.

When discussing factor trees, ask different groups of students to use the same number but use a different factor pair as the starting point. Take feedback on the prime factors to illustrate that they will always be the same (EP) (1.4).

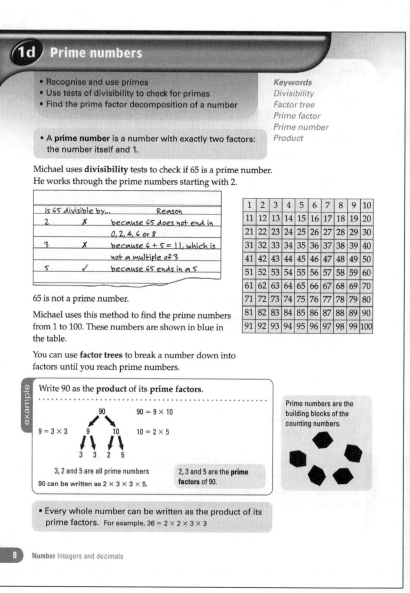

Plenary

Refer back to the lesson objectives and ask students to identify for themselves what they have done during the lesson that shows they understand and can use their knowledge of prime numbers, tests of divisibility and factors. Introduce this initially as an opportunity to think for themselves, then to discuss with a friend and finally to share as a whole group to encourage engagement with moving learning forward (RL).

Simplification

Some students may need further consolidation of tests of divisibility from the previous spread. A useful method of reinforcement is to ask students to design an additional 'text book example' of their own to show how they would explain to another student.

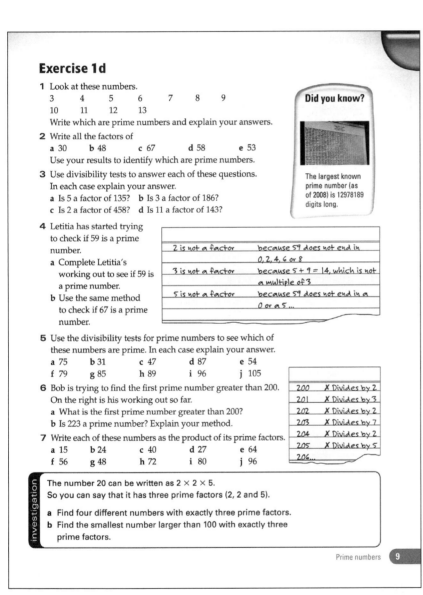

Exercise 1d

1 Look at these numbers.

3   4   5   6   7   8   9
10   11   12   13

Write which are prime numbers and explain your answers.

2 Write all the factors of
a 30   b 48   c 67   d 58   e 53
Use your results to identify which are prime numbers.

3 Use divisibility tests to answer each of these questions. In each case explain your answer.
a Is 5 a factor of 135?   b Is 3 a factor of 186?
c Is 2 a factor of 458?   d Is 11 a factor of 143?

4 Letitia has started trying to check if 59 is a prime number.
a Complete Letitia's working out to see if 59 is a prime number.
b Use the same method to check if 67 is a prime number.

2 is not a factor	because 59 does not end in 0, 2, 4, 6 or 8
3 is not a factor	because 5 + 9 = 14, which is not a multiple of 3
5 is not a factor	because 59 does not end in a 0 or a 5 ...

Did you know?

The largest known prime number (as of 2008) is 12978189 digits long.

5 Use the divisibility tests for prime numbers to see which of these numbers are prime. In each case explain your answer.
a 75   b 31   c 47   d 87   e 54
f 79   g 85   h 89   i 96   j 105

6 Bob is trying to find the first prime number greater than 200. On the right is his working out so far.
a What is the first prime number greater than 200?
b Is 223 a prime number? Explain your method.

200	X Divides by 2
201	X Divides by 3
202	X Divides by 2
203	X Divides by 7
204	X Divides by 2
205	X Divides by 5
206...	

7 Write each of these numbers as the product of its prime factors.
a 15   b 24   c 40   d 27   e 64
f 56   g 48   h 72   i 80   j 96

investigation

The number 20 can be written as $2 \times 2 \times 5$.
So you can say that it has three prime factors (2, 2 and 5).

a Find four different numbers with exactly three prime factors.
b Find the smallest number larger than 100 with exactly three prime factors.

Prime numbers   **9**

Extension

Challenge students to design an interesting question (with answer!) of their own along the lines of the **investigation** and to swap with another student to offer a solution. This offers an opportunity to develop the process skill of communicating and reflecting through discussing and reflecting on different approaches to solving a numerical problem.

Exercise 1d commentary

This exercise gives an opportunity to develop classroom dialogue via an emphasis on explanation and justification.

Question 1 – Encourage students to memorise the early primes.

Question 2 – Encourage the use of divisibility tests; emphasise that primes have only two factors (1.3).

Question 3 – Insist on explanations (1.2, 1.3).

Question 4 – Relates to the text. Ask why you don't need to test 4 or 6 (1.2, 1.3, 1.4).

Question 5 – Develops question **4**. Discuss efficient approaches: 85 is clearly a multiple of 5, therefore no need test for divisibility by 2 and 3. Why does this not undermine using a systematic approach? (IE) (1.3, 1.4)

Question 6 – Develops question **1**. Again place an emphasis on method. (1.2, 1.3)

Question 7 –Point out that one can start the tree with any factor pair, a unique answer is guaranteed (1.3).

Investigation – There is no short cut way to do part **b** (SM) (1.3).

Assessment Criteria – UAM, level 6: solve problems and carry through substantial tasks by breaking them into smaller, more manageable tasks, using a range of efficient techniques.

Links

Erastothenes lived from about 276 BC to 194 BC. He was born in Cyrene, North Africa (now Libya) and became not only a mathematician but also a poet, athlete, geographer and astronomer. One of his achievements was to use geometry to estimate the circumference of the Earth. There is more information about Erastothenes at http://en.wikipedia.org/wiki/Erast othenes (TW).

- Use multiples, factors, common factors, highest common factors, lowest common multiples and primes
(L5)

Useful resources
Mini whiteboards

Starter – Prime calculations!

Ask students questions involving prime numbers less than 20. For example,

Two prime numbers that have a difference of 9? (2, 11)
Two prime numbers that have a total of 22? (5, 17)
Two prime numbers with a product of 65? (5, 13)
Sum of first four prime numbers? (17)

Make a list of prime numbers on the board if necessary.

Teaching notes

Help students to make connections by linking back to the earlier spreads in this chapter and establish through questioning that they are certain of the definitions of factor and multiple (1.5). Suggest a number and ask, for example, for half of the students to suggest a factor of that number and the other half to suggest a multiple of the same number. Discuss the responses. Mini whiteboards could be used for easy whole class involvement and assessment (EP, RL).

Develop this idea, giving half the group one number and the other half another number, enter student suggestions into two Venn diagrams, one for factors and one for multiples, as a way to find the HCF or LCM of pairs of numbers. Ensure that students are confident with this before moving on to discuss how prime factors can be used to find these in a more efficient way: a single Venn diagram with far fewer entries will be needed. A diagrammatic approach will help to support some students who struggle.

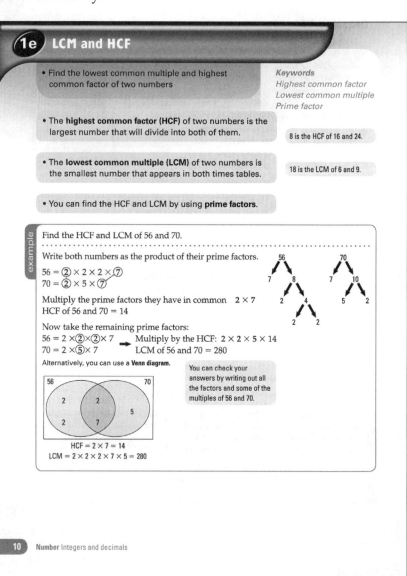

1e LCM and HCF

- Find the lowest common multiple and highest common factor of two numbers

Keywords
Highest common factor
Lowest common multiple
Prime factor

- The **highest common factor (HCF)** of two numbers is the largest number that will divide into both of them.

8 is the HCF of 16 and 24.

- The **lowest common multiple (LCM)** of two numbers is the smallest number that appears in both times tables.

18 is the LCM of 6 and 9.

- You can find the HCF and LCM by using **prime factors**.

Find the HCF and LCM of 56 and 70.

Write both numbers as the product of their prime factors.
56 = ②× 2 × 2 ×⑦
70 = ②× 5 ×⑦

Multiply the prime factors they have in common 2 × 7
HCF of 56 and 70 = 14

Now take the remaining prime factors:
56 = 2 ×②×②× 7 ➝ Multiply by the HCF: 2 × 2 × 5 × 14
70 = 2 ×⑤× 7   LCM of 56 and 70 = 280

Alternatively, you can use a **Venn diagram**.

You can check your answers by writing out all the factors and some of the multiples of 56 and 70.

HCF = 2 × 7 = 14
LCM = 2 × 2 × 2 × 7 × 5 = 280

10   **Number** Integers and decimals

Plenary

As a very brief assessment overview, give two numbers and ask students to draw a Venn diagram themselves, using the student book example as a model, to demonstrate how they have used prime factors to find the HCF and LCM. Additional support is provided by offering a selection of pairs of numbers and inviting students to share their responses with others with a different number pair.

Simplification

This exercise is supportive of a range of abilities of student as the more able will use prime factors readily whilst the weakest will still be able to achieve using a method that may be more familiar and can be encouraged to make connections at their own pace.

For students who are experiencing difficulty with this work, give additional practice with smaller numbers and encourage them to become confident with the table method as illustrated in question **1** before introducing the use of prime factors.

All questions (1.3)

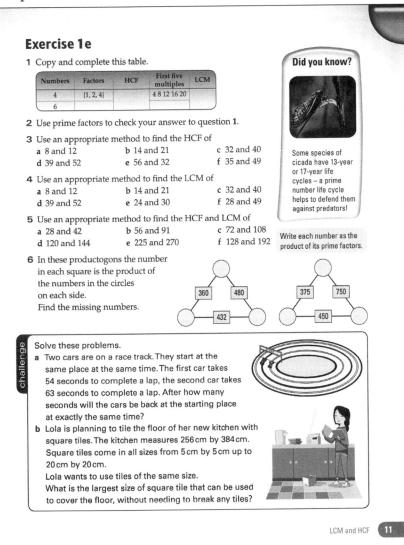

Extension

Ask pairs to work together to create their own productogon with answer, using question **6** as a model, which they can then swap with another pair (CT).

Exercise 1e commentary

Question 1 – This is the simplest way to find the HCF and LCM.

Question 2 – A complementary and more efficient approach to question **1**.

Questions 3 and **4** – The approaches used in questions **1** and **2** are both options, though moving to the prime factor method should be encouraged.

Question 5 – Less confident students could use the approach of question **1**.

Question 6 – For students who have struggled with the previous questions, it may be necessary to revisit the word 'product'.

Challenge – A good opportunity for paired work to establish what mathematical skills to use. This exemplifies the use of the process skill of representing, involving students in identifying the type of problem and the operations needed to reach a solution (IE). (1.1)

Assessment Criteria – UAM, level 5: identify and obtain necessary information to carry through a task and solve mathematical problems.

Links

Periodical cicadas live for years underground as grubs, emerge from their burrows for a few weeks to breed and then die. The population is synchronised so that they all become adults and emerge together. This can result in huge populations, up to 1.5M/acre. The males make a distinctive drumming or singing noise. There is more information about cicadas at http://magicicada.org/magicicada.php and at http://biology.clc.uc.edu/steincarter/cicadas.htm

- Use squares, positive and negative square roots and cubes (L6)
- Use index notation for small positive integer powers (L5)

Useful resources
Multilink cubes

Starter – Prime bingo

Ask students to draw a 3 × 3 grid and enter 9 prime numbers less than 100.

Give calculations yielding prime numbers, for example, half of 82, 7 × 5 + 2, 39 ÷ 3.

The winner is the first student to cross out all their primes.

The game can be differentiated by the choice of calculations.

Teaching notes

Make links with students' prior knowledge of number patterns as a way of introducing this work. Refer to the visual representations of square and cube numbers in the student book to reinforce concepts. This offers the opportunity to develop the process skill of analysing – using appropriate mathematical procedures, visualising images to support mental methods (1.2).

When developing the use of index notation, it is important to remember that a common misconception amongst students is that, for example, 3^2 means 3×2, rather than 3 multiplied by itself. Again, referring to the use of the visual images in the student book will provide support in overcoming this.

The first example offers the opportunity to discuss what happens when two negative numbers are multiplied together.

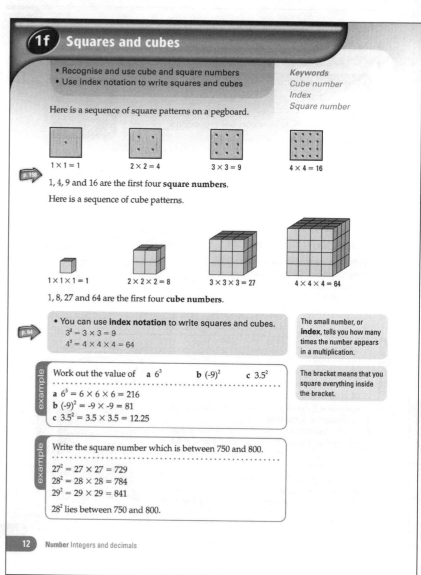

1f Squares and cubes

- Recognise and use cube and square numbers
- Use index notation to write squares and cubes

Keywords
Cube number
Index
Square number

Here is a sequence of square patterns on a pegboard.

$1 \times 1 = 1$ $2 \times 2 = 4$ $3 \times 3 = 9$ $4 \times 4 = 16$

1, 4, 9 and 16 are the first four **square numbers**.

Here is a sequence of cube patterns.

$1 \times 1 \times 1 = 1$ $2 \times 2 \times 2 = 8$ $3 \times 3 \times 3 = 27$ $4 \times 4 \times 4 = 64$

1, 8, 27 and 64 are the first four **cube numbers**.

- You can use **index notation** to write squares and cubes.
  $3^2 = 3 \times 3 = 9$
  $4^3 = 4 \times 4 \times 4 = 64$

The small number, or **index**, tells you how many times the number appears in a multiplication.

example
Work out the value of **a** 6^3 **b** $(-9)^2$ **c** 3.5^2

a $6^3 = 6 \times 6 \times 6 = 216$
b $(-9)^2 = -9 \times -9 = 81$
c $3.5^2 = 3.5 \times 3.5 = 12.25$

The bracket means that you square everything inside the bracket.

example
Write the square number which is between 750 and 800.

$27^2 = 27 \times 27 = 729$
$28^2 = 28 \times 28 = 784$
$29^2 = 29 \times 29 = 841$

28^2 lies between 750 and 800.

12   Number Integers and decimals

Plenary

Ask students to write a definition of a square and a cube number and to discuss how they would explain to someone who didn't understand, what the small number (index) means (CT).

Simplification

Allow students to develop their understanding of square and cube numbers using a kinaesthetic approach, building the numbers with cubes.

All questions (1.3)

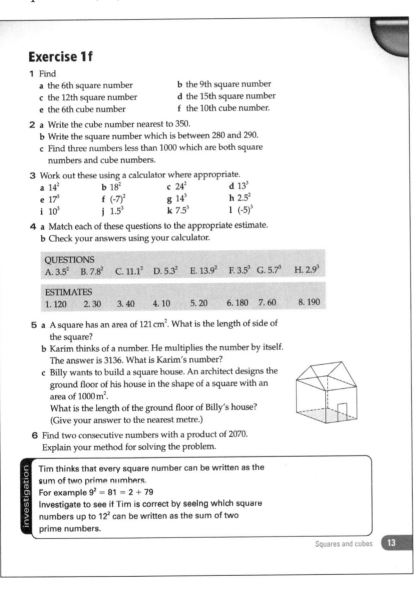

Exercise 1f

1 Find
  a the 6th square number
  b the 9th square number
  c the 12th square number
  d the 15th square number
  e the 6th cube number
  f the 10th cube number.

2 a Write the cube number nearest to 350.
  b Write the square number which is between 280 and 290.
  c Find three numbers less than 1000 which are both square numbers and cube numbers.

3 Work out these using a calculator where appropriate.
  a 14^2 **b** 18^2 **c** 24^2 **d** 13^3
  e 17^3 **f** $(-7)^2$ **g** 14^3 **h** 2.5^2
  i 10^3 **j** 1.5^3 **k** 7.5^3 **l** $(-5)^3$

4 a Match each of these questions to the appropriate estimate.
  b Check your answers using your calculator.

QUESTIONS
A. 3.5^2 B. 7.8^2 C. 11.1^2 D. 5.3^2 E. 13.9^2 F. 3.5^3 G. 5.7^3 H. 2.9^3

ESTIMATES
1. 120   2. 30   3. 40   4. 10   5. 20   6. 180   7. 60   8. 190

5 a A square has an area of 121 cm^2. What is the length of side of the square?
  b Karim thinks of a number. He multiplies the number by itself. The answer is 3136. What is Karim's number?
  c Billy wants to build a square house. An architect designs the ground floor of his house in the shape of a square with an area of 1000 m^2.
    What is the length of the ground floor of Billy's house?
    (Give your answer to the nearest metre.)

6 Find two consecutive numbers with a product of 2070.
  Explain your method for solving the problem.

> **investigation**
> Tim thinks that every square number can be written as the sum of two prime numbers.
> For example $9^2 = 81 = 2 + 79$
> Investigate to see if Tim is correct by seeing which square numbers up to 12^2 can be written as the sum of two prime numbers.

Extension

Ask students to investigate the relationship between square numbers and odd numbers. Encourage them to use visual images to explain their thinking (SM).

Exercise 1f commentary

Question 1 – Students may need support with parts **e** and **f**.

Question 2 – Similar to the second example; some students may need a calculator.

Question 3 – Similar to the second example. Check that students can correctly use their calculator. It will be useful to make a discussion point of the negative numbers. What happens when you multiply a negative number by itself? What happens when you multiply a negative number by itself three times?

Question 4 – Check that students try to match as many pairs as they can without defaulting to using a calculator.

Question 5 – Inverse problems requiring the mathematis to be identified: students may use 'guess work' or the 'square-root' key (1.1).

Question 6 – Suggest trying numbers close to the square root.

Investigation – Here, work on prime numbers is revisited. For some students, a brief reminder of the meaning of 'sum' may be appropriate. For the square of an even number this is a special case of the (presently unresolved) Goldbach conjecture (1.4).

Links

The sugar cube was invented in 1841 by Jakub Kryštof Rad in Dačice, Bohemia (now the Czech Republic). At this time, sugar was produced in a large, solid cone shape and had to be cut for the customer. Rad invented the sugar cube after his wife cut her finger slicing the sugar. Sugar cubes first appeared in shops in Vienna in 1843. A granite memorial of a sugar cube stands in the Town Square in Dačice.

1a

1 Put these numbers in order from smallest to largest.

a -5	4	-3	2	-1
b 1.5	-2.4	-3.1	-0.9	-2
c 3.2	-1.4	-2.6	2.7	-1.1
d -1.5	0.15	-2.1	-1.6	-0.5

2 Calculate

a 8 + -4	**b** 6 + -9	**c** 4 + -8	**d** -3 + -7	**e** -11 + -2
f 9 − -2	**g** 3 − -12	**h** 1 − -5	**i** -8 − -4	**j** -6 − -11
k -9 − -3	**l** -6 − -12	**m** -13 + -21	**n** 12 + -15	**o** 17 − -12
p 11 + -21	**q** 13 − -24	**r** 28 + -23	**s** 35 − -21	**t** -17 + -31

1b

3 Calculate

a 2 × -5	**b** -4 × 3	**c** -6 × -4	**d** -8 × -5	**e** -9 × 6
f -7 × -9	**g** 9 × -9	**h** -12 × -3	**i** -14 × 5	**j** -15 × -6
k -40 ÷ -8	**l** -35 ÷ 7	**m** -36 ÷ -9	**n** -56 ÷ -8	**o** 45 ÷ -9
p -13 × 5	**q** -25 × 4	**r** -11 × -15	**s** -84 ÷ -7	**t** -96 ÷ 8

4 Copy and complete these calculations.

a $8 \times \square = -72$ **b** $-54 \div \square = -6$ **c** $-7 \times \square = 49$

d $-104 \div \square = 8$ **e** $\square \times -12 = -48$ **f** $-6 \times \square = -72$

g $\square \div -4 = 64$ **h** $\square \div -9 = -14$

1c

5 Write all the factors of

a 30	**b** 48	**c** 65	**d** 72	**e** 96	**f** 100
g 130	**h** 108	**i** 120	**j** 132	**k** 144	**l** 150

6 Use the divisibility tests to answer each of these questions. In each case explain your answer.

a Is 2 a factor of 98?   **b** Is 3 a factor of 93?

c Is 4 a factor of 112?   **d** Is 5 a factor of 157?

e Is 6 a factor of 184?   **f** Is 7 a factor of 135?

g Is 8 a factor of 196?   **h** Is 9 a factor of 289?

i Is 12 a factor of 200?   **j** Is 11 a factor of 385?

1d

7 Use the divisibility tests for prime numbers to see which of these numbers are prime. In each case explain your answer.

a 35	**b** 38	**c** 37	**d** 47	**e** 51	**f** 53
g 75	**h** 79	**i** 76	**j** 85	**k** 93	**l** 91

8 Work out the value of
   a $3 \times 3 \times 5$ **b** $2 \times 3 \times 5$ **c** $2 \times 3 \times 3$ **d** $3 \times 3 \times 5 \times 5$
   e $2 \times 3 \times 3 \times 3$ **f** $3 \times 5 \times 5$ **g** $7 \times 7 \times 7$ **h** $3 \times 5 \times 7$
   i $5 \times 5 \times 7$ **j** $3 \times 5 \times 7 \times 11$

9 Write each of these numbers as the product of its prime factors.
   a 18 **b** 28 **c** 45 **d** 57 **e** 63 **f** 76
   g 88 **h** 92 **i** 108 **j** 115 **k** 130 **l** 132
   m 144 **n** 160 **o** 170 **p** 175 **q** 188 **r** 240

10 Use an appropriate method to find the HCF of
   a 6 and 10 **b** 12 and 16 **c** 18 and 27
   d 15 and 20 **e** 24 and 32 **f** 25 and 30
   g 28 and 40 **h** 56 and 80

11 Use an appropriate method to find the LCM of
   a 6 and 10 **b** 12 and 16 **c** 18 and 27
   d 15 and 20 **e** 16 and 24 **f** 24 and 30
   g 28 and 32 **h** 50 and 56

12 In these productogons the number in each square is the
   product of the numbers in the circles on each side of it.
   Find the missing numbers in each of these productogons.

   Write each number as the product of its prime factors.

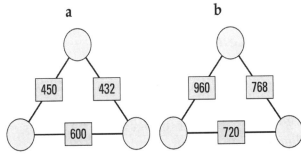

13 Work out these using a calculator where appropriate.
   a 7^3 **b** 13^2 **c** 21^2 **d** 9^3 **e** 15^3 **f** $(-3)^2$
   g 24^3 **h** 0.5^2 **i** 0.1^2 **j** 0.1^3 **k** $(-2)^2$ **l** $(-2)^3$

1 Summary

Assessment criteria

- Recognise and describe number relationships including multiple, factor and square   Level 4
- Order negative numbers in context   Level 5
- Solve simple problems involving ordering, adding, subtracting negative numbers in context   Level 5
- Identify and obtain necessary information to carry through a task and solve mathematical problems   Level 5
- Solve problems and carry through substantial tasks, using a range of efficient techniques   Level 6

Question commentary

Example	The example illustrates a typical problem about negative numbers. Emphasise the use of a number line in parts **a** and **b**. In part **c**, emphasise that two numbers of the same sign cannot multiply to make a negative number. Ask students to find alternative answers to those given. Ask probing questions such as "Does addition always make numbers bigger?" and "Does subtraction always make numbers smaller?"	**a** $-3 + 1 = -2$ **b** $4 - 6 = -2$ **c** $2 \times -1 = -2$

Past question	The question asks students to identify whether multiples and factors of given even numbers could be odd or even. Encourage thorough reading of the question to avoid confusion of factor and multiple. Some students may find the justification difficult in part **b**. Encourage a thorough approach by asking students to list all the factors of 20 and then identify examples of odd and even factors.	**Answer**
		 Level 5
		 2 a My number must be even
		     If a number is a multiple of 4, it must be even
		   b My number could be odd or even
		     1 is odd and is a factor of 20
		     2 is even and is a factor of 20

Development and links

The work on factors and multiples is developed in Chapter 4 when students work with equivalent fractions. Squares and cubes are used in work on area in Chapter 2 and volume in Chapter 15 and extended to square and cube roots in Chapter 10. Students will use negative numbers when drawing graphs in Chapter 7.

Factors and multiples are used in fraction work and in algebra to factorise equations and formulae. This has links with work with scientific formulae in science and multiples of fractional quantities in food technology and resistant materials. Negative numbers are important in science and geography when working with concepts such as temperature, electric charge and height above or below sea level.

Objectives

Level

- Choose and use units of measurement to measure, estimate, calculate and solve problems in a range of contexts...................... **4**
- Know rough metric equivalents of imperial measures in common use, such as miles, pounds(lb) and pints **5**
- Calculate areas of compound shapes **5**
- Derive and use the formulae for the area of a triangle, parallelogram and trapezium.. **5, 6**

MPA

1.1	2b
1.2	2a, d
1.3	2a, b, c, d, e
1.4	2b, d
1.5	2d

Introduction

In this chapter, students review their knowledge of metric units of measurement, read and interpret scales on measuring instruments and consider rough metric equivalents of imperial units of length, mass and capacity. They extend to calculating the area and perimeter of a rectangle and using the rectangle to deduce the formula for the area of a triangle, a parallelogram and a trapezium.

There are many historical systems of units, but the SI (Systeme International d'Unites) of metric units is now recognised and used in most countries of the World. The student book discusses the importance of using units when measuring distances. Students will often forget to give units with an answer and some may not realise that 2 km is considerably shorter than 2 miles. It should be emphasised that without units, a measurement is ambiguous and incomplete. There is more information about systems of measurement at http://en.wikipedia.org/wiki/Units_of_measurement (EP).

PLTS

IE	2a, c, d, e
CT	2a, c, e
RL	2a
TW	2a, b
SM	2d, e
EP	2I, a, b, d

Extra Resources

2 Start of chapter presentation
2b Starter: Metric unit matching
2b Worked solution: Q3e
2b Consolidation sheet
2d Animation: Area
2d Worked solution: Q3c
2e Consolidation sheet

   Assessment: chapter 2

Fast-track

2a, b, e

- Convert one metric unit to another (L5)
- Choose and use units of measurement to measure and estimate in a range of contexts (L5)

Useful resources
1m ruler, 1 kg weight, (empty) 1 l bottle
Box of 50 or 100 paper clips

Starter – Powers of ten

Write 10.7 on the board.
Ask students what answer you will get if you multiply the number by 1000, 100, 10.
Repeat using different starting numbers.
Can be extended by using division.

Teaching notes

Ask three or four questions in context, for example, how much liquid is in a 2 l bottle about a quarter full? How heavy is a large text book (over 1 kg)? What is the length of the teacher's desk? Lead into a discussion of metric measures and common conversions: the desk is 1.5 m = 150 cm, *etc*. It is now appropriate to supply a table for the different conversions and measures or the families of measures: tonne, kilogram, gram, milligram, *etc*. (1.3).

Questions **1** and **2** lend themselves to whole class discussion.

A worked example, such as given, of how to convert between units will provide students with a template for setting out their answers to questions **3** and **4**.

Wider issues are relevant to questions **5** and **6** where more than one solution might exist. Can a student convince the class that they have all the solutions or why they have the best solution: are four small bottles better than one large one? (EP)

The **problem** question allows a discussion of what does weigh 500 kg: a large motor bike, 8-12 students or 3-4 head teachers.

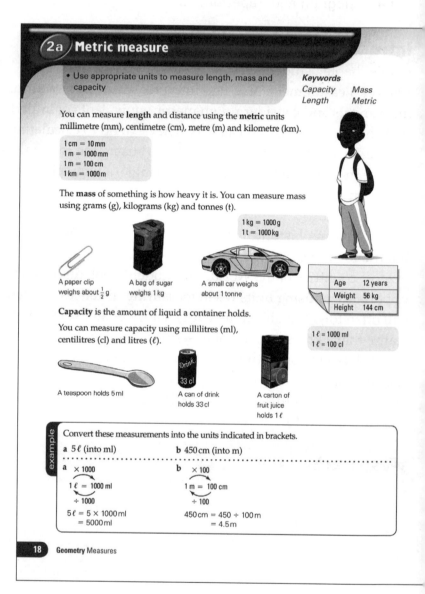

2a Metric measure

- Use appropriate units to measure length, mass and capacity

Keywords
Capacity   Mass
Length   Metric

You can measure **length** and distance using the **metric** units millimetre (mm), centimetre (cm), metre (m) and kilometre (km).

1 cm = 10 mm
1 m = 1000 mm
1 m = 100 cm
1 km = 1000 m

The **mass** of something is how heavy it is. You can measure mass using grams (g), kilograms (kg) and tonnes (t).

1 kg = 1000 g
1 t = 1000 kg

A paper clip weighs about $\frac{1}{2}$ g

A bag of sugar weighs 1 kg

A small car weighs about 1 tonne

Age	12 years
Weight	56 kg
Height	144 cm

Capacity is the amount of liquid a container holds.

You can measure capacity using millilitres (ml), centilitres (cl) and litres (ℓ).

1 ℓ = 1000 ml
1 ℓ = 100 cl

A teaspoon holds 5 ml

A can of drink holds 33 cl

A carton of fruit juice holds 1 ℓ

example

Convert these measurements into the units indicated in brackets.

a 5 ℓ (into ml)   b 450 cm (into m)

a × 1000
1 ℓ = 1000 ml
÷ 1000
5 ℓ = 5 × 1000 ml
= 5000 ml

b × 100
1 m = 100 cm
÷ 100
450 cm = 450 ÷ 100 m
= 4.5 m

18 **Geometry** Measures

Plenary

Give students working in pairs – 5 lengths (heights), 5 weights and 5 animals (blue whale, giraffe, large python, lion, Labrador dog). Can they match the appropriate animal with their weight and length.

Simplification

This type of measures work mixes a great number of different topics – where students find the work challenging it is appropriate to practice questions involving only one of length or mass or capacity. With question **2** ask students to look at the measurement column and agree which are measures of length.

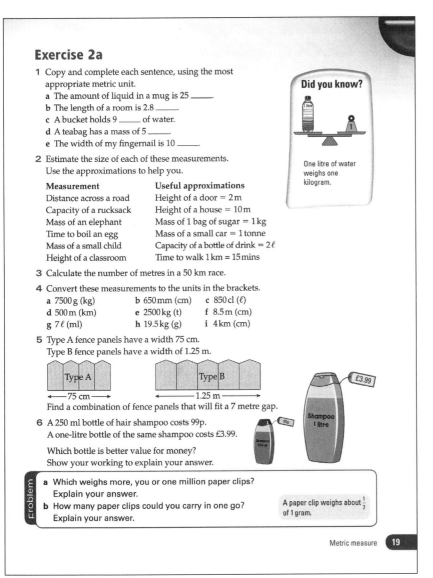

Exercise 2a

1 Copy and complete each sentence, using the most appropriate metric unit.
   a The amount of liquid in a mug is 25 _____.
   b The length of a room is 2.8 _____.
   c A bucket holds 9 _____ of water.
   d A teabag has a mass of 5 _____.
   e The width of my fingernail is 10 _____.

2 Estimate the size of each of these measurements. Use the approximations to help you.

Measurement	Useful approximations
Distance across a road	Height of a door = 2 m
Capacity of a rucksack	Height of a house = 10 m
Mass of an elephant	Mass of 1 bag of sugar = 1 kg
Time to boil an egg	Mass of a small car = 1 tonne
Mass of a small child	Capacity of a bottle of drink = 2ℓ
Height of a classroom	Time to walk 1 km = 15 mins

3 Calculate the number of metres in a 50 km race.

4 Convert these measurements to the units in the brackets.
   a 7500 g (kg)   b 650 mm (cm)   c 850 cl (ℓ)
   d 500 m (km)   e 2500 kg (t)   f 8.5 m (cm)
   g 7 ℓ (ml)   h 19.5 kg (g)   i 4 km (cm)

5 Type A fence panels have a width 75 cm. Type B fence panels have a width of 1.25 m.

Find a combination of fence panels that will fit a 7 metre gap.

6 A 250 ml bottle of hair shampoo costs 99p. A one-litre bottle of the same shampoo costs £3.99.

Which bottle is better value for money? Show your working to explain your answer.

Problem
a Which weighs more, you or one million paper clips? Explain your answer.
b How many paper clips could you carry in one go? Explain your answer.

A paper clip weighs about $\frac{1}{2}$ of 1 gram.

Did you know?
One litre of water weighs one kilogram.

Metric measure **19**

Extension

It is important that competent students are able to explain their mathematical thinking and these students should be asked to justify their choices and answers for questions **5, 6** and the **problem**. What if the fence lengths had been 80 cm and 1.5 metres – how would this have affected the best answer to the problem? (RL)

Exercise 2a commentary

Question 1 and **2** – Students could be encouraged to work in pairs to identify relevant information, agree on appropriate units and good approximations (TW, IE) (1.2, 1.3).

Question 3 and **4f** – **i** involve conversion from large to small units whilst parts **4a** – **e** require conversions from small to large units (1.3).

Question 5 – Students must use a consistent set of units and can be encouraged to find more than one solution (1.3).

Question 6 – It will be easiest to compare the prices of $1 \, l = 4 \times \frac{1}{4} \, l$ of shampoo. The suitability of the solution as well as value for money may be introduced as a consideration.

Problem – A larger number is involved in the conversion; for the comparison only an estimate of the student's weight is necessary. In part **b** a consideration should be the paper clip's bulk as well as weight (CT).

Assessment Criteria – SSM, level 4: choose and use appropriate units.

Links

Measurement of length was originally based on the human body. The ancient Egyptians used a unit called a cubit, which was the length of an arm from the elbow to the fingertips. As everybody's arm was a different length, the Egyptians developed the standard Royal cubit and preserved this length as a black granite rod. Other measuring sticks were made the same length as this rod. There is a picture of a cubit rod at http://www.globalegyptianmuseum. org/detail.aspx?id=4424 (TW)

- Know rough metric equivalents of imperial measures in common use (L5)
- Read and interpret scales on a range of measuring instruments (L5)

Useful resources
Rulers with cm and in.
A 1m rule also showing inches (builder/dressmaker's tape)
Scales with both kg and lb/oz weights

Starter – Metric pairs

Write the following measurements on the board:
0.01 g, 0.1 g, 0.1 kg, 1 g, 1 kg, 1t, 10 mg, 10 g, 100 mg, 100 g, 1000 mg, 1000 g, 1000 kg, 10 000 mg.
Ask students to find the equivalent pairs (1000 mg = 1 g).
Can be extended by asking students to make their own equivalent pairs for capacity or length.

Teaching notes

Ask students to name units used in everyday life: miles on roads, litre bottles and pint glasses, *etc*. Make lists, grouping the units for length, mass and capacity and within each group split them into metric and imperial units. Invite students to match comparable units: inch ~ centimetre, yard ~ metre, mile ~ kilometre, *etc*. and give examples of objects with these sizes. A ruler/tape with inches and centimetres can be used to confirm the examples. Likewise scales can be used to confirm that 1 kg = 2 lb + 3 oz (1.1).

Introduce the formal conversion factors and show how these can be used to convert between units by multiplication or division.

A wider discussion may involve which system of units is better (1.4). Metric is more systematic but what is $\frac{1}{3}$, $\frac{1}{6}$ or a $\frac{1}{9}$ of a metre or a yard? What is the role of the European Union in promoting the metric system? How is international business organised, for example America uses (a version) of imperial units (EP).

A demonstration of how to read a scale should focus on establishing what a sub-interval is worth and not to assume that it is a tenth of the interval unit.

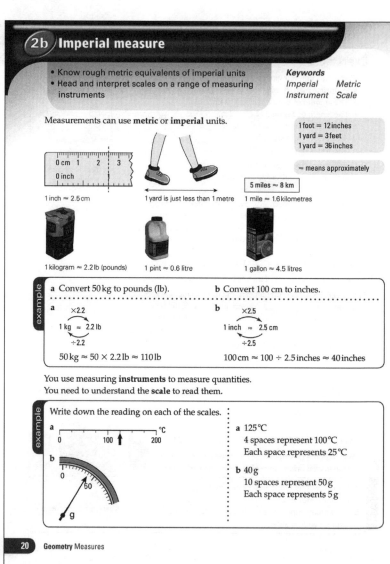

2b Imperial measure

- Know rough metric equivalents of imperial units
- Read and interpret scales on a range of measuring instruments

Keywords
Imperial   Metric
Instrument   Scale

Measurements can use **metric** or **imperial** units.

1 foot = 12 inches
1 yard = 3 feet
1 yard = 36 inches

≈ means approximately

1 inch ≈ 2.5 cm   1 yard is just less than 1 metre
5 miles ≈ 8 km
1 mile ≈ 1.6 kilometres

1 kilogram ≈ 2.2 lb (pounds)   1 pint ≈ 0.6 litre   1 gallon ≈ 4.5 litres

example

a Convert 50 kg to pounds (lb).   b Convert 100 cm to inches.

a   ×2.2
1 kg ≈ 2.2 lb
        ÷2.2
50 kg ≈ 50 × 2.2 lb ≈ 110 lb

b   ×2.5
1 inch ≈ 2.5 cm
        ÷2.5
100 cm ≈ 100 ÷ 2.5 inches ≈ 40 inches

You use measuring **instruments** to measure quantities.
You need to understand the **scale** to read them.

example

Write down the reading on each of the scales.

a
0   100   200   °C

b
0   50   g

a 125 °C
4 spaces represent 100 °C
Each space represents 25 °C

b 40 g
10 spaces represent 50 g
Each space represents 5 g

20 **Geometry** Measures

Plenary

The most widely used imperial measures are for length. Miles are used exclusively on UK roads and students will probably only be able to give their height in feet and inches. Ask all students to convert their height from feet and inches to metres. On average are the boys taller than the girls? Ask students to estimate the average, then in small groups work out the averages (TW).

Simplification

Many students find this difficult because they just try to apply the different rules without a context to check their answers against. For question **2** real life examples with these measurements may help. Also measure items in the classroom in both imperial (ft and in and lb and oz) and in metric (m and kg). For example, weigh a 2 l bottle in kilograms (2 kg) and then in pounds and ounces (4 lb 6 or 7oz); also measure length of books, tables, *etc*. (1.1).

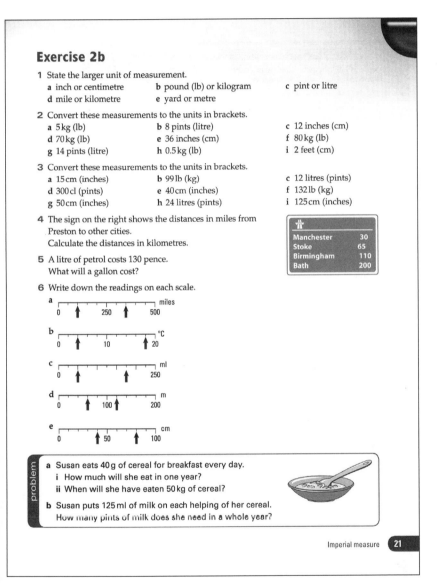

Exercise 2b

1 State the larger unit of measurement.
 a inch or centimetre   b pound (lb) or kilogram   c pint or litre
 d mile or kilometre   e yard or metre

2 Convert these measurements to the units in brackets.
 a 5 kg (lb)   b 8 pints (litre)   c 12 inches (cm)
 d 70 kg (lb)   e 36 inches (cm)   f 80 kg (lb)
 g 14 pints (litre)   h 0.5 kg (lb)   i 2 feet (cm)

3 Convert these measurements to the units in brackets.
 a 15 cm (inches)   b 99 lb (kg)   c 12 litres (pints)
 d 300 cl (pints)   e 40 cm (inches)   f 132 lb (kg)
 g 50 cm (inches)   h 24 litres (pints)   i 125 cm (inches)

4 The sign on the right shows the distances in miles from Preston to other cities.
 Calculate the distances in kilometres.

Manchester	30
Stoke	65
Birmingham	110
Bath	200

5 A litre of petrol costs 130 pence.
 What will a gallon cost?

6 Write down the readings on each scale.

a   miles
 0   250   500

b   °C
 0   10   20

c   ml
 0   250

d   m
 0   100   200

e   cm
 0   50   100

problem
 a Susan eats 40 g of cereal for breakfast every day.
  i How much will she eat in one year?
  ii When will she have eaten 50 kg of cereal?
 b Susan puts 125 ml of milk on each helping of her cereal.
  How many pints of milk does she need in a whole year?

Imperial measure   **21**

Extension

For those who have mastered the idea of conversion, develop more problem solving activities. At a market the cost of potatoes is given as 40p a pound and also £1 per kilogram – is this acceptable? A leg of lamb weighs 2.4 kg and costs £10, how much is this per pound?

Exercise 2b commentary

Question 1 – Helps develop an intuition for comparable units.

Question 2 and **3** – Metric—imperial conversions as in the first example. Question **2** generally involves multiplications and question **3** divisions. Part **2f** requires a preliminary conversion 2 ft = 24 in (1.3).

Question 4 and **5** – Real life situations requiring conversion of units (1.3).

Question 6 – Students will need to be clear how much a sub-division is worth: 50 miles, 2°C, *etc*. (1.3)

Problem – Students could first make an estimate based on 400 days before multiplying by 365; answers should be converted to appropriate units.

Assessment Criteria – SSM, level 5: solve problems involving the conversion of units and make sensible estimates of a range of measures in relation to everyday situations. SSM, level 5: read and interpret scales on a range of measuring instruments, explaining what each labeled division represents.

Links

The United States has its own system of weights and measures which is largely similar to the imperial system. Yards, feet, inches and pounds are all in everyday use in the US, however, the US pint and US gallon are both smaller than the imperial pint and gallon. There is more information about the differences between the two systems at http://home.clara.net/brianp/usa.html and at http://en.wikipedia.org/wiki/Imperial_units (TW).

- Know and use the formula for the area of a rectangle   (L5)
- Calculate the perimeter and area of shapes made from rectangles   (L5)

Useful resources
Squared paper

Starter – Estimation

Draw lines on the board.

Ask students to estimate the length of the lines in cm.

Ask how much this would be in inches.

Use a scoring system for the estimations, for example, within 10% score 3 points, within 20% score 1 point.

Bonus points for correct metric to imperial conversion.

Teaching notes

It is important for students to be aware of the units used and convert where necessary.

A good introduction to the lesson would be to give students the area of a rectangle, say 36 cm², and ask them what the dimensions could be. Why are there more than two possible answers? Do the lengths have to be whole numbers? Which lengths and widths give the largest and smallest perimeters? This could form the basis of a paired activity. It also relates directly to question **2** (CT).

When dealing with composite shapes, strongly recommend that students make their own sketch first. This is especially important if the original drawing is not to scale. To find the perimeter, carefully identify any missing lengths and work them out before adding up all the lengths. To find the area divide the shape into simpler rectangles. It may help to label these as A and B, work out their areas and then compute A + B.

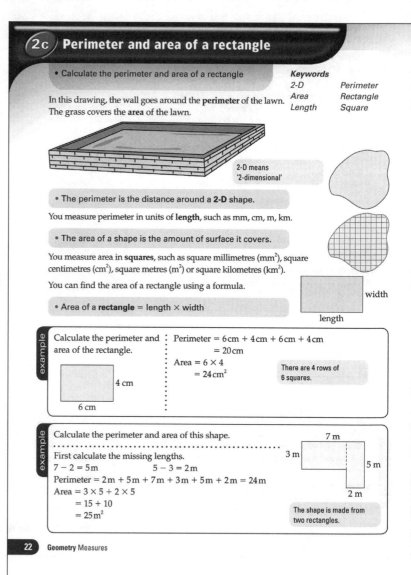

2c   Perimeter and area of a rectangle

- Calculate the perimeter and area of a rectangle

Keywords
2-D   Perimeter
Area   Rectangle
Length   Square

In this drawing, the wall goes around the **perimeter** of the lawn. The grass covers the **area** of the lawn.

2-D means '2-dimensional'

- The perimeter is the distance around a **2-D** shape.

You measure perimeter in units of **length**, such as mm, cm, m, km.

- The area of a shape is the amount of surface it covers.

You measure area in **squares**, such as square millimetres (mm²), square centimetres (cm²), square metres (m²) or square kilometres (km²).

You can find the area of a rectangle using a formula.

- Area of a **rectangle** = length × width

example   Calculate the perimeter and area of the rectangle.

4 cm

6 cm

Perimeter = 6 cm + 4 cm + 6 cm + 4 cm
         = 20 cm
Area = 6 × 4
     = 24 cm²

There are 4 rows of 6 squares.

example   Calculate the perimeter and area of this shape.

First calculate the missing lengths.
7 − 2 = 5 m   5 − 3 = 2 m
Perimeter = 2 m + 5 m + 7 m + 3 m + 5 m + 2 m = 24 m
Area = 3 × 5 + 2 × 5
     = 15 + 10
     = 25 m²

7 m
3 m
5 m
2 m

The shape is made from two rectangles.

22   **Geometry** Measures

Plenary

Ask students to work in pairs on a problem like - A rectangular towel is folded in half four times and the dimensions are now 20 × 30 cm. What is the area of the towel and what could the dimensions of the towel be? (IE)

Simplification

Using an accurate scale diagram on grid paper and counting squares can be used to give confidence in the numerical approach to finding perimeters and areas.

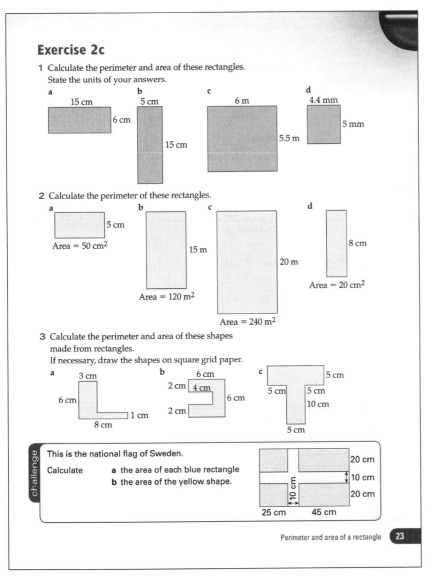

Exercise 2c

1 Calculate the perimeter and area of these rectangles. State the units of your answers.

a 15 cm, 6 cm

b 5 cm, 15 cm

c 6 m, 5.5 m

d 4.4 mm, 5 mm

2 Calculate the perimeter of these rectangles.

a 5 cm, Area = 50 cm²

b 15 m, Area = 120 m²

c 20 m, Area = 240 m²

d 8 cm, Area = 20 cm²

3 Calculate the perimeter and area of these shapes made from rectangles.
If necessary, draw the shapes on square grid paper.

a 3 cm, 6 cm, 1 cm, 8 cm

b 6 cm, 2 cm, 4 cm, 6 cm, 2 cm

c 5 cm, 5 cm, 5 cm, 10 cm, 5 cm

challenge

This is the national flag of Sweden.

Calculate **a** the area of each blue rectangle
          b the area of the yellow shape.

20 cm, 10 cm, 20 cm, 10 cm, 25 cm, 45 cm

Perimeter and area of a rectangle   **23**

Extension

Challenge students with a problem like – A rectangle has a perimeter of 28 cm, what different areas might this rectangle have? You are told that the area of this rectangle is 30 cm² to the nearest square cm. What could the dimensions be? (remember the perimeter is still 28 cm.) Students may need a calculator as this will bring in decimal lengths.

Exercise 2c commentary

Question 1 – Areas and perimeters, similar to first example (1.3).

Question 2 – An 'inverse' problem, an example of which could be discussed in the lesson (1.3).

Question 3 – Composite shapes as illustrated in the second example. There is more than one way to decompose the shapes; it is also possible to use subtraction (1.3).

Challenge – There are several ways to find the area of the yellow cross by dissection or subtracting four blue rectangles from the flag.

Assessment criteria – SSM, level 4: find perimeters of simple shapes and find areas by counting squares. SSM, level 5: understand and use the formula for the area of a rectangle and distinguish area from perimeter.

Links

An area of 10 000 m² is called a hectare and is a common unit used to measure an area of land. Measure or estimate the size of the classroom. What fraction is this of a hectare? Estimate how many hectares the school field or other local open space covers. The O2 arena in Greenwich has a ground area of over 80 000 m². How many hectares is this? There is a picture of the O2 arena at http://en.wikipedia.org/wiki/Image:Canary.wharf.and.dome.london.arp.jpg (1.3)

- Derive and use the formula for the area of a triangle (L5)
- Calculate areas of compound shapes (L5)

Useful resources
Squared Paper
Mini whiteboards

Starter – Calculating length

Give the area and width of rectangles and ask students for the lengths, for example,

  Area = 39 cm² and width = 3 cm
  Area = 48 cm² and width = 6 cm
  Area = 125 cm² and width = 25 cm

Can be extended by using numbers that will generate decimal lengths.

Teaching notes

A natural introduction would be to demonstrate why the area of a triangle is given by half the area of the 'bounding' rectangle. The case of an 'overhanging' triangle should be treated with some care and an emphasis placed on deciding what is the perpendicular height of the triangle. Doing a worked example, as supplied, will show students how to set out their written answers.

Question **2** could be used as the basis of a paired activity to find multiple solutions. Classroom discussion can then be used to see what features groups of answers have in common and why the formula gives the same answer for each triangle.

It may be helpful to provide a worked example involving a composite shape, such as those in question **5** and the **challenge**. Emphasise being methodical and careful setting out of workings: make your own sketch, dissect it into more basic shapes, apply formula to calculate these sub-areas and sum the results and add units.

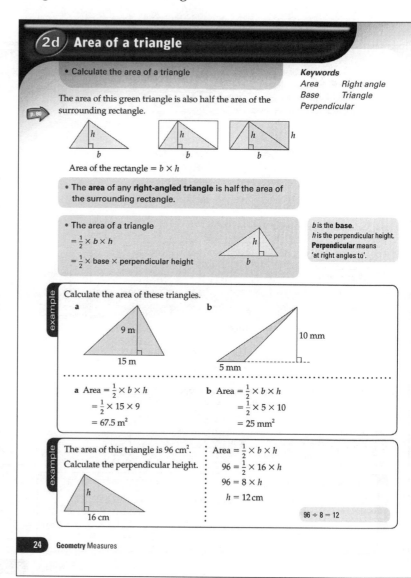

Plenary

Tell all students that you have drawn a triangle of area 18 cm² - they can ask questions but only with a yes or no answer. Their target is to work out what your triangle looks like and sketch it (white boards are very useful here). Questions could be: are the height and base of the triangle equal? Is the height greater than the base? *etc.*

Simplification

As in question **1**, accurately drawing shapes on squared paper will allow a 'square counting' approach to be used. This can be used to give confidence in the use of the formula. This approach will also work for the more complex shapes in question **5**.

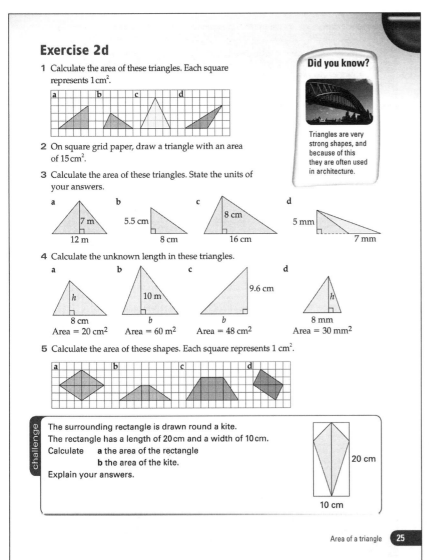

Extension

To challenge the more able student, draw a 3, 4, 5 right angled triangle with the base as 5 cm. Show the right angle and draw in the height h. Ask them to find the area $\left(\frac{1}{2} \times 3 \times 4\right)$ cm² and then to find the given height. You can add a similar question with a 5, 12, 13 triangle (SM).

Exercise 2d commentary

Question 1 – The question can be answered using the formula or by counting squares; students could be asked to use both methods and confirm agreement. If counting squares, encourage students to group part squares to make whole squares.

Question 2 – Students could be asked to find several solutions and note what they have in common.

Question 3 – Similar to the first example. Part **d** may cause confusion over what is the height of the triangle. Make sure that students know how to set out their working (1.3).

Question 4 – Similar to the second example. Students may need reminding to include the factor $\frac{1}{2}$ and should be encouraged to check their answers using the formula (1.3).

Question 5 – Encourage students to draw the shapes and break them down into triangles and rectangles.

Challenge – The challenge is to give a clear explanation for the area of the kite being half of that of the rectangle. It may help to further split the kite into four right-angle triangles (IE) (1.2, 1.3, 1.4).

Assessment Criteria – SSM, level 6: deduce and use the formula for the area of a triangle.

Links

The area of the sail of a yacht affects its performance, the greater the area, the greater the force that the wind can exert on the yacht. However if the sail is too large, the boat will become unstable. There is more information about how sails work at http://www.seed.slb.com/en/scictr/watch/sailing/index.htm (EP).

- Derive and use the formulae for the area of a parallelogram and trapezium

(L6)

Useful resources
Large cardboard triangles,
parallelograms and trapeziums
Pre-prepared tanagrams.
Scissors, glue, squared paper

Starter – Calculating height

Give the area and base of triangles and ask students for the heights, for example,

Area = 18 cm² and base = 3 cm (12 cm)
Area = 24 cm² and base = 6 cm (8 cm)
Area = 25 cm² and base = 10 cm (5 cm)

Can be extended by using numbers that will generate decimal heights.

Teaching notes

As an introduction to the lesson, students could be shown and challenged to understand the formulae for area of parallelograms and trapeziums – clear explanations are provided in the student books.

Using large cardboard shapes allows the derivation of the formulae to be illustrated. For example, two identical triangles could be fitted together to make a parallelogram or a triangle cut off the end of a parallelogram and the pieces rearranged to make a rectangle. Likewise two identical trapeziums could be fitted together to make a parallelogram or a trapezium could be cut down the middle, parallel to the two parallel sides, and the pieces fitted together to make a long-narrow parallelogram.

Once the formulae are understood, examples can be provided to show how they are applied and how the answers should be set out. For the parallelogram, emphasise that it is the perpendicular height and not the 'slope height' which is relevant for the area. For the trapezium, it may help students to think of the $\frac{1}{2}(a+b)$ term as the average length of the two parallel sides.

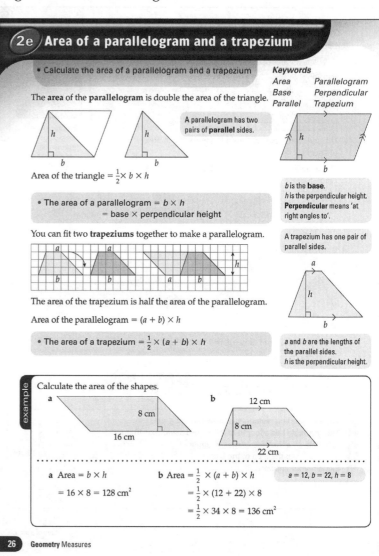

Plenary

Give students the area 10 cm² - in pairs can they sketch out a rectangle, triangle, parallelogram, trapezium and square of this area (allow 3.1 – 3.2 cm for the side of the square). Compare answers, are they unique? Only the square has one solution, how can students find the exact measurements required for the square? (CT)

Simplification

The use of squared paper to count squares in an accurate drawing can be used to develop confidence in the application of the formulae.

Finding the area of trapeziums using the formula is challenging – use isosceles trapeziums and allow students to break them into a rectangle and two equal right angled triangles to find the area of each and total.

Exercise 2e commentary

Question 1 – Similar to the first example; check that units are given (1.3).

Question 2 – Similar to the second example; check that students are correctly applying the formula.

Question 3 – 'Inverse' problems with parallelograms (1.3).

Activity – Students will need squared paper, scissors and possibly glue to stick solutions into their books. Emphasise the need for clear workings and explanations. This may be done as a paired activity (IE).

Assessment Criteria – SSM, level 6: deduce and use the formula for the area of a parallelogram.

Links

Sanders parallelogram is an optical illusion published in 1926 by the German psychologist Friedrich Sander. Sander's parallelogram can be found together with a selection of other optical illusions at http://library.thinkquest.org/05aug/01744/slanted_box.htm and at http://www.wyrmcorp.com/galleries/illusions/geometry.shtml

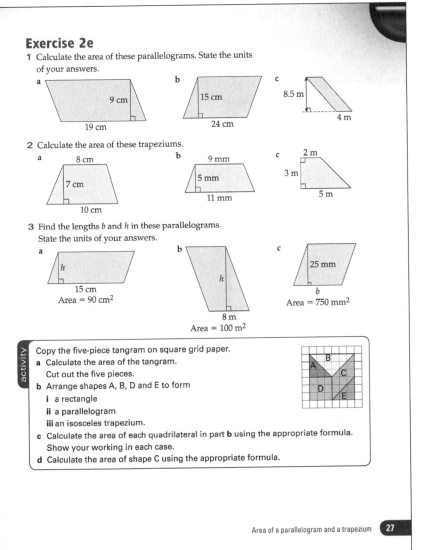

Exercise 2e

1 Calculate the area of these parallelograms. State the units of your answers.

   a   9 cm   19 cm

   b   15 cm   24 cm

   c   8.5 m   4 m

2 Calculate the area of these trapeziums.

   a   8 cm   7 cm   10 cm

   b   9 mm   5 mm   11 mm

   c   2 m   3 m   5 m

3 Find the lengths b and h in these parallelograms. State the units of your answers.

   a h 15 cm   Area = 90 cm²

   b h 8 m   Area = 100 m²

   c   25 mm b Area = 750 mm²

Copy the five-piece tangram on square grid paper.
a Calculate the area of the tangram.
   Cut out the five pieces.
b Arrange shapes A, B, D and E to form
   i a rectangle
   ii a parallelogram
   iii an isosceles trapezium.
c Calculate the area of each quadrilateral in part **b** using the appropriate formula.
   Show your working in each case.
d Calculate the area of shape C using the appropriate formula.

Area of a parallelogram and a trapezium   27

Extension

An accuracy challenge – give students two side lengths for a parallelogram, for example, 8 cm and 5 cm and an area of 12 cm² - can they construct this parallelogram (or draw it accurately on squared paper)? (SM)

2a

1 Copy and complete each sentence, using the most appropriate metric unit of length.

 a A banana weighs 150 ___.

 b I can walk 1 ___ in 15 minutes.

 c The weight of a football is just over 400 ___.

 d A ruler is 30 ___ long.

 e A small flask holds 300 ___ of liquid.

2 Convert these measurements to the units indicated in brackets.

 a 380 mm (cm)   **b** 4.5 kg (g)   **c** 6.5 ℓ (cl)

 d 3500 mm (m)   **e** 2500 kg (t)

2b

3 Convert these measurements to the units indicated in brackets.

 a 8 inches (cm)   **b** 132 lbs (kg)   **c** 40 km (miles)

 d 5 pints (ℓ)   **e** 10 gallons (ℓ)

4 Write down each reading on the scales.

2c

5 Copy and complete the table for the rectangles.

	Length	Width	Perimeter	Area
a	15 cm	10 cm		
b	25 cm	20 cm		
c	9 cm			63 cm^2
d	4.5 cm			18 cm^2
e	5.5 cm		20 cm	
f	7.5 cm		26 cm	
g			28 cm	48 cm^2

6 Calculate the perimeter and area of these shapes made from rectangles.

a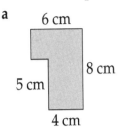

b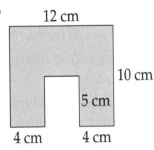

c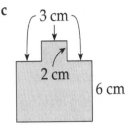

7 Calculate the area of the shaded region.

a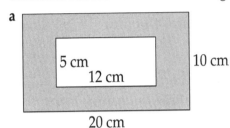

b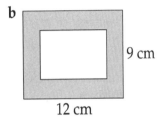

The shaded border is 2 cm wide.

8 On square grid paper, draw
  a two different rectangles, each with an area of 6 cm^2
  b two different right-angled triangles, each with an area of 6 cm^2
  c two different triangles, neither of them right-angled each with an area of 6 cm^2.

9 Calculate the area of these parallelograms.

a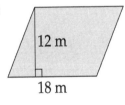

b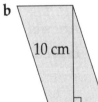

c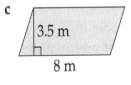

10 Calculate the area of these trapeziums.

a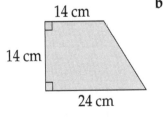

b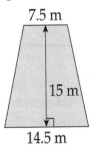

c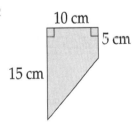

2 Summary

Assessment criteria

- Choose and use appropriate units — Level 4
- Find perimeters of simple shapes and find areas by counting squares — Level 4
- Read and interpret scales on a range of measuring instruments, explaining what each labelled division represents — Level 5
- Solve problems involving the conversion of units and make sensible estimates of a range of measures in relation to everyday situations — Level 5
- Understand and use the formula for the area of a rectangle and distinguish area from perimeter — Level 5
- Deduce and use formulae for the area of a triangle and parallelogram — Level 6

Question commentary

Example	The example asks students to identify a quadrilateral and find its area given the areas of the four constituent parallelograms. Encourage students to think of the area of a parallelogram in terms of the area of a rectangle and to use the terms *base* and *height* to prevent confusion with the length of the sloping sides. Ask probing questions such as "Why do you have to multiply the base by the perpendicular height?" Emphasise that units must be given with the final answer.	**a** parallelogram **b** $4 \times (18 \times 8) =$ $4 \times 144 = 576\,\text{cm}^2$

Past question	The question requires students to convert three lengths into different metric units. Some students will forget that there are 100 cm in a metre but 1000 m in a kilometre and give the same numerical value for each answer. Emphasise the meaning of the prefixes *milli-* (a thousandth), *centi-* (a hundredth) and *kilo-* (One thousand). Encourage visualisation of known lengths such as the divisions on a 30 cm ruler.	**Answer**
		Level 5
		2 a 12 cm
		  b 1.2 m
		  c 0.12 km

Development and links

Shape work is developed further in Chapter 6 to include the properties of triangles, quadrilaterals and other polygons. Students will consider changes in area due to enlargement in Chapter 9 and use measures extensively in work on scale drawings in Chapter 9 and constructions in Chapter 15.

Students will encounter a range of SI units in science, where the skill of reading measuring scales accurately is especially important. Measures are used in all areas of the curriculum, for example in geography for measuring distances and areas of land, in food technology for measuring ingredients and volumes and in design technology for measuring quantities of materials.

Objectives

Level

- Know that if the probability of an event occurring is p then the probability of it not occurring is $1 - p$ **5**

- Compare estimated experimental probabilities with theoretical probabilities **5**

- Recognise that if an experiment is repeated the outcome may, and usually will, be different **5**

- Recognise that increasing the number of times an experiment is repeated generally leads to better estimates of probability **5**

- Use diagrams and tables to record in a systematic way all possible mutually exclusive outcomes for single events and for two successive events **6**

MPA	
1.1	3a, b, c, CS
1.2	3b, d, CS
1.3	3a, c, d, CS
1.4	3c, d, CS
1.5	3a, b, c, d, CS

PLTS	
IE	3a, c, d, CS
CT	3a, b, c, d, CS
RL	3a, b, c, d, CS
TW	3a, b, c, CS
SM	3c, CS
EP	3I a, b, c, d, CS

Introduction

The focus of this chapter is on the difference between theoretical and experimental probability. Students will develop their knowledge of theoretical probability by calculating mutually exclusive probabilities and listing outcomes for two successive events using tree diagrams and sample space diagrams. They will calculate experimental probabilities and compare with theoretical probabilities for the same experiment, considering the reasons for any differences.

The student book discusses product warranties and the probability of a product breaking down during the warranty period. Although a process may be random and unpredictable, probabilities give an idea of the patterns that are likely to occur. Theoretical probabilities are only valid when all primitive outcomes are equally likely. Experiments are used to check that all outcomes are equally likely, or to estimate probabilities when outcomes are not equally likely. Students may have seen demonstrations of durability tests, for example some kitchen manufacturers display a demonstration of testing drawer durability and shoe retailers often have a shoe in water to demonstrate water resistance testing (EP).

Fast-track

All spreads

Extra Resources

3 Start of chapter presentation
3a Worked solution: Q3a
3a Consolidation sheet
3b Worked solution: Q4c
3b Consolidation sheet
3c Starter: Complimentary probability matching
3c Consolidation sheet
3d Consolidation sheet
Maths life: Traditional games and pastimes

Assessment: chapter 3

- Use diagrams and tables to record in a systematic way all possible mutually exclusive outcomes for single events and for two successive events   (L6)

Useful resources

Starter – Ice cream

Ask students how many different combinations of ice cream they could make choosing two different flavours from the following six flavours: vanilla, strawberry, toffee, mint choc, pistachio and banana. (15)

What if they had seven flavours to choose from? (21)

Teaching notes

Ask students, what is the probability of getting a head and a tail when two coins are tossed? Expect answers of $\frac{1}{3}$ and $\frac{1}{2}$ (HT and TH are different and both contribute). Ask students how they reached their answers. Guide them to the need to list all possible outcomes and count the number that contain a head and a tail so that they can apply the formula in spread 3b. Can students suggest ways to systematically list all the outcomes from one or two events? (RL, EP)

For the two coins, show how to construct the sample space diagram (also known as a two-way table) and the tree diagram. In the tree diagram emphasise the need to label the 'First' and 'Second' set of branches and to label individual 'H' and 'T' outcomes at the end of the branches. When many branches are involved it may help to draw the outside branches first and then fill in the middle branches.

Counting arguments work for calculating probabilities with these diagrams if each outcome is equally likely.

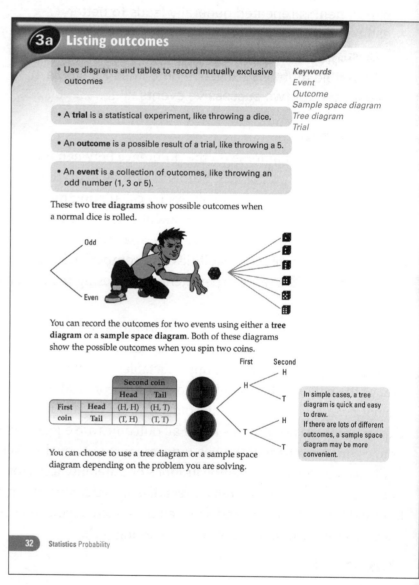

Plenary

Develop question **2**: suppose Danni likes to give her cat meat one day and fish the next – can the students show this in either a tree diagram or a sample space diagram?

Simplification

Students' understanding of what the various diagrams show can be clarified by giving them four labelled examples and asking them to explain what is shown. For example,

Tree diagram: one dice – show more than 4 and 4 or less.
Sample space diagram: results of two football games – show win, draw and loss.
Sample space diagram: morning and afternoon weather – show sunny, cloudy and rainy.
Tree diagram: driving test result – show pass or fail

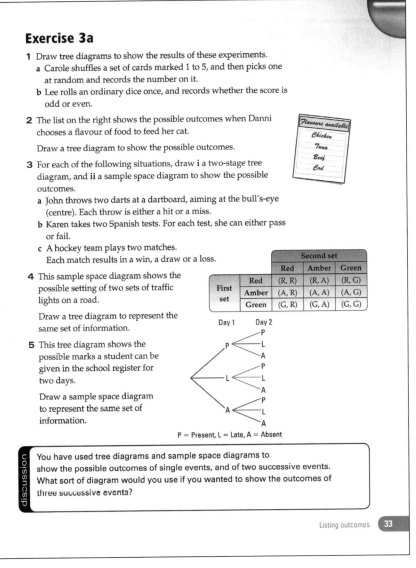

Exercise 3a

1 Draw tree diagrams to show the results of these experiments.
  a Carole shuffles a set of cards marked 1 to 5, and then picks one at random and records the number on it.
  b Lee rolls an ordinary dice once, and records whether the score is odd or even.

2 The list on the right shows the possible outcomes when Danni chooses a flavour of food to feed her cat.

Draw a tree diagram to show the possible outcomes.

Flavours available
Chicken
Tuna
Beef
Cod

3 For each of the following situations, draw **i** a two-stage tree diagram, and **ii** a sample space diagram to show the possible outcomes.
  a John throws two darts at a dartboard, aiming at the bull's-eye (centre). Each throw is either a hit or a miss.
  b Karen takes two Spanish tests. For each test, she can either pass or fail.
  c A hockey team plays two matches. Each match results in a win, a draw or a loss.

4 This sample space diagram shows the possible setting of two sets of traffic lights on a road.

		Second set		
		Red	Amber	Green
First set	Red	(R, R)	(R, A)	(R, G)
	Amber	(A, R)	(A, A)	(A, G)
	Green	(G, R)	(G, A)	(G, G)

Draw a tree diagram to represent the same set of information.

5 This tree diagram shows the possible marks a student can be given in the school register for two days.

Draw a sample space diagram to represent the same set of information.

P = Present, L = Late, A = Absent

> **discussion**
> You have used tree diagrams and sample space diagrams to show the possible outcomes of single events, and of two successive events. What sort of diagram would you use if you wanted to show the outcomes of three successive events?

Listing outcomes **33**

Extension

Students could be encouraged to use simple counting arguments to calculate probabilities based on the list of outcomes at the end of a tree diagram or in a sample space diagram. For example, the probability of obtaining a factor of twelve on the roll of a fair dice or the probability of two heads or one head and one tail when two fair coins are tossed (IE).

Exercise 3a commentary

Question 1 and **2** – Drawing one-stage tree diagrams.

Question 3 – Drawing two-stage tree and sample space diagrams. Check that students correctly label their diagrams (1.1, 1.3).

Question 4 and **5** – Converting between sample space and two-stage tree diagrams (1.3).

Discussion – This could be done in pairs. Students could be asked to draw actual diagrams to show the outcomes of tossing one, two or three coins. A 'cuboidal' sample space drawing might be possible using perspective or colours but is unlikely to be practical (TW, CT) (1.1, 1.5).

Assessment Criteria – HD, level 6: find and record all mutually exclusive outcomes for single events and two successive events in a systematic way.

Links

The traffic light was invented before the motor car. In 1868 a gas-powered lantern was used to control the horse-drawn and pedestrian traffic at a junction outside the Houses of Parliament in London. The lantern had rotating red and green lamps which were turned manually to face the oncoming traffic. The lantern exploded on January 2, 1869 injuring its operator. There is more information about traffic lights at http://www.bbc.co.uk/dna/h2g2/A9559407

- Understand and use the probability scale from 0 to 1 (L5)
- Find and justify probabilities based on equally likely outcomes in simple contexts (L5)
- Know that if the probability of an event occurring is p then the probability of it not occurring is 1 − p (L5)

Useful resources
A large horizontal probability line on display (marked 0.1, 10%, $\frac{1}{10}$ etc.).
Statement cards

Starter – Dice bingo

Ask students to draw a 3 × 3 grid and enter nine numbers from 2 to 12 inclusive, duplicates allowed. Throw two dice. Students add the scores and cross out the total if they have it in their grid (only one number at a time). The winner is the first student to cross out all their numbers

Teaching notes

As an introduction to probability, students could be given statements on large cards to read out and then stand at the appropriate point on the probability line. For example, 'I will get my maths correct', 'I will arrive at school on time tomorrow', 'the head of year will come in the room this lesson', 'there will be some one away in class tomorrow', 'it will be sunny at the end of school', 'there will be a fire alarm in school today', *etc.*

Probability is described in several ways and students need to be aware of this. In words, only 'certain', 'equally likely/evens' and 'impossible/certain not to happen' have definite meanings, words such as 'likely', *etc.* are subjective. Using numbers gives definite meaning but they can be expressed in one of three ways: decimals, percentages or fractions. Students may need reminding how to simplify fractions and recognise common FDP equivalences (RL).

Working through a problem, such as that in the example, will show students how to apply the probability formula and to calculate the probability of a complementary event.

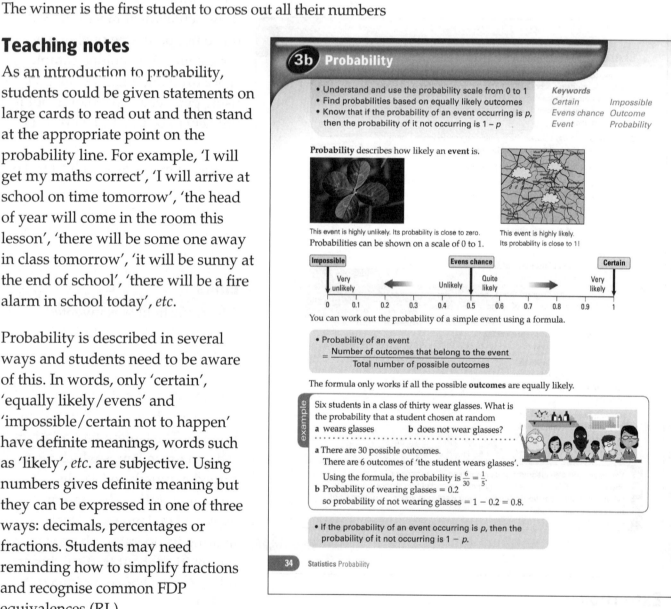

3b Probability

- Understand and use the probability scale from 0 to 1
- Find probabilities based on equally likely outcomes
- Know that if the probability of an event occurring is *p*, then the probability of it not occurring is 1 – *p*

Keywords
Certain   Impossible
Evens chance Outcome
Event   Probability

Probability describes how likely an **event** is.

This event is highly unlikely. Its probability is close to zero.

This event is highly likely. Its probability is close to 1!

Probabilities can be shown on a scale of 0 to 1.

Impossible — Very unlikely — Unlikely — Evens chance — Quite likely — Very likely — Certain

0   0.1   0.2   0.3   0.4   0.5   0.6   0.7   0.8   0.9   1

You can work out the probability of a simple event using a formula.

- Probability of an event
  $= \dfrac{\text{Number of outcomes that belong to the event}}{\text{Total number of possible outcomes}}$

The formula only works if all the possible **outcomes** are equally likely.

example

Six students in a class of thirty wear glasses. What is the probability that a student chosen at random
a wears glasses   b does not wear glasses?

a There are 30 possible outcomes.
  There are 6 outcomes of 'the student wears glasses'.
  Using the formula, the probability is $\frac{6}{30} = \frac{1}{5}$.
b Probability of wearing glasses = 0.2
  so probability of not wearing glasses = 1 − 0.2 = 0.8.

- If the probability of an event occurring is *p*, then the probability of it not occurring is 1 − *p*.

34   Statistics Probability

Plenary

Give pairs of students a small piece of card showing a percentage or fraction written on, for example, $\frac{1}{4}$ or 20%. Ask each pair to write a statement that they feel would fit the probability they have been given. Allow four or five pairs to get together to discuss and agree their statements and place on a probability line (TW).

Simplification

In questions **2** and **5**, use only fractions: $0.3 = \frac{3}{10}$ and $45\% = \frac{9}{20}$.

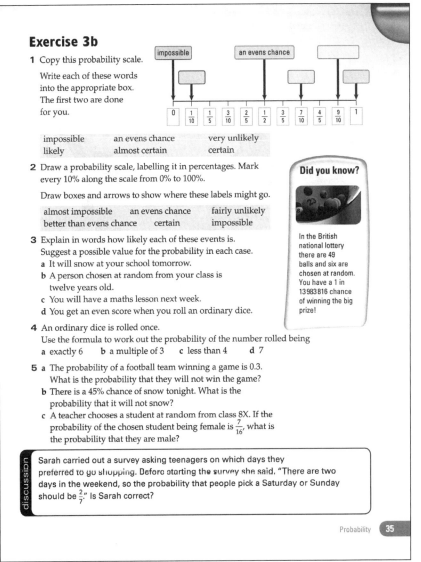

Exercise 3b

1 Copy this probability scale.

Write each of these words into the appropriate box. The first two are done for you.

| impossible | | an evens chance | | | |

Scale: 0, $\frac{1}{10}$, $\frac{1}{5}$, $\frac{3}{10}$, $\frac{2}{5}$, $\frac{1}{2}$, $\frac{3}{5}$, $\frac{7}{10}$, $\frac{4}{5}$, $\frac{9}{10}$, 1

| impossible | an evens chance | very unlikely |
| likely | almost certain | certain |

2 Draw a probability scale, labelling it in percentages. Mark every 10% along the scale from 0% to 100%.

Draw boxes and arrows to show where these labels might go.

| almost impossible | an evens chance | fairly unlikely |
| better than evens chance | certain | impossible |

3 Explain in words how likely each of these events is. Suggest a possible value for the probability in each case.
 a It will snow at your school tomorrow.
 b A person chosen at random from your class is twelve years old.
 c You will have a maths lesson next week.
 d You get an even score when you roll an ordinary dice.

4 An ordinary dice is rolled once. Use the formula to work out the probability of the number rolled being
 a exactly 6   **b** a multiple of 3   **c** less than 4   **d** 7

5 a The probability of a football team winning a game is 0.3. What is the probability that they will not win the game?
 b There is a 45% chance of snow tonight. What is the probability that it will not snow?
 c A teacher chooses a student at random from class 8X. If the probability of the chosen student being female is $\frac{7}{16}$, what is the probability that they are male?

discussion
Sarah carried out a survey asking teenagers on which days they preferred to go shopping. Before starting the survey she said, "There are two days in the weekend, so the probability that people pick a Saturday or Sunday should be $\frac{2}{7}$." Is Sarah correct?

Did you know?

In the British national lottery there are 49 balls and six are chosen at random. You have a 1 in 13 983 816 chance of winning the big prize!

Probability **35**

Extension

Ask students (in pairs) to write a set of probability questions for an octahedral dice (a dice with eight faces numbered from 1 to 8) with each face equally likely to be thrown. Ask pairs to exchange questions, discuss the answers and the difficulty.

Exercise 3b commentary

Questions 1 and **2** – Students could be asked to work in pairs to discuss the most appropriate placement. If the statements are written on pieces of paper it will be easier to rearrange them and help avoid unduly messy answers (EP).

Question 3 – Students could be asked to explain how they reached their answers (EP) (1.2).

Question 4 – Similar to the example; requires students to use fractions, decimals and percentages.

Question 5 – Applications of P(not A) = 1 − P(A).

Discussion – This could be done in small groups. Ask students to give two or three arguments to explain whether $\frac{2}{7}$ is a sensible answer. Can they think of a practical way to measure the probability? (CT) (1.1, 1.5)

Assessment Criteria – HD, level 5: understand and use the probability scale from 0 to 1.

Links

Snow can still fall even when the air temperature is above freezing. As snowflakes fall, they gain heat from the surrounding air by conduction but they also lose heat by evaporation. If the flakes lose more heat than they gain, the flakes will remain frozen. The rate of evaporation is greatest in dry air, so whether the snowflakes melt or not depends on both the temperature of the surrounding air and on the relative humidity. There is a snow probability calculator at http://www.sciencebits.com/SnowProbCalc&calc=yes

- Estimate probabilities by collecting data from a simple experiment (L5)
- Recognise that increasing the number of times an experiment is repeated generally leads to better estimates of probability (L5)

Useful resources
Calculator
Coins/two-sided tokens

Starter – Higher or lower

Using either a set of playing cards or a set of numbered cards, show students the first card and ask whether they think the next card will be higher or lower. Repeat several times.

Can be extended by being more specific, for example, the chance of the next card being a square number. (If using playing cards, the naming and numerical value of face cards may need to be explained.)

Teaching notes

Give groups of students three coins and ask them what do they think is the probability of getting two heads and one tail when they are tossed. Ask them to repeatedly toss the coins and create a tally chart for the possible outcomes (HHH, HHT, TTH, TTT). Show how each group's results can be used to calculate an estimate for P(HHT) (= $\frac{3}{8}$ = 0.375) using the formula. Do they all agree? How could they get a better estimate? Combine all the groups' results to obtain a better estimate (RL).

Probabilities can be given as fractions, decimals or percentages. You should clarify which form you prefer for an answer (decimals/percentages make comparisons easier) and if necessary how to convert between them, possibly using a calculator.

The **discussion** question can be organised as a **think, pair and share** activity. Give students two minutes to **think** about the problem themselves – what information would Tara need? Then give students, in **pairs**, time to agree and list the data they would require. Then put two pairs together to **share** their ideas and agree how best Tara could work out the probability (CT, TW) (1.5).

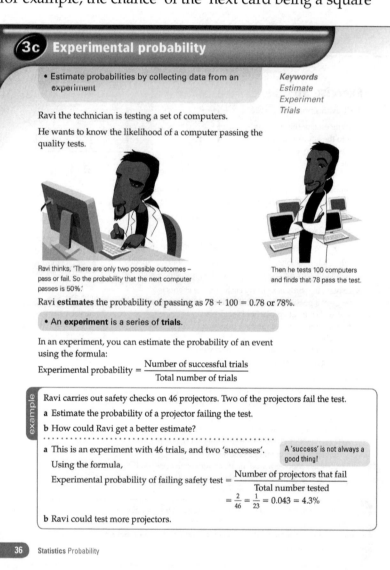

3c Experimental probability

- Estimate probabilities by collecting data from an experiment

Keywords
Estimate
Experiment
Trials

Ravi the technician is testing a set of computers.

He wants to know the likelihood of a computer passing the quality tests.

Ravi thinks, 'There are only two possible outcomes – pass or fail. So the probability that the next computer passes is 50%.'

Then he tests 100 computers and finds that 78 pass the test.

Ravi **estimates** the probability of passing as $78 \div 100 = 0.78$ or 78%.

- An **experiment** is a series of **trials**.

In an experiment, you can estimate the probability of an event using the formula:

$$\text{Experimental probability} = \frac{\text{Number of successful trials}}{\text{Total number of trials}}$$

example

Ravi carries out safety checks on 46 projectors. Two of the projectors fail the test.
a Estimate the probability of a projector failing the test.
b How could Ravi get a better estimate?
....
a This is an experiment with 46 trials, and two 'successes'.

Using the formula,

A 'success' is not always a good thing!

$$\text{Experimental probability of failing safety test} = \frac{\text{Number of projectors that fail}}{\text{Total number tested}}$$
$$= \frac{2}{46} = \frac{1}{23} = 0.043 = 4.3\%$$

b Ravi could test more projectors.

36   Statistics Probability

Plenary

Display the table for question **2** on a whiteboard. On one day the buses are on strike: four students who normally catch a bus don't get to school, three cycle, two walk and the rest come by car. In pairs, ask students to work out the different probabilities for coming to school. Compare these percentages with those when buses were running. Why have they all increased? What is their total?

Simplification

A few students will find changing fractions into percentages a challenge, allow them to leave their answers as fractions. If there is time, work with them (or allow a competent student time) to show them how to change their fraction answers into percentages.

Exercise 3c

1 A scientist checks some trees to see whether they are infected with a disease. She checks 28 trees, and 5 of them are infected. Estimate the probability that a tree chosen at random has the disease.

2 The table shows the results of a survey about Year 8 students' journeys to school.

Transport	Cycle	Walk	Bus	Car
Number	18	29	14	8

Estimate the probability that a student chosen at random from Year 8 would have cycled to school.

3 This table shows the number of points scored by twenty competitors in an athletics competition.

15	14	11	8	11	17	7	16	16	8
18	10	4	17	13	6	15	3	9	12

Use the data in the table to estimate the probability that a competitor chosen at random would have scored more than ten points.

4 Andy and Ben both carry out experiments to estimate the probability of a particular kind of seed germinating. The table shows their results.

	Andy	Ben
Germinated	32	58
Failed	18	42

a Estimate the probability of a seed germinating, using Andy's results.
b Now estimate the probability using Ben's results.
c Whose results do you think should be more reliable – Andy's or Ben's?

discussion
Tara wants to estimate the probability of it snowing in London next Christmas Day. Explain how she could do this.

Experimental probability **37**

Extension

Ask students to construct a three-stage tree diagram for the results of tossing three coins and hence calculate P(HHT) from the introductory example.

Develop question **3**: the athletes receive a gold, silver, or bronze, medal or a certificate depending on how many points they score. What would be suitable boundaries for each of these awards? Note, there should be less gold medals than silver and bronze with most receiving a certificate (IE, SM).

Exercise 3c commentary

Question 1 – Similar to the example. Check how students set out their answer, should it be given as a percentage?

Question 2 – Requires students to add up the total number of pupil journeys.

Question 3 – Requires students to identify and count 'successful' outcomes from the list.

Question 4 – Requires students to extract the appropriate information from a two-way table and appreciate the significance of the total number of trials. Ask the class if knowing the conditions used in the two experiments would affect their answers (CT) (1.3, 1.4).

Discussion – Will it snow on Christmas day in London next year? Students are asked to consider what information they need to be able to work out this as a probability (EP) (1.1, 1.5).

Assessment Criteria – HD, level 5: understand that different outcomes may result from repeating an experiment.

Links

Dutch Elm disease has affected millions of trees in Britain since the late 1960s. It is caused by a fungus which is spread by the elm bark beetle. By the late 1990s, over 25 million of the UK's 30 million elm trees had died due to the disease. What fraction of the trees died?

- Compare estimated experimental probabilities with theoretical probabilities (L5)
- Recognise that if an experiment is repeated the outcome may, and usually will, be different (L5)

Useful resources
Calculator
A4 envelope
Small coloured cards
  (red, blue, green, yellow)
Stiff cardboard pentagons

Starter – Probability jumble

Write a list of anagrams on the board and ask students to unscramble them.
Possible anagrams are
ANCHEC, COMETOU, TEENV, LIRAT, NAMDOR, SLIPSOMBIE, KELLIY, TRAINCE
(chance, outcome, event, trial, random, impossible, likely, certain)
Can be extended by asking students to make a probability word search.

Teaching notes

As students enter the lesson ask them, without looking, to pick one of four coloured cards from an envelope, remember the colour, then replace it. Record the results on a tally chart (use student helpers if necessary) and use them to calculate the probabilities of obtaining each colour. This set of data is analogous to that discussed in the first example; can students explain the similarities and any differences (is the spinner biased?).

Ensure that students appreciate that results obtained by repeating exactly the same experiment may differ from one another and from the theoretical probability. It is also important that they can clearly explain why, for example, an experiment may be biased.

Question **1** can be used to make sure that students explain their thinking in well-constructed statements. Allow a few minutes to complete the question then ask students to exchange their work with their neighbour – who should write a positive comment and a developmental comment (how the work could be better explained) in pencil on the book. Allow paired discussion after this with a small time for feedback (1.5).

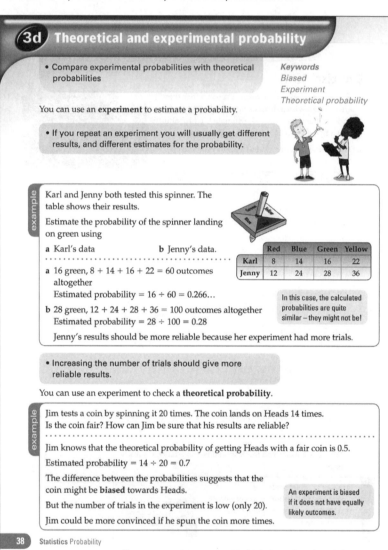

Plenary

In an envelope have 2 red, 2 blue and 1 green card. Split the class into teams and explain that the envelope contains five cards. Draw out a card, show it and replace it. After 5, 7, 9, 11, 13 and 15 repeats, ask the teams to guess the colours of the five cards. Is any team's final guess correct? What strategies were used for 'best' guesses at each stage? Why is after five cards not a good time to have a guess? (CT) (1.2)

The exercise can be repeated, say, using 1 red, 3 blue and 1 green or 2 red and 3 green cards.

Simplification

Question **3** can be made into a practical activity using an envelope with coloured cards (4 red, 3 green, 2 blue, 1 yellow). Discuss with students what they may expect and then allow them to complete the experiment (if there are two groups then the data can be put together). Discuss the outcomes.

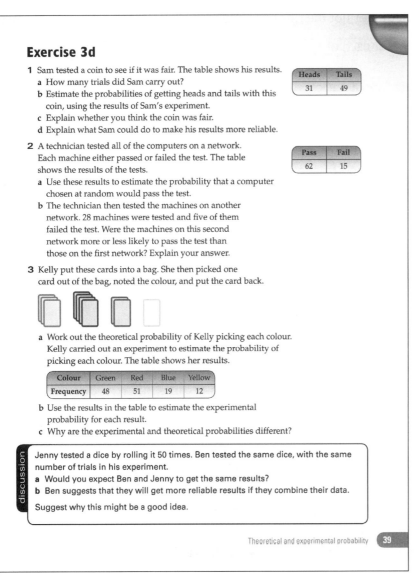

Exercise 3d

1 Sam tested a coin to see if it was fair. The table shows his results.

Heads	Tails
31	49

   a How many trials did Sam carry out?
   b Estimate the probabilities of getting heads and tails with this coin, using the results of Sam's experiment.
   c Explain whether you think the coin was fair.
   d Explain what Sam could do to make his results more reliable.

2 A technician tested all of the computers on a network. Each machine either passed or failed the test. The table shows the results of the tests.

Pass	Fail
62	15

   a Use these results to estimate the probability that a computer chosen at random would pass the test.
   b The technician then tested the machines on another network. 28 machines were tested and five of them failed the test. Were the machines on this second network more or less likely to pass the test than those on the first network? Explain your answer.

3 Kelly put these cards into a bag. She then picked one card out of the bag, noted the colour, and put the card back.

   a Work out the theoretical probability of Kelly picking each colour. Kelly carried out an experiment to estimate the probability of picking each colour. The table shows her results.

Colour	Green	Red	Blue	Yellow
Frequency	48	51	19	12

   b Use the results in the table to estimate the experimental probability for each result.
   c Why are the experimental and theoretical probabilities different?

> **discussion**
>
> Jenny tested a dice by rolling it 50 times. Ben tested the same dice, with the same number of trials in his experiment.
> **a** Would you expect Ben and Jenny to get the same results?
> **b** Ben suggests that they will get more reliable results if they combine their data.
>
> Suggest why this might be a good idea.

Theoretical and experimental probability **39**

Extension

Make a biased spinner, for example, by cutting across two neighbouring sections of a five sided spinner to make into one. After working out the theoretical probabilities, estimate how many of each score should be obtained in 40 spins. Do 40 spins and work out the experimental probabilities. Discuss how the answers compare (IE).

Exercise 3d commentary

Question 1 – A structured question similar to the second example, that encourages students to assess the reliability of their results (RL). (1.3)

Question 2 – Requires students to think about an estimate's reliability/accuracy.

Question 3 – Requires students to realise that experimental results naturally vary (1.4).

Discussion – Suggest working in pairs. Ask pairs to agree and list three reasons why it would be a good idea to combine results. These reasons can be shared with another pair and the best idea fed back to the whole class (EP) (1.5).

Assessment Criteria – HD, level 5: in probability, select methods based on equally likely outcomes and experimental evidence, as appropriate.

Links

Dice that are deliberately biased are called crooked or loaded dice. Dice can be loaded by adding a small amount of metal to one side or by manufacturing the dice with a hollow gap inside so that one side is lighter than the others. One way of testing for a loaded die is to drop it several times into a glass of water. If it is hollow it will float with the hollow side uppermost; if it is weighted, it will sink with the same number always facing down.

3 Consolidation

1 Draw a tree diagram for each of these situations.

  a An ordinary dice is rolled, and the score is noted; then a coin is spun, and the result is recorded.

  b An experimenter rolls an ordinary dice, and notes whether the score is odd or even; then a coin is spun, and the result is recorded.

  c A coin is spun, and the result is written down; then the experimenter rolls a dice and records whether or not the result is a multiple of 3.

2 Draw a sample space diagram for each of these situations.

  a A player turns up two cards from a pack of playing cards, and notes the suits.

Hearts   Diamonds   Clubs   Spades

  b A player picks two letters from a bag of letter tiles. Each tile has a vowel or a consonant.

  c A shopper picks two cans of cat food from a shelf. The flavours available are chicken, beef and fish.

3 A computer is used to choose a random whole number between 1 and 100 (inclusive).

Find the probability that the chosen number is

  a exactly 13

  b even

  c a multiple of 10

  d a multiple of 7

4 Here are the scores obtained by a sample of people who carried out a safety test.

| 15 | 17 | 13 | 18 | 19 | 20 | 19 | 13 | 14 | 19 |
| 20 | 11 | 12 | 16 | 18 | 17 | 20 | 14 | 18 | 13 |

  a A score of 15 or more is needed to pass the test. Estimate the experimental probability of passing the test.

  b If 500 people took the test, how many would you expect to pass? Explain your answer.

3d

5 Max put 25 blue counters, 50 red counters and 25 yellow counters into a bag.
He shook the bag, picked a counter without looking, recorded the colour and returned the counter to the bag. He did this 200 times altogether.

  a Calculate the theoretical probability of choosing each colour.
  b Max actually obtained 61 blues, 108 reds and 31 yellows. Estimate the experimental probability of obtaining each colour.
  c Max thought that the theoretical and experimental probabilities were different. Give an example of a factor that could have caused this difference.

- Try out and compare mathematical representations   (L5)
- Evaluate the efficiency of alternative strategies and approaches   (L5)

Useful resources
Dice
Counters

Background

Games differ in the extent to which they depend on strategy/skill and luck. Ludo and snakes and ladders are games of pure chance - they rely on the roll of a dice to make them play differently. Noughts and crosses and chess are games of pure skill. Some games combine both skill and chance.

This case study takes two traditional games that involve both chance and strategies that can tip the balance of play. 'Scissors, paper, stone' is likely to be familiar whilst 'Shut the Box' might be less well known (it is sometimes played as a child's game to help with addition and number bonds). It uses dice in a way that complicates the underlying probabilities (EP).

Teaching notes

Ask students to think about games and the role that chance and skill play in them. How many examples can they give of games of pure skill, pure chance or a combination of the two?

Focus on 'Scissors, Paper, Stone' from the student book: ensure that all students know how to play and let them try a few turns with a partner.

Look at the questions on the pale blue notepad. Ask the students to note all the pairs of moves that can be made in the game. When they have had a few minutes, check that they have found all nine possible pairs. Did anyone have a system for finding all the pairs? (IE) (1.1)

Discuss possible ways of making sure that all the pairs have been found, such as fixing one player on scissors

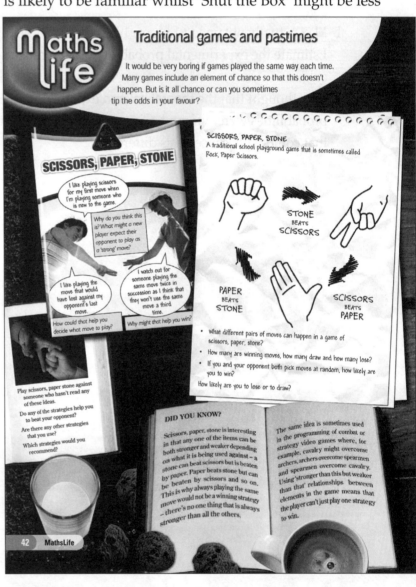

while the second player plays all three possible moves, then moving player one on to paper and repeating the procedure and finally moving player one on to stone.

Move on to consider winning, losing and drawing moves and how they all have a probability 1/3. Discuss how this means that random play would be likely to result in an overall draw.

Ask students to describe how they decide what to play next, is it random or do they have strategies? Discuss how making random moves when you are choosing what to play next is actually quite difficult. For example, most people playing in a way that they meant to be random would be very unlikely to play stone, stone, stone, stone. However, if they used a dice and allocated 1 and 6 to scissors, 3 and 4 to paper and 2 and 5 to stone, they might not be quite so surprised to get stone, stone, stone, stone.

Teaching notes continued

Look at the strategies on the clipboard; give students a few minutes to talk about them with a partner and to note down why they might help you beat an opponent. Discuss as a class how well they think the strategies might work and why. Do they have other strategies? (EP) (1.5)

Move on to consider 'Shut the Box': it might be a good idea to play a game or two to familiarise students with the rules. To encourage debate organise students to work in pairs (TW). Ask them to work through the questions on the yellow notepad. Next ask them to systematically list all the ways of obtaining the scores from 1 to 12 with either one or two dice. For example, to obtain 4 with one die roll 4, with two dice roll (1, 3) (2, 2) or (3, 1). As a class use this table to establish probabilities and revisit the questions. Does knowing the probabilities affect how you might play the game? (RL).

Students now know the probabilities of getting particular scores. Challenge them to list the possible numbers they can cover given a particular score from 1 to 12. For example, a score of 6 allows you to cover (6), (5, 1) (4, 2) or (3, 2, 1); there are 59 possibilities in total (SM) (1.3). Use the results to answer the following questions (CT) (1.2):

How many combinations include a nine (5) or a one (25)?

Is it be harder to cover a nine or a one on your first turn?

Which numbers would it be best to cover first? Why?

Do you have any strategies for how to play?

After discussing the answers encourage the students to play a few more games of 'Shut the Box' to test their strategies. Suggest that they also reverse their strategy to test if it really does make a difference. For example, rather than play 'unlikely' numbers (higher numbers) first switch to playing likely numbers (lower numbers) first (1.4).

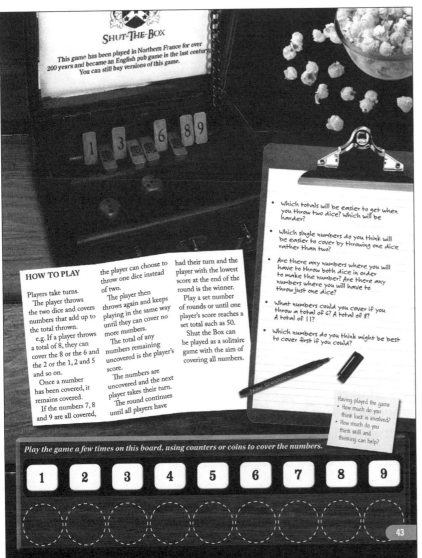

HOW TO PLAY

Players take turns. The player throws the two dice and covers numbers that add up to the total thrown.

e.g. If a player throws a total of 8, they can cover the 8 or the 6 and the 2 or the 1, 2 and 5 and so on.

Once a number has been covered, it remains covered.

If the numbers 7, 8 and 9 are all covered, the player can choose to throw one dice instead of two.

The player then throws again and keeps playing in the same way until they can cover no more numbers.

The total of any numbers remaining uncovered is the player's score.

The numbers are uncovered and the next player takes their turn. The round continues until all players have had their turn and the player with the lowest score at the end of the round is the winner.

Play a set number of rounds or until one player's score reaches a set total such as 50.

Shut the Box can be played as a solitaire game with the aim of covering all numbers.

- Which totals will be easier to get when you throw two dice? Which will be harder?
- Which single numbers do you think will be easier to cover by throwing one dice rather than two?
- Are there any numbers where you will have to throw both dice in order to make the number? Are there any numbers where you will have to throw just one dice?
- What numbers could you cover if you throw a total of 6? A total of 8? A total of 11?
- Which numbers do you think might be best to cover first if you could?

Having played the game
- How much do you think luck is involved?
- How much do you think skill and thinking can help?

Play the game a few times on this board, using counters or coins to cover the numbers.

| 1 | 2 | 3 | 4 | 5 | 6 | 7 | 8 | 9 |

43

Extension

The actual probabilities involved when playing 'Shut the Box' are quite difficult to calculate as it is an evolving situation. For example, if on the first go, you cover the 9 then you would cut out just 5 of the 59 possible plays, but if you played the 1 you would cut out 25. So the probabilities of subsequent turns depend on choices that precede them. Students could play a simplified version of 'Shut the Box' that had the numbers 2 to 12 inclusive, always used two dice and initially only let you cover one number on any turn. That would play according to the normal distribution of the outcomes of two dice. Then they could maybe play using the difference of two dice. What numbers would they need on the 'box'? Which ones should they cover first?

3 Summary

Assessment criteria

- In probability, select methods based on equally likely outcomes and experimental evidence, as appropriate   Level 5

- Understand and use the probability scale from 0 to 1.   Level 5

- Understand that different outcomes may result from repeating an experiment   Level 5

- Find and record all mutually exclusive outcomes for single events and two successive events in a systematic way.   Level 6

Question commentary

Example	
The example asks students to extract information and determine probability from a table. The answer to part **b** can be given as a fraction or a decimal. This offers the opportunity to revise the equivalence of fractions and decimals. Part **c** requires students to identify that using more trials is likely to give a more reliable result.	**a** $100 - 60 = 40$ **b** $\frac{30}{50} = \frac{3}{5}$ or 0.6 **c** larger sample, $100 > 50$
Past question	
The question is an example of a typical question about probability. In part **a**, students are required to find a simple probability. Part **b** is more difficult. Encourage the student to break the problem down into smaller steps. Ask questions such as "What is the probability that the counter will be white" and "How many white counters are in the bag?" Some students may give the answer as 4, the total number of black counters in the bag. Encourage thorough reading of the question and emphasise the word *more* in the final sentence.	**Answer** **Level 5** **2 a** $\frac{1}{3}$ 　**b** 3 more

Development and links

The topic of probability is developed further in Year 9.

Students will compare experimental results with theoretical results in science. They will use probability and sample space diagrams when studying inherited characteristics such as eye colour in biology. Probability is important in PSHE where students learn to make risk/reward assessments for events in their own lives. All forms of gambling are based on probability including buying a Lottery ticket and speculating on the Stock Exchange.

4 Fractions, decimals and percentages

Number

Objectives

Introduction

In this chapter, students will consolidate and extend skills in working with decimals, fractions and percentages. They will use knowledge of lowest common multiples and highest common factors gained in chapter 1 to find equivalent fractions and add and subtract fractions with different denominators. They will order decimals and fractions, find fractions and percentages of a quantity using mental, written and calculator methods and convert between fractions, decimals and percentages.

The student book discusses price reductions in shops expressed as percentages. Understanding and being able to work with fractions, decimals and percentages is vital as students encounter them on a daily basis: in shops, in recipes, in measures, when dealing with money and when dividing pizza!

Fast-track

4c, d

Level

MPA

1.1	4b, c, d, f
1.2	4b
1.3	4a, b, c, d, e, f
1.4	4b, c, e
1.5	4a, b

PLTS

IE	4a, d, e
CT	4a, d, e, f
RL	4a, b, c, d, e, f
TW	4a, b, e
SM	4b, f
EP	4d, e, f

Extra Resources

4	Start of chapter presentation
4a	Consolidation sheet
4b	Animation: Fractions and decimals
4b	Consolidation sheet
4c	Starter Factors and multiples T or F
4c	Animation: Adding fractions
4c	Worked solution: Q4b
4c	Consolidation sheet
4d	Consolidation sheet
4e	Starter: Fraction-decimal matching
4e	Worked solution: Q3h
4e	Consolidation sheet
4f	Animation: Fractions, decimals and percentages
4f	Worked solution: Q4c
4e	Consolidation sheet
	Assessment: chapter 4

- Understand and use decimal notation and place value (L4)   *Useful resources*
- Order decimals   (L4)

• •

Starter – Decimal grid

Write 9 decimal numbers in a 3 × 3 grid. Ask students to find

the sum of the top row

the product of the top left number and bottom right number

the difference between the middle left and middle right numbers *etc.*

Can be differentiated by the number of decimal places in the chosen numbers.

Teaching notes

Use an initial classroom discussion to assess students' prior knowledge of place value in whole numbers. Then invite students to explain how to write 45 hundredths as a decimal. Emphasise that $\frac{45}{100} = \frac{40}{100} + \frac{5}{100} = \frac{4}{10} + \frac{5}{100}$ and that the 4 goes in the tenths column and the 5 in the hundredths column: the diagram supplied should help to make this clear.

The comparison of decimals causes difficulty through KS3 and into KS4, so it is important that the basic understanding is secure. Using a diagram with columns labelled tenths (t), hundredths (h), thousandths (th) *etc.* will help students correctly order decimals. An *aide-memoire* for the inequality signs is 'the narrower end always points to the smaller number'. Ensure students have the opportunity to discuss the common misconception that a number with more digits is bigger, or a larger digit is always worth more than a smaller digit ($0.10 > 0.09$) linking this to the use of columns.

The second example and **group work** activity, involving class intervals, can both be done as whole class activities and provide the opportunity to clear up any misconceptions and consolidate understanding in a real life context.

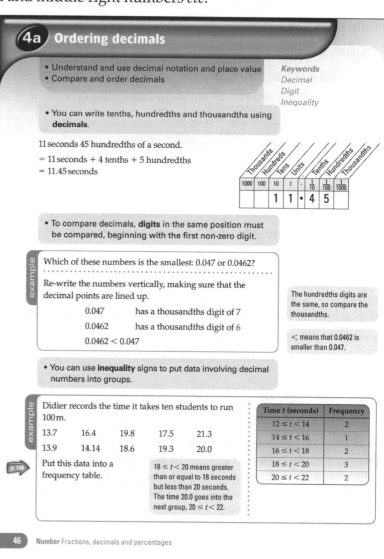

Plenary

Invite students to challenge each other by asking them to generate a set of decimals to be either ordered or grouped (according to their ability) and to swap with a friend and check that they agree or debate where they disagree. This activity offers an opportunity for students to support their conclusions, using reasoned arguments (IE) (1.5).

Simplification

Some students will be challenged with the ordering task and may well need additional practise to consolidate place value in whole numbers – it will be helpful to use more examples such as asking how eleven units should be written.

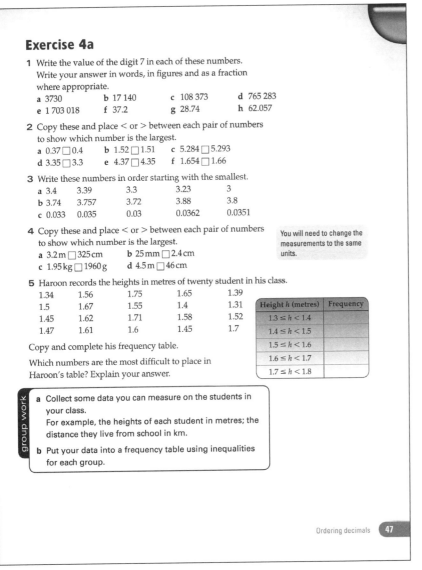

Exercise 4a

1 Write the value of the digit 7 in each of these numbers.
  Write your answer in words, in figures and as a fraction where appropriate.
  a 3730   b 17 140   c 108 373   d 765 283
  e 1 703 018   f 37.2   g 28.74   h 62.057

2 Copy these and place < or > between each pair of numbers to show which number is the largest.
  a 0.37 ☐ 0.4   b 1.52 ☐ 1.51   c 5.284 ☐ 5.293
  d 3.35 ☐ 3.3   e 4.37 ☐ 4.35   f 1.654 ☐ 1.66

3 Write these numbers in order starting with the smallest.
  a 3.4   3.39   3.3   3.23   3
  b 3.74   3.757   3.72   3.88   3.8
  c 0.033   0.035   0.03   0.0362   0.0351

4 Copy these and place < or > between each pair of numbers to show which number is the largest.

  You will need to change the measurements to the same units.

  a 3.2 m ☐ 325 cm   b 25 mm ☐ 2.4 cm
  c 1.95 kg ☐ 1960 g   d 4.5 m ☐ 46 cm

5 Haroon records the heights in metres of twenty student in his class.

  1.34   1.56   1.75   1.65   1.39
  1.5   1.67   1.55   1.4   1.31
  1.45   1.62   1.71   1.58   1.52
  1.47   1.61   1.6   1.45   1.7

Height h (metres)	Frequency
$1.3 \leq h < 1.4$	
$1.4 \leq h < 1.5$	
$1.5 \leq h < 1.6$	
$1.6 \leq h < 1.7$	
$1.7 \leq h < 1.8$	

  Copy and complete his frequency table.

  Which numbers are the most difficult to place in Haroon's table? Explain your answer.

  group work

  a Collect some data you can measure on the students in your class.
    For example, the heights of each student in metres; the distance they live from school in km.

  b Put your data into a frequency table using inequalities for each group.

Ordering decimals   **47**

Extension

Encourage more able students to spend additional time on grouping data that includes decimal values – perhaps making links to how this may be used in other areas of the curriculum, particularly in science experiments.

Students could also be challenged to find a number halfway between two decimals: 6.7 and 6.8, 2.18 and 2.19, 3.7 and 3.76.

Exercise 4a commentary

The exercise allows opportunities for paired discussion to encourage the development of mathematical language and to open up debate where there are misconceptions.

Question 1 – Could be done as a whole class oral activity to promote dialogue and debate. The meaning of 'as a fraction where appropriate' may need clarifying.

Question 2 – Similar to example one. Students could work in pairs and be asked to explain how they know which answer is largest (1.3).

Question 3 – Suggest students check with and justify answers to a partner (RL).

Question 4 – Ask students to explain why it is important to use consistent units (1.3).

Question 5 – Similar to example two. Ask students to explain why some numbers are harder to place than others.

Group work – Emphasise the link between students' maths and real life scenarios. Students could be encouraged to suggest their own ideas for data collection (CT, TW).

Assessment Criteria – NNS, level 4: order decimals to three decimal places. UAM, level 5: draw simple conclusions of their own and give an explanation of their reasoning.

Links

Bring in some sheet music for the class to use. The inequality signs are similar to the musical symbols *crescendo* (becoming louder) and *decrescendo* or *diminuendo* (becoming softer). Ask the class to find examples of these symbols on the music.

- Convert terminating decimals to fractions (L5) **Useful resources**
- Use division to convert a fraction to a decimal (L5)
- Order fractions by converting them to decimals (L5)

Starter – Sums and products

Challenge students to find two numbers

with a sum of 8.9 and a product of 7.2 etc (0.9, 8)
with a sum of 1.1 and a product of 0.24 (0.3, 0.8)
with a sum of 0.75 and a product of 0.035 (0.05, 0.7)
with a sum of 1.3 and a product of 0.42 (0.6, 0.7).

Teaching notes

Prepare students for this work by making links to their prior learning with a brief recapitulation of the associated vocabulary. Ensure students know and understand terms such as numerator, denominator and discuss terminating and recurring decimals.

In the first example, attention should be drawn to the need to cancel down fractions to their simplest form. Students could be invited to generate examples (for which they have an answer) that can be set as challenges for the class to simplify (CT). When converting from fractions to decimals, students should be encouraged to discuss whether mental, written or calculator methods are most efficient (Key Process for number). They should be discouraged from saying 'calculator' as a default response.

The second example covers an important skill used throughout KS3 and KS4 and is a Key Indicator for level 6. A careful approach should be adopted as it brings together a number of skills. Rather than converting to decimals, an alternative approach is to rewrite the fractions with a common denominator.

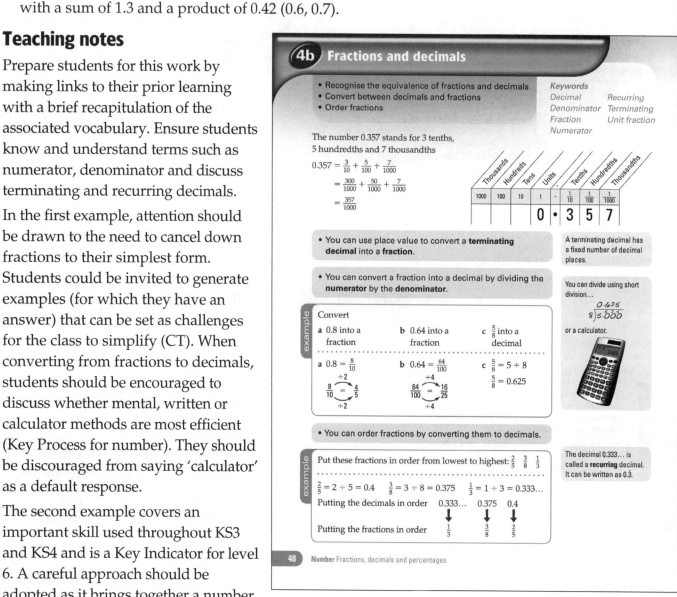

Plenary

Provide an example for ordering that includes both fractions and decimals. Ask students to check the reasonableness of their answer by sketching a diagram of the number line.

Simplification

Provide simple examples to support understanding of how to cancel down fractions and how to carry out short division, as used in converting fractions to decimals.

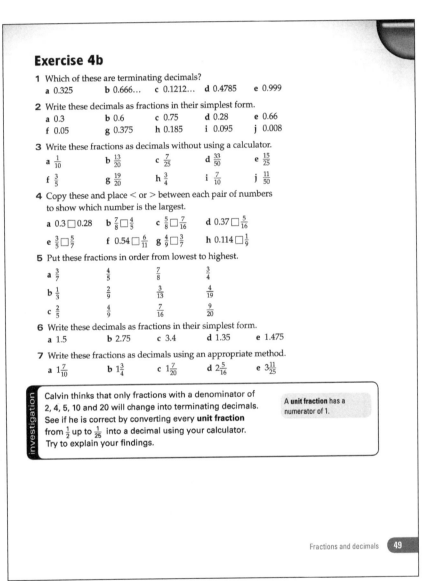

Exercise 4b

1 Which of these are terminating decimals?
  a 0.325 b 0.666... c 0.1212... d 0.4785 e 0.999

2 Write these decimals as fractions in their simplest form.
  a 0.3 b 0.6 c 0.75 d 0.28 e 0.66
  f 0.05 g 0.375 h 0.185 i 0.095 j 0.008

3 Write these fractions as decimals without using a calculator.
  a $\frac{1}{10}$ b $\frac{13}{20}$ c $\frac{7}{25}$ d $\frac{33}{50}$ e $\frac{15}{25}$
  f $\frac{3}{5}$ g $\frac{19}{20}$ h $\frac{3}{4}$ i $\frac{7}{10}$ j $\frac{11}{50}$

4 Copy these and place < or > between each pair of numbers to show which number is the largest.
  a 0.3 □ 0.28 b $\frac{7}{8}$ □ $\frac{4}{5}$ c $\frac{5}{8}$ □ $\frac{7}{16}$ d 0.37 □ $\frac{5}{16}$
  e $\frac{3}{5}$ □ $\frac{5}{7}$ f 0.54 □ $\frac{6}{11}$ g $\frac{4}{9}$ □ $\frac{3}{7}$ h 0.114 □ $\frac{1}{9}$

5 Put these fractions in order from lowest to highest.
  a $\frac{3}{7}$ $\frac{4}{5}$ $\frac{7}{8}$ $\frac{3}{4}$
  b $\frac{1}{3}$ $\frac{2}{9}$ $\frac{3}{13}$ $\frac{4}{19}$
  c $\frac{2}{5}$ $\frac{4}{9}$ $\frac{7}{16}$ $\frac{9}{20}$

6 Write these decimals as fractions in their simplest form.
  a 1.5 b 2.75 c 3.4 d 1.35 e 1.475

7 Write these fractions as decimals using an appropriate method.
  a $1\frac{7}{10}$ b $1\frac{3}{4}$ c $1\frac{7}{20}$ d $2\frac{5}{16}$ e $3\frac{11}{25}$

investigation

Calvin thinks that only fractions with a denominator of 2, 4, 5, 10 and 20 will change into terminating decimals. See if he is correct by converting every **unit fraction** from $\frac{1}{2}$ up to $\frac{1}{25}$ into a decimal using your calculator. Try to explain your findings.

A **unit fraction** has a numerator of 1.

Fractions and decimals 49

Extension

Give examples that include mixed numbers or that include the equivalence of percentages for comparison and ordering.

Exercise 4b commentary

Question 1 – The meaning of terminating and recurring may need to be reiterated.

Question 2 – Similar to the first example. The second step, simplifying the fraction, may need additional support (1.3).

Question 3 – Similar to the first example. Consider getting students to work in pairs to support one another and explain their working, perhaps working on alternate parts (RL) (1.3, 1.5).

Question 4 – A question to focus on as it requires an important skill and combines several ideas from this spread and spread **4a** (1.3).

Question 5 – Further consolidation.

Question 6 and 7 – How to handle mixed numbers may need explanation/discussion (1.3).

Investigation – This is naturally done in pairs or small groups. Emphasise the need for students to clearly explain their findings (TW, SM, RL) (1.2, 1.2, 1.4).

Assessment Criteria – NNS, level 5: use equivalence between fractions and order fractions and decimals.

Links

The Ancient Egyptians represented all their fractions as Egyptian fractions. An Egyptian fraction is the sum of a number of unit fractions, where all the unit fractions are different. All positive rational numbers can be represented by Egyptian fractions. More information about Egyptian fractions, including a calculator to convert a fraction to an Egyptian fraction can be found at http://www.mcs.surrey.ac.uk/Personal/R.Knott/Fractions/egyptian.html (TW).

- Identify equivalent fractions (L5)
- Add and subtract fractions by writing them with a common denominator (L6)

Useful resources
Squared paper

Starter – Dice fractions

Ask students to draw six boxes representing the numerators and denominators of three fractions side by side.

Throw a dice six times. After each throw, ask students to place the score in one of their boxes.

Students score points if the first fraction is bigger than the second fraction which in turn is bigger than the third fraction.

Teaching notes

A precursor to being able to add/subtract fractions is the ability to find equivalent fractions and to identify a suitable (lowest) common denominator. Taking time to make sure students can multiply up / cancel down fractions and find lowest common denominators will allow them to focus on the essentials of adding two fractions.

To help understand the mechanics of adding fractions the visual approach used in the first example and question **3** is likely to be very valuable. Students should be encouraged to suggest their own examples using diagrams.

A common mistake is to add both numerator and denominator. Tackle this misconception with a simple example: clearly $\frac{1}{2} + \frac{1}{2} = 1$ but the 'wrong' method would give $\frac{1+1}{2+2} = \frac{1}{2}$ which leads to the absurdity $1 = \frac{1}{2}$. Another more careless mistake is to forget whether the calculation calls for an addition or a subtraction (RL).

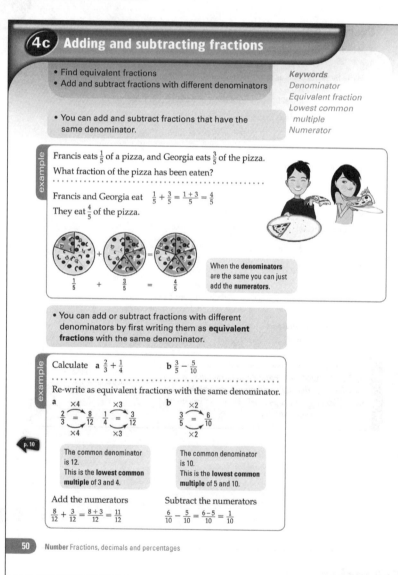

Plenary

Provide a word problem similar to the last question in the exercise and ask students to identify the key information and the mathematics required to answer it. The strengthening of literacy skills in this way is important for developing students' understanding and capacity to be functional in mathematics.

Simplification

Provide examples showing how to find lowest common multiples and develop this into the addition of simple fractions requiring a common denominator. Encourage the use of diagrams such as in question **3** to support understanding of the method being used to add and subtract fractions.

Exercise 4c

1 Find the missing number in each of these pairs of equivalent fractions.

a $\frac{1}{4} = \frac{\square}{12}$ b $\frac{2}{5} = \frac{\square}{25}$ c $\frac{3}{7} = \frac{\square}{28}$ d $\frac{4}{9} = \frac{\square}{63}$

e $\frac{5}{8} = \frac{45}{\square}$ f $\frac{6}{11} = \frac{\square}{88}$ g $\frac{7}{12} = \frac{\square}{36}$ h $\frac{8}{15} = \frac{\square}{150}$

2 Calculate each of these, giving your answer as a fraction in its simplest form.

a $\frac{2}{7} + \frac{3}{7}$ b $\frac{1}{8} + \frac{5}{8}$ c $\frac{4}{5} - \frac{1}{5}$ d $\frac{7}{8} - \frac{5}{8}$

e $\frac{3}{11} + \frac{5}{11}$ f $\frac{9}{13} - \frac{6}{13}$ g $\frac{5}{3} - \frac{1}{3}$ h $\frac{8}{5} - \frac{4}{5}$

3 Copy the grids and use them to show how to add each of these pairs of fractions.

a $\frac{1}{2} + \frac{1}{5}$ ▦ + ▦ = ▦

b $\frac{2}{3} + \frac{1}{4}$ ▦ + ▦ = ▦

c $\frac{2}{5} + \frac{1}{3}$ ▦ + ▦ = ▦

4 Calculate each of these additions and subtractions, giving your answer as a fraction in its simplest form.

a $\frac{1}{3} + \frac{1}{4}$ b $\frac{2}{3} + \frac{1}{5}$ c $\frac{1}{6} + \frac{1}{5}$ d $\frac{2}{5} + \frac{1}{3}$

e $\frac{5}{8} + \frac{1}{3}$ f $\frac{3}{10} + \frac{1}{3}$ g $\frac{8}{9} - \frac{3}{5}$ h $\frac{9}{11} - \frac{2}{3}$

5 Kyle owns lots of computer games. Exactly $\frac{2}{5}$ of his games are sports games and $\frac{1}{4}$ of his games are action games. The rest of his games are adventure games. What fraction of Kyle's computer games are adventure games?

Jameela and Ursula are working out $\frac{2}{9} + \frac{4}{7}$

Ursula says   'The answer is $\frac{6}{16}$.'

Jameela says   'The answer must be more than a half.'

a Explain what Ursula has done wrong.
b Explain how Jameela knows the answer is more than a half.
c Work out the correct answer.

Extension

Set questions involving addition and subtraction of mixed numbers or three fractions.

Exercise 4c commentary

Question 1 – Links to work on cancelling down. Ask students to 'Explain how you did this' (1.3)

Question 2 – Similar to the first example. Check students do not add, rather than subtract, in all cases.

Question 3 – This approach provides both understanding and a good 'fall back' method.

Question 4 – Similar to the second example. Encourage students to discus what is the best denominator to use (RL) (1.3).

Question 5 – An example of more functional maths. Literacy skills, rather than mathematical competency, may be an issue in this word problem. Suggest students work in pairs in order to help identify key information and the maths to use (1.1).

Challenge – Focuses on a common misconception. Students responses to this (or similar) questions could be used as the basis for a poster display.

Assessment Criteria – UAM, level 5: show understanding of situations by describing them mathematically using symbols words and diagrams. Calculating, level 6: add and subtract fractions by writing them with a common denominator.

Links

In chemistry, a fraction is a mixture of liquids with similar boiling points. The fractions in crude oil have individual names, (for example, diesel, kerosene, petrol) and have different properties and uses. They are separated using a fractionating column. There is more information at http://www.bbc.co.uk/schools/gcsebitesize/science/edexcel/oneearth/fuelsrev2.shtml

- Calculate fractions of quantities (fraction answers)   (L6)
- Multiply an integer by a fraction   (L6)
- Express a smaller whole number as a fraction of a larger one   (L5)

Useful resources
Mini white boards
World Atlas

Starter – Fraction sort

Write the following fractions on the board.

$$\frac{14}{49}, \frac{2}{12}, \frac{7}{42}, \frac{18}{30}, \frac{2}{7}, \frac{4}{24}, \frac{3}{5}, \frac{6}{21}, \frac{1}{6}, \frac{24}{40}, \frac{4}{14}, \frac{15}{25}$$

Ask students to sort the fractions into 3 sets of equivalent fractions.
Can be extended by asking students to make up their own
fraction sort puzzle.

Teaching notes

The spread places an emphasis on
using mental calculation strategies.
Using mini white boards challenge the
class to write down the answer to
simple 'fraction of an amount'
problems, beginning with unit
fractions. Ask successful students to
share the methods they used to do the
calculations until you have built up a
battery of suitable tricks (EP). For
example, $\frac{1}{5} = 2 \times \frac{1}{10}$, $\frac{1}{4} = \frac{1}{2} \times \frac{1}{2}$, $\frac{1}{6} = \frac{1}{2} \times \frac{1}{3}$, *etc.* An alternative way to
approach the same topic is to use a
spider diagram with, say, 450 in the
middle and various fractions of this
number surrounding it.

As necessary, it may help to
recapitulate how to convert improper
fractions into mixed numbers, as in the
first example, and how to cancel down
a fraction, as in the second example.
Students can also be reminded of the
connection between fractions and
division.

It may also help to give an example
which can be done using mental,
written and calculator methods. Ask
students which is the best method in a
given case and how the answers can be
checked.

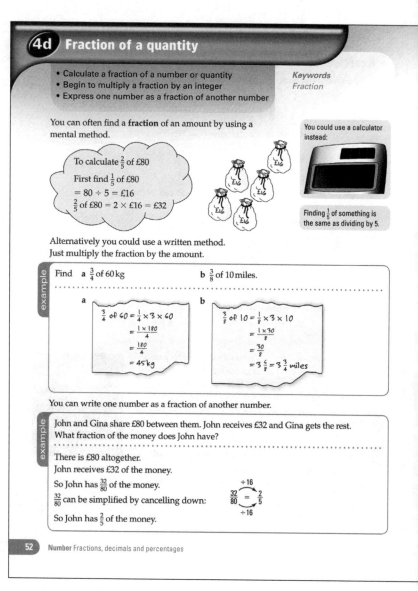

Plenary

Ask students to identify two things they can do now that
they were unable to do at the start of the unit and one thing
that they are still unclear about. Use this to inform further
teaching. This activity also encourages students to review
their own progress and act on the outcome (RL).

Simplification

Provide simple examples designed to develop and reinforce students' confidence in applying mental strategies, beginning with unit fractions.

Encourage students to use mental, written and calculator methods as part of their strategy for checking answers. Also emphasise the functional aspects of maths being used in the real life contexts.

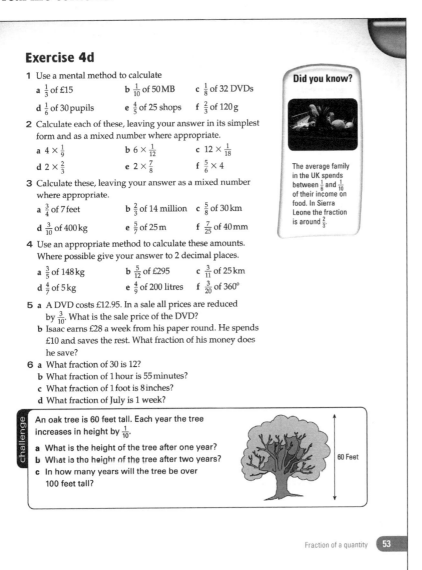

Exercise 4d

1 Use a mental method to calculate

  a $\frac{1}{3}$ of £15   **b** $\frac{1}{10}$ of 50 MB   **c** $\frac{1}{8}$ of 32 DVDs

  d $\frac{1}{6}$ of 30 pupils   **e** $\frac{4}{5}$ of 25 shops   **f** $\frac{2}{3}$ of 120 g

2 Calculate each of these, leaving your answer in its simplest form and as a mixed number where appropriate.

  a $4 \times \frac{1}{9}$ **b** $6 \times \frac{1}{12}$ **c** $12 \times \frac{1}{18}$

  d $2 \times \frac{2}{3}$ **e** $2 \times \frac{7}{8}$ **f** $\frac{5}{6} \times 4$

3 Calculate these, leaving your answer as a mixed number where appropriate.

  a $\frac{3}{4}$ of 7 feet   **b** $\frac{2}{3}$ of 14 million   **c** $\frac{5}{8}$ of 30 km

  d $\frac{3}{10}$ of 400 kg   **e** $\frac{5}{7}$ of 25 m   **f** $\frac{7}{25}$ of 40 mm

4 Use an appropriate method to calculate these amounts. Where possible give your answer to 2 decimal places.

  a $\frac{3}{5}$ of 148 kg   **b** $\frac{5}{12}$ of £295   **c** $\frac{3}{11}$ of 25 km

  d $\frac{4}{7}$ of 5 kg   **e** $\frac{4}{9}$ of 200 litres   **f** $\frac{3}{20}$ of 360°

5 **a** A DVD costs £12.95. In a sale all prices are reduced by $\frac{3}{10}$. What is the sale price of the DVD?

  b Isaac earns £28 a week from his paper round. He spends £10 and saves the rest. What fraction of his money does he save?

6 **a** What fraction of 30 is 12?

  b What fraction of 1 hour is 55 minutes?

  c What fraction of 1 foot is 8 inches?

  d What fraction of July is 1 week?

Did you know?

The average family in the UK spends between $\frac{1}{6}$ and $\frac{1}{10}$ of their income on food. In Sierra Leone the fraction is around $\frac{2}{3}$.

Challenge

An oak tree is 60 feet tall. Each year the tree increases in height by $\frac{1}{10}$.

  a What is the height of the tree after one year?

  b What is the height of the tree after two years?

  c In how many years will the tree be over 100 feet tall?

60 Feet

Fraction of a quantity   **53**

Extension

More able students can be given a variety of examples and be asked to identify and then justify which calculation method they would use to find a solution.

Exercise 4d commentary

Question 1 – Similar to the first example. In parts **e** and **f** beware of students still using unit fractions.

Question 2 and **3** – Similar to the first example. A discussion of how to change an improper fraction to a mixed number and then simplifying may be helpful here.

Question 4 – Here students may need to revisit converting a fraction to a decimal and to discuss rounding to a given number of decimal places (1.3).

Question 5 – Students should be encouraged to pause to identify the important information and what they need to calculate (1.1).

Question 6 – Similar to the second example. Ask able students to generate their own real life questions with which to challenge a partner (CT).

Challenge – This could be done as a paired activity. In part **b** check that students know to calculate $\frac{1}{10}$ of the new height 66 ft (not 60 ft).

Assessment Criteria – Calculating, level 6: calculate fractions of quantities (fraction answers), multiply an integer by a fraction.

Links

Use an atlas to show Sierra Leone on the West Coast of Africa. The country has a lush tropical climate and is rich in diamonds but is still recovering from a bitter civil war which ended in 2002. During the war, over 2M people left their homes. If the population of Sierra Leone is around 6M, what fraction is this? There is more information about Sierra Leone at https://www.cia.gov/library/publications/the-world-factbook/geos/sl.html

- Recognise the equivalence of percentages, fractions and decimals (L5)
- Express one given number as a percentage of another (L6)
- Calculate percentages (L5)

Useful resources

Starter – Half-way fractions

Ask students to find the fraction that is half-way between

$\frac{1}{3}$ and $\frac{2}{3}$, $\frac{3}{5}$ and $\frac{7}{10}$, $\frac{3}{8}$ and $\frac{1}{2}$.

Answers: $\frac{1}{2}$, $\frac{13}{20}$, $\frac{7}{16}$

Can be extended by asking for other fractions in the given ranges.

Teaching notes

To assess and develop students' mental methods draw a spider diagram with centre labelled 100% and surrounding bubbles labelled 10%, 20%, 5%, 1%, etc. Supply a value for the central bubble, say 280, and invite students to find the values that should go in the surrounding bubbles (28, 56, 14, 2.8). The surrounding percentages can be made harder as students gain in confidence: 43%, 2.5%, 1.5%, 17.5%, etc. This will allow students to take responsibility, showing confidence in themselves and their contribution. (TW).

The examples given can be used to provide templates for how written solutions should be laid out.

All of these questions can be approached in different ways, based on fraction-decimal-percentage equivalences, and computed using mental, written and calculator methods. Students should be encouraged to try different approaches and once confidence is gained the efficiency of the various approaches discussed.(IE, EP)

It may be necessary to devote time to discussing questions like question **5** and the **investigation** where language may be an issue and it is necessary to identify relevant information.

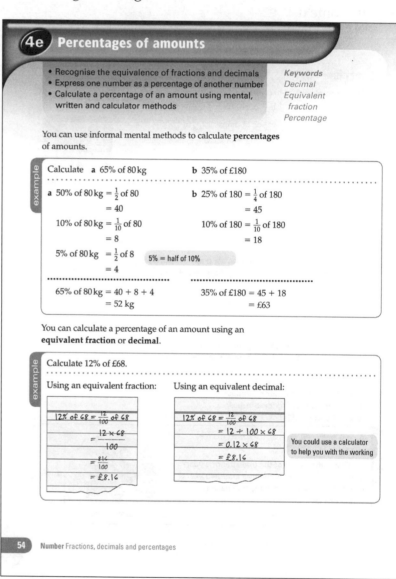

Plenary

Ask students to write a word problem that involves calculating a percentage in an everyday situation. Use question **5** from the exercise as a model. Take a few examples that students have generated and consider using them as starter activities next lesson (CT).

Simplification

Students who struggle with calculating percentages could be given more opportunity to develop mental methods. Mental strategies are essential for full understanding and are an excellent checking strategy for all abilities.

Questions can be answered using a variety of methods: encourage students to compare methods and gain confidence in their ability to use a range of strategies.

Exercise 4e

1 Calculate these percentages using a mental method.
   a 50% of £70   b 10% of 45 kg
   c 1% of 1500 m   d 25% of 256 MB

2 Calculate these percentages using a mental method.
   a 20% of £40   b 5% of 60 DVDs   c 2% of 150 MB
   d 40% of £75   e 60% of £700   f 15% of 180°
   g 11% of £5500   h 30% of 250 N   i 8% of 240 ml
   j 35% of £20 000   k 65% of 440 yards   l 95% of 400 kJ

3 Calculate these using a mental, written or calculator method, giving
   your answers to 2 decimal places where appropriate.
   a 18% of £40   b 7% of 71 kg   c 11% of 58 km
   d 16% of 85 euros   e 3% of 75 mm   f 24% of 55 kB
   g 29% of 18 litres   h 35% of 92 mph   i 46% of 46 m
   j 49% of 90 MB   k 63% of 15 cm   l 77% of 90°

4 Calculate these using a calculator. Show all the steps of your working
   out, and give your answer to 2 decimal places where appropriate.
   a 12% of £148   b 35% of 96 kg   c 52% of 512 MB
   d 86% of 355 km   e 4% of 185 mm   f 55% of 420 ml
   g 2.5% of £800   h 47% of 925 g   i 12.5% of 48 N
   j 41% of £8000   k 73% of 840 kJ   l 110% of 5 million

5 a Naheeda scores 45% in her English exam. The maximum score on
     the exam is 120. How many marks did Naheeda score on the exam?
   b Gavin starts to download an 8 GB file from the internet.
     He downloads 65% of the file in 10 minutes. How much
     of the file has he downloaded?
   c The label on the back of a 150 g packet of crisps says that
     the crisps are 6% fat. How much fat is that in grams?

investigation

a Sheena eats 240 g of baked beans. The beans contain

   Sugar   5%
   Fat   0.2%
   Protein   4.9%
   Carbohydrates 13%

Calculate how much sugar, fat, protein and carbohydrate
Sheena has eaten.

b Investigate the labels on the back of other things that you eat.

Extension

More able students could be challenged to find as many different ways of finding a mental solution to a given problem as possible (for example, how many different ways are there of finding 65%?) and could be offered further challenge by being asked to solve increasingly difficult word problems, always being required to explain their thinking.

Exercise 4e commentary

Question 1 – Ask students to write a sentence explaining how they obtained their answers: '50% means half, so I divided by 2'.

Question 2 – Similar to the first example. Encourage pairs of students to compare their methods and invent new ones: $5\% = \frac{1}{10} \times 50\%$ versus $\frac{1}{2} \times 10\%$ or $5 \times 1\%$, *etc.*

Question 3 – Similar to the second example. Consider pairing students, alternately one answers a part using a mental method whilst the other uses a written method, they then compare answers and resolve any discrepancies (RL) (1.3, 1.4).

Question 4 – Students may need reminding how to round and why it is appropriate (1.4).

Question 5 – Students may need guidance extracting the relevant information and deciding what to calculate.

Investigation – This could form the basis of a group-based small project in which students are required to report their findings to the class via a presentation or display (TW, EP, IE).

Links

Crisps are made from slices of potato fried in oil. The amount of fat absorbed during frying depends on the kind of oil used and the temperature. Manufacturers have lowered the amount of saturated fat in the crisps. by using different oils A typical 35 g bag of crisps still contains about 2.5 teaspoons of oil. How many teaspoons of oil will you consume if you eat one bag of crisps every day for a month?

- Recognise the equivalence of percentages, fractions and decimals (L5)
- Express one given number as a percentage of another (L5)

Useful resources
Newspapers and magazines

Starter – Percentage bingo

Ask students to draw a 3 × 3 grid and enter nine amounts from the following list:

£2 £3 £4 £5 £6 £7 £8 £9 £10 £11 £12 £13 £14 £15 £16 £17

Give questions, for example, 36% of £25, 68% of £25, 44% of £25.

Winner is the first student to cross out all nine amounts.
(Hint: 36% of 25 = 25% of 36.)

See also the plenary spread **4e**.

Teaching notes

Students are required to synthesise a large amount of information on various methods and approaches (SM).

Begin by giving a number of worked examples that illustrate the different techniques, their inter-connections, relative merits and how to check calculations. Then ask the class to supply 'helpful hints' on the important points they need to know. This could be developed as a homework activity to produce a poster showing a model worked example that is annotated with hints and explanations (CT).

As the lesson progresses, it may be useful to collate a list of problematic questions and identify common mistakes. Classroom discussion could then be used to understand where the errors lie and to agree correct approaches (RL).

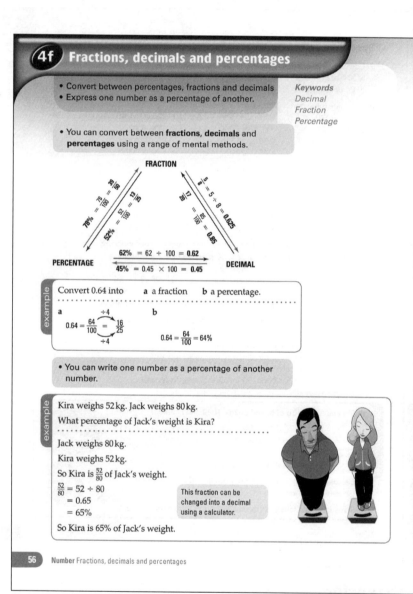

Plenary

Invite students to refer back to the exercise and take a vote on what was found to be the hardest question (or part question). Model a solution to this and then set a similar challenge to 'try again'.

Simplification

As a confidence building exercise ask students to complete a fraction-decimal-percentage table containing entries which involve straightforward conversions. They could also choose their own values.

All questions (1.3).

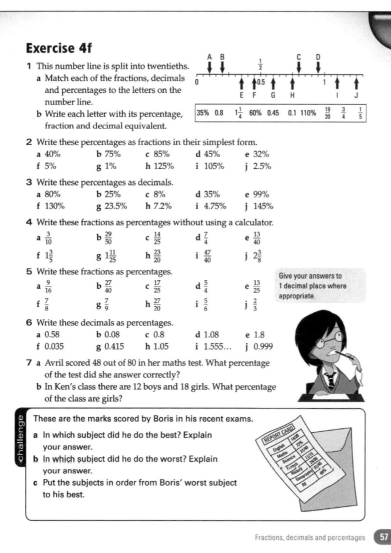

Exercise 4f

1 This number line is split into twentieths.
 a Match each of the fractions, decimals and percentages to the letters on the number line.
 b Write each letter with its percentage, fraction and decimal equivalent.

$$35\% \quad 0.8 \quad 1\tfrac{1}{4} \quad 60\% \quad 0.45 \quad 0.1 \quad 110\% \quad \tfrac{19}{20} \quad \tfrac{3}{4} \quad \tfrac{1}{5}$$

2 Write these percentages as fractions in their simplest form.
 a 40% b 75% c 85% d 45% e 32%
 f 5% g 1% h 125% i 105% j 2.5%

3 Write these percentages as decimals.
 a 80% b 25% c 8% d 35% e 99%
 f 130% g 23.5% h 7.2% i 4.75% j 145%

4 Write these fractions as percentages without using a calculator.
 a $\frac{3}{10}$ b $\frac{29}{50}$ c $\frac{14}{25}$ d $\frac{7}{4}$ e $\frac{13}{40}$
 f $1\frac{3}{5}$ g $1\frac{11}{25}$ h $\frac{23}{20}$ i $\frac{47}{40}$ j $2\frac{3}{8}$

5 Write these fractions as percentages.
 a $\frac{9}{16}$ b $\frac{27}{40}$ c $\frac{17}{25}$ d $\frac{5}{4}$ e $\frac{13}{25}$
 f $\frac{7}{8}$ g $\frac{7}{9}$ h $\frac{27}{20}$ i $\frac{5}{6}$ j $\frac{2}{3}$

Give your answers to 1 decimal place where appropriate.

6 Write these decimals as percentages.
 a 0.58 b 0.08 c 0.8 d 1.08 e 1.8
 f 0.035 g 0.415 h 1.05 i 1.555... j 0.999

7 a Avril scored 48 out of 80 in her maths test. What percentage of the test did she answer correctly?
 b In Ken's class there are 12 boys and 18 girls. What percentage of the class are girls?

challenge
These are the marks scored by Boris in his recent exams.
 a In which subject did he do the best? Explain your answer.
 b In which subject did he do the worst? Explain your answer.
 c Put the subjects in order from Boris' worst subject to his best.

Fractions, decimals and percentages 57

Extension

More able students could create a PowerPoint presentation showing how to convert between fractions, decimals and percentages, that could be made available on the school website for students having difficulty with this topic (EP).

Exercise 4f commentary

Question 1 – It may help students to understand what to do if the first two numbers are matched as a class (1.1).

Question 2 – Remind students to cancel. Parts **h** and **i** may require discussion of percentages ≥ 100%.

Question 3 – In parts **g–i** students may need to discuss 'fractional' percentages.

Question 4 and **5** – Allow students to skip parts that initially appear too difficult before returning to them for paired and then whole class discussion.

Question 6 – Similar to the second example. Parts **d**, **e**, **h** and **i** may require discussion of fractions >1. Ask students how they might 'sensibly' round some answers.

Question 7 – Similar to the second example.

Challenge – This exemplifies a practical and very relevant use of percentages. Encourage students to use their own most recent test results in a similar way (in their own time to avoid a potentially threatening situation) (EP)

Assessment Criteria – NNS, level 6: use the equivalence of fractions, decimals and percentages to compare proportions.

Links

Bring in some newspapers or magazines. Ask students to find any article or advertisement where a decimal, percentage or fraction is used. Which format is used most frequently? Would the article or advertisement have the same affect if, for example, a decimal was used in place of a percentage, or a percentage in place of a fraction? Are there any examples where a conversion has been used?

4a

1 Write these numbers in order starting with the smallest.

a 4.5	4.48	4.4	4.34	4
b 5.96	5.979	5.94	6	5.9
c 0.066	0.068	0.06	0.0695	0.0684
d 2.8	2.771	2.16	2.776	2.77

2 Copy these and place $<$ or $>$ between each pair of numbers to show which number is the larger.

a $0.46 \square 0.5$ **b** $1.61 \square 1.6$ **c** $4.375 \square 4.384$ **d** $5.24 \square 5.2$

e $7.13 \square 7.14$ **f** $2.753 \square 2.76$ **g** $8.0444 \square 8.044$ **h** $6.999 \square 7.1$

4b

3 Change these fractions into decimals using division. Use an appropriate method.

a $\frac{7}{8}$ **b** $\frac{7}{16}$ **c** $\frac{7}{20}$ **d** $\frac{1}{6}$ **e** $\frac{5}{9}$

4 Put these fractions in order from lowest to highest.

a $\frac{2}{9}$ $\frac{1}{4}$ $\frac{3}{10}$ $\frac{4}{13}$

b $\frac{2}{3}$ $\frac{3}{4}$ $\frac{13}{18}$ $\frac{11}{15}$

c $\frac{3}{5}$ $\frac{5}{9}$ $\frac{9}{16}$ $\frac{13}{20}$

4c

5 Find the missing number in each of these pairs of equivalent fractions.

a $\frac{2}{3} = \frac{\square}{15}$ **b** $\frac{3}{7} = \frac{\square}{21}$ **c** $\frac{4}{9} = \frac{\square}{36}$ **d** $\frac{5}{7} = \frac{\square}{49}$

e $\frac{4}{11} = \frac{44}{\square}$ **f** $\frac{3}{13} = \frac{\square}{39}$ **g** $\frac{5}{12} = \frac{\square}{72}$ **h** $\frac{7}{16} = \frac{\square}{80}$

6 Calculate each of these additions and subtractions, giving your answer as a fraction in its simplest form.

a $\frac{2}{5} + \frac{1}{4}$ **b** $\frac{3}{7} + \frac{1}{5}$ **c** $\frac{1}{3} + \frac{1}{5}$ **d** $\frac{3}{4} + \frac{1}{9}$

e $\frac{3}{15} + \frac{9}{20}$ **f** $\frac{5}{6} - \frac{4}{9}$ **g** $\frac{5}{8} + \frac{7}{12}$ **h** $\frac{6}{7} - \frac{2}{21}$

7 Use an appropriate method to calculate these amounts. Where appropriate give your answer to 2 decimal places.

a $\frac{2}{7}$ of 236 g   b $\frac{7}{16}$ of £500   c $\frac{4}{7}$ of 18 km   d $\frac{8}{5}$ of 47 miles

e $\frac{5}{12}$ of 48 hours   f $\frac{3}{5}$ of $25   g $\frac{7}{9}$ of 25 tonnes   h $\frac{2}{5}$ of 360°

8 **a** In Karla's class there are 14 boys and 21 girls. What fraction of the class are boys?

b Ronald has 12 music CDs, 10 of which are Country and Western music. What fraction of Ronald's CDs are not Country and Western music?

9 Calculate these using an appropriate method, giving your answers to 2 decimal places where appropriate.

a 7% of £50   b 12% of 45 kg   c 31% of 18 km

d 57% of 39 euros   e 29% of £87   f 41% of 63 kg

10 Copy and complete this table.

Fraction	Decimal	Percentage
$\frac{17}{25}$		
	0.61	
		42%
$\frac{9}{40}$		

4 Summary

Assessment criteria

- Order decimals to three decimal places — Level 4
- Use equivalence between fractions and order fractions and decimals — Level 5
- Show understanding of situations by describing them mathematically using symbols, words and diagrams — Level 5
- Draw simple conclusions of their own and give an explanation of their reasoning — Level 5
- Add and subtract fractions by writing them with a common denominator — Level 6
- Calculate fractions of quantities (fraction answers), multiply an integer by a fraction — Level 6
- Use the equivalence of fractions, decimals and percentages to compare proportions — Level 6

Question commentary

Example

The example illustrates a level 6 question on adding and subtracting fractions. Some students will still try to add the numerators and denominators without finding equivalent fractions with a common denominator. Use of a number line, divided into 21sts for part **b**, may help students to work out where they would expect their answers to be. Emphasise that fractional answers should always be expressed in their simplest form.

a $\frac{1}{2} + \frac{2}{5} = \frac{(5 + 4)}{10} = \frac{9}{10}$

b $\frac{5}{7} - \frac{1}{21} = \frac{(15 - 1)}{21} = \frac{14}{21} = \frac{2}{3}$

Past question

The question asks students to find some equivalent fractions and percentages. Some students may find the word format of the question confusing. Ask how 3 out of 4 is written mathematically. Encourage students to write down the fractions in the question in mathematical form before converting to percentages. In part **b**, any answers resulting in a fraction equivalent to $\frac{1}{20}$ are acceptable.

Answer

Level 5

2 a 7, 50%

    b 5 out of 100, 1 out of 20 are two possibilities

Development and links

The skills gained in this chapter are important for the work on ratio and proportion in Chapter 12, where the topic will be extended to calculating percentage increases and decreases. Fractions and percentages are also needed for the work on statistics, especially for pie charts in Chapter 11, and handling probabilities.

Fractions, decimals and percentages are used throughout the curriculum, especially in subjects where students work with ratio and proportion. Students will use fractions, decimals and percentages for measure in design technology, for recipes and recipe conversion in food technology, for pie charts and statistical analysis in geography, and for comparing concentrations of liquids in science. Outside school, students will use percentages to compare interest rates on bank and building society accounts, to calculate VAT and to calculate price increases and reductions.

5 Expressions and formulae Algebra

Objectives

- Understand that algebraic operations, including the use of brackets, follow the rules of arithmetic 5
- Use index notation for small positive integer powers 5
- Simplify or transform linear expressions by collecting like terms 5
- Multiply a single term over a bracket 5
- Use formulae from mathematics and other subjects 5
- Substitute integers into simple formulae, including examples that lead to an equation to solve 5
- Derive simple formulae 5

Introduction

The focus of this chapter is on using and manipulating symbols in expressions and formulae and builds on the idea of using symbols to replace numbers encountered in Year 7. Students will substitute values for symbols in expressions and formulae, simplify expressions by collecting like terms and using index notation, expand brackets and construct and use simple formulae.

The student book discusses the mathematician Diophantus and his work on algebra. Algebra is a universal language used for hundreds of years and links across the curriculum to ICT, and the sciences. Translating mathematical problems into the language of algebra makes them quicker to write and easier to solve. There is more information about Diophantus and his work, including Diophantus' Riddle at http://en.wikipedia.org/wiki/Diophantus and at http://www.mathsisgoodforyou.com/people/diophantus.htm (TW).

Fast-track

All spreads

Level

MPA

1.1	5a, d, f
1.2	5a, d, f
1.3	5a, b, c, d, e, f
1.4	
1.5	5b

PLTS

IE	5b, c, d, e
CT	5a, c, d, f
RL	5a, b, c, d
TW	5I, a, e
SM	5d
EP	5a, e, f

Extra Resources

5 Start of chapter presentation

5a Starter: BIDMAS multichoice

5a Worked solution: Q3

5c Worked solution: Q5a

5c Consolidation sheet

5d Animation: Multiplying brackets

5d Animation: Expanding brackets

5d Consolidation sheet

5e Worked solution: Q5a

5e Consolidation sheet

5f Starter: Collecting like terms T or F

Assessment: chapter 5

- Understand that algebraic operations follow the rules of arithmetic (L5)
- Substitute positive integers into linear expressions and formulae (L5)

Useful resources
Mini whiteboards

Starter – What is my number?

I am greater than 3^3 and…

I am less than 4^3 and…

I am a multiple of 7 and…

I have exactly 3 factors. (49)

Can be extended by asking students which clue could have been omitted and asking them to make up their own puzzles (CT).

Teaching notes

As a precursor to constructing and substituting into formulae it is important to ensure that students are comfortable with the notation: x for $1x$, $2x$ rather than $x2$ for $x + x = 2 \times x = x \times 2$ (1.3).

Questions **2** onwards involve generating expressions from a description. This is a skill that large numbers of students struggle with and it is well worth focusing here. After providing an example, organise students into small groups and ask them to write their own sentences. Then in turn, the first student reads out his or her sentence and challenges the second to supply the correct formula, the second then reads his or her their sentence and challenges the third *etc*. Ensure all students agree on whether the answers are correct (EP, RL).

The second example involves substitution into a 'two operation' expression. This assumes that students understand the order of precedence. Using mini whiteboards, a few numerical BIDMAS examples should allow you to asses if this is the case. Further examples can be used to check that students understand how to substitute into formulae.

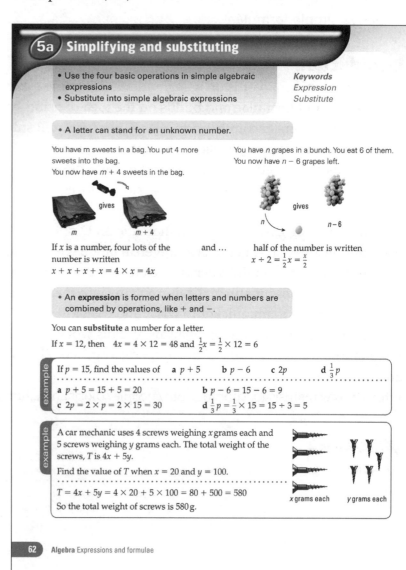

5a Simplifying and substituting

- Use the four basic operations in simple algebraic expressions
- Substitute into simple algebraic expressions

Keywords
Expression
Substitute

- A letter can stand for an unknown number.

You have m sweets in a bag. You put 4 more sweets into the bag. You now have $m + 4$ sweets in the bag.

You have n grapes in a bunch. You eat 6 of them. You now have $n - 6$ grapes left.

If x is a number, four lots of the number is written
$x + x + x + x = 4 \times x = 4x$

and … half of the number is written
$x \div 2 = \frac{1}{2}x = \frac{x}{2}$

- An **expression** is formed when letters and numbers are combined by operations, like + and −.

You can **substitute** a number for a letter.
If $x = 12$, then $4x = 4 \times 12 = 48$ and $\frac{1}{2}x = \frac{1}{2} \times 12 = 6$

example

If $p = 15$, find the values of **a** $p + 5$ **b** $p - 6$ **c** $2p$ **d** $\frac{1}{3}p$

a $p + 5 = 15 + 5 = 20$ **b** $p - 6 = 15 - 6 = 9$

c $2p = 2 \times p = 2 \times 15 = 30$ **d** $\frac{1}{3}p = \frac{1}{3} \times 15 = 15 \div 3 = 5$

example

A car mechanic uses 4 screws weighing x grams each and 5 screws weighing y grams each. The total weight of the screws, T is $4x + 5y$.

Find the value of T when $x = 20$ and $y = 100$.

$T = 4x + 5y = 4 \times 20 + 5 \times 100 = 80 + 500 = 580$

So the total weight of screws is 580 g.

x grams each y grams each

62   Algebra Expressions and formulae

Plenary

Use the best two or three examples generated by the class in the activity outlined above to share as a whole class and, through this task, encourage students to collaborate with others to work towards common goals (TW).

Simplification

Provide more simple examples such as those in questions **1** and **2**.

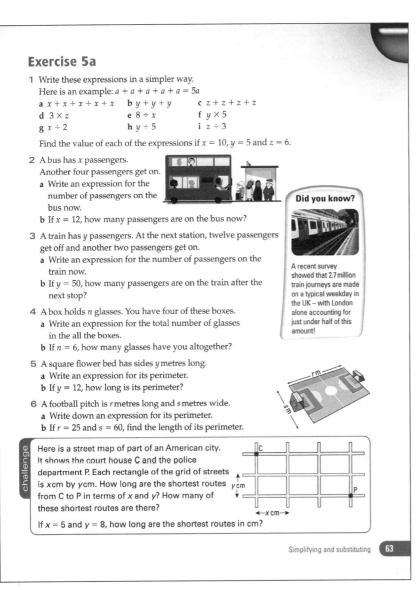

Exercise 5a

1 Write these expressions in a simpler way.
Here is an example: $a + a + a + a + a = 5a$
  a $x + x + x + x + x$ b $y + y + y$ c $z + z + z + z$
  d $3 \times z$ e $8 \div x$ f $y \times 5$
  g $x \div 2$ h $y \div 5$ i $z \div 3$

Find the value of each of the expressions if $x = 10$, $y = 5$ and $z = 6$.

2 A bus has x passengers.
Another four passengers get on.
  a Write an expression for the number of passengers on the bus now.
  b If $x = 12$, how many passengers are on the bus now?

3 A train has y passengers. At the next station, twelve passengers get off and another two passengers get on.
  a Write an expression for the number of passengers on the train now.
  b If $y = 50$, how many passengers are on the train after the next stop?

4 A box holds n glasses. You have four of these boxes.
  a Write an expression for the total number of glasses in the all the boxes.
  b If $n = 6$, how many glasses have you altogether?

5 A square flower bed has sides y metres long.
  a Write an expression for its perimeter.
  b If $y = 12$, how long is its perimeter?

6 A football pitch is r metres long and s metres wide.
  a Write down an expression for its perimeter.
  b If $r = 25$ and $s = 60$, find the length of its perimeter.

Did you know?
A recent survey showed that 2.7 million train journeys are made on a typical weekday in the UK – with London alone accounting for just under half of this amount!

challenge
Here is a street map of part of an American city. It shows the court house C and the police department P. Each rectangle of the grid of streets is x cm by y cm. How long are the shortest routes from C to P in terms of x and y? How many of these shortest routes are there?
If $x = 5$ and $y = 8$, how long are the shortest routes in cm?

Simplifying and substituting **63**

Extension

Set substitution questions involving negative numbers or possibly decimals.

Exercise 5a commentary

Question 1 – Similar to the first example. A point for discussion is the difference between $3 \times z = z \times 3$ and $3 \div z \neq z - 3$. Compare with $y + 3 = 3 + y$ but $y - 3 \neq 3 - y$ (in general) (1.3).

Question 3 – A development of question **2**. In part **a** the expression should be simplified.

Question 4 – Check that students use multiplication and not repeated addition.

Question 5 – Some students may need to be reminded of the method for finding the perimeter.

Question 6 – Similar to the second example. Look for a simplified expression (1.1).

Challenge – Some explanation of what is required may be necessary here (1.1, 1.2).

Assessment Criteria – UAM, level 5: show understanding of situations by describing them mathematically using symbols, words and diagrams.

Links

The city of Milton Keynes in the UK was designed as a new town and construction began in 1967. The town is laid out in a grid system with ten horizontal roads (Ways) at 1 km intervals and eleven vertical roads (Streets), also at 1 km intervals. The roads are numbered with H or V numbers for horizontal or vertical, for example, H6 Childs Way. There is more information about the Milton Keynes grid system at http://www.road-to-nowhere.co.uk/features/milton_keynes.html

• Use index notation for small positive integer powers   (L5)   **Useful resources**
Mini whiteboards

Starter – ABC

Each letter of the alphabet represents a number:
$a = 1$, $b = 2$, $c = 3$, $d = 4$, $e = 5$, etc.
Challenge students to decode the following message:

$(f+g)$ $(t-s)$ $(2j)$ $\left(\dfrac{p}{b}\right)$ $(e+n)$ c^2 $(3f+1)$ $\left(\dfrac{r}{c}\right)$ $(3g)$ (d^2-b)

(Coded message = maths is fun: $f + g = 6 + 7 = 13 = m$, $t - s = 1 = a$, etc.)

Can be extended by asking students to code their own message.

Teaching notes

Ensure that students understand the notation and do not confuse, say, 2^3 and 2×3. This is a potentially persistent problem which should be tackled early. The meaning of the notation can be emphasised by providing examples of how expressions can be expanded out as a product or collected as a power (RL) (1.3).

The **research** question places an emphasis on numerical values. This gives an opportunity to demonstrate how quickly numbers become very big when using powers and how indices allow very large numbers to be written succinctly. This will lay the foundation for future work on standard form and can also be linked to units which arise in science.

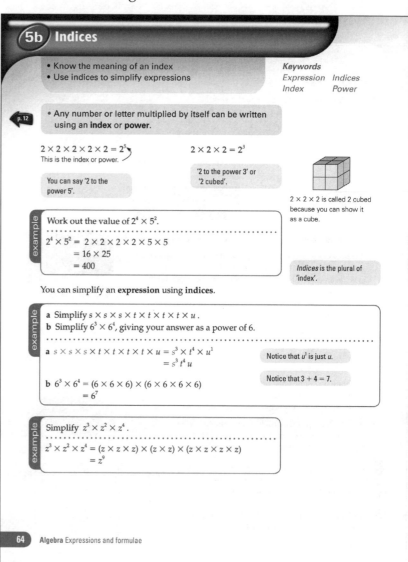

5b Indices

• Know the meaning of an index
• Use indices to simplify expressions

Keywords
Expression   Indices
Index   Power

p. 12

• Any number or letter multiplied by itself can be written using an **index** or **power**.

$2 \times 2 \times 2 \times 2 \times 2 = 2^5$
This is the index or power.

You can say '2 to the power 5'.

$2 \times 2 \times 2 = 2^3$
'2 to the power 3' or '2 cubed'.

$2 \times 2 \times 2$ is called 2 cubed because you can show it as a cube.

example
Work out the value of $2^4 \times 5^2$.

$2^4 \times 5^2 = 2 \times 2 \times 2 \times 2 \times 5 \times 5$
$= 16 \times 25$
$= 400$

Indices is the plural of 'index'.

You can simplify an **expression** using **indices**.

example
a Simplify $s \times s \times s \times t \times t \times t \times t \times u$.
b Simplify $6^3 \times 6^4$, giving your answer as a power of 6.

a $s \times s \times s \times t \times t \times t \times t \times u = s^3 \times t^4 \times u^1$
$= s^3 t^4 u$

Notice that u^1 is just u.

b $6^3 \times 6^4 = (6 \times 6 \times 6) \times (6 \times 6 \times 6 \times 6)$
$= 6^7$

Notice that $3 + 4 = 7$.

example
Simplify $z^3 \times z^2 \times z^4$.

$z^3 \times z^2 \times z^4 = (z \times z \times z) \times (z \times z) \times (z \times z \times z \times z)$
$= z^9$

64   **Algebra** Expressions and formulae

Plenary

Revisit some of the more challenging questions using the mini whiteboards to assess understanding and clear up any confusions.

Simplification

Supply more examples like those in questions **1** and **2**.

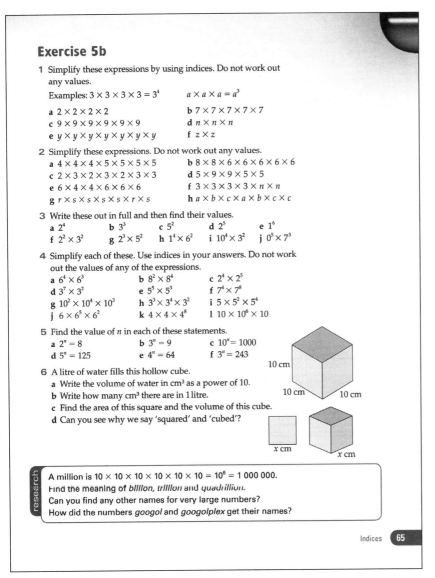

Exercise 5b

1 Simplify these expressions by using indices. Do not work out any values.

Examples: $3 \times 3 \times 3 \times 3 = 3^4$ $a \times a \times a = a^3$

  a $2 \times 2 \times 2 \times 2$

  b $7 \times 7 \times 7 \times 7 \times 7$

  c $9 \times 9 \times 9 \times 9 \times 9 \times 9$

  d $n \times n \times n$

  e $y \times y \times y \times y \times y \times y \times y$

  f $z \times z$

2 Simplify these expressions. Do not work out any values.

  a $4 \times 4 \times 4 \times 5 \times 5 \times 5 \times 5$

  b $8 \times 8 \times 6 \times 6 \times 6 \times 6 \times 6$

  c $2 \times 3 \times 2 \times 3 \times 2 \times 3 \times 3$

  d $5 \times 9 \times 9 \times 5 \times 5$

  e $6 \times 4 \times 4 \times 6 \times 6 \times 6$

  f $3 \times 3 \times 3 \times 3 \times n \times n$

  g $r \times s \times s \times s \times r \times s$

  h $a \times b \times c \times a \times b \times c \times c$

3 Write these out in full and then find their values.

  a 2^4 **b** 3^3 **c** 5^2 **d** 2^5 **e** 1^6

  f $2^2 \times 3^2$ **g** $2^3 \times 5^2$ **h** $1^4 \times 6^2$ **i** $10^4 \times 3^2$ **j** $0^5 \times 7^3$

4 Simplify each of these. Use indices in your answers. Do not work out the values of any of the expressions.

  a $6^4 \times 6^3$ **b** $8^2 \times 8^4$ **c** $2^4 \times 2^5$

  d $3^7 \times 3^2$ **e** $5^5 \times 5^3$ **f** $7^4 \times 7^8$

  g $10^2 \times 10^4 \times 10^3$ **h** $3^3 \times 3^4 \times 3^2$ **i** $5 \times 5^2 \times 5^4$

  j $6 \times 6^5 \times 6^2$ **k** $4 \times 4 \times 4^8$ **l** $10 \times 10^6 \times 10$

5 Find the value of n in each of these statements.

  a $2^n = 8$ **b** $3^n = 9$ **c** $10^n = 1000$

  d $5^n = 125$ **e** $4^n = 64$ **f** $3^n = 243$

6 A litre of water fills this hollow cube.

  a Write the volume of water in cm³ as a power of 10.

  b Write how many cm³ there are in 1 litre.

  c Find the area of this square and the volume of this cube.

  d Can you see why we say 'squared' and 'cubed'?

> **research**
> A million is $10 \times 10 \times 10 \times 10 \times 10 \times 10 = 10^6 = 1\,000\,000$.
> Find the meaning of *billion*, *trillion* and *quadrillion*.
> Can you find any other names for very large numbers?
> How did the numbers *googol* and *googolplex* get their names?

Indices **65**

Extension

Introduce harder versions of question **4** that use a variable or combinations of numbers and variable(s).

Exercise 5b commentary

Question 1 and **2** – Similar to the first example. Could be done as a whole class with mini whiteboards. Question **2** will require re-ordering of the individual terms (1.3).

Question 3 – In parts **e** and **h** check that 1^6 is not interpreted as 6 *etc*.

Question 4 – It may be helpful for students to rewrite the expressions in full (1.3).

Question 5 – It may help to write the RHS of each question as the product of (prime) factors.

Question 6 – A chance to expand on how indices naturally arise when discussing real life situations (1.5).

Research – It is useful to encourage students to investigate big numbers as this will be solid groundwork when they move on to use standard form much later on (IE).

Links

The traditional nursery rhyme below is a riddle that can be investigated using indices.

As I was going to St. Ives (1)
I met a man with 7 wives
Each wife had 7 sacks
Each sack had 7 cats
Each cat had 7 kits
Kits, cats, sacks, wives
How many were going to St. Ives?
$\left(1 + 1 + 7 + 7^2 + 7^3 + 7^4 = 2802\right)$

Of course the answer to the riddle is actually 1 as everyone else is on their way back!

- Simplify or transform [linear] expressions by collecting like terms (L5)

Useful resources
Coloured beads

Starter – Power products

Draw a 4 × 4 table on the board. Label the columns 3^2, 2^5, 2^3 and 3^4.

Label the rows 2^2, 2^4, 3^5 and 3^2.

Ask students to fill in the table with the products, for example, the top row in the table would read $2^2 \times 3^2$, 2^7, 2^5, $2^2 \times 3^4$.

Can be differentiated by the choice of powers.

Teaching notes

Start by challenging students to write definitions for the four keywords (Expression, Like terms, Simplify, Term); allow pairs of students to confer. Then share ideas as a class before agreeing on a best definition.

Sources of error are likely to include failing to take into account the $+/-$ signs in front of terms or failing to distinguish between terms with different powers of the same variable . Carefully explained examples should help students know what to look for and how to proceed (RL).

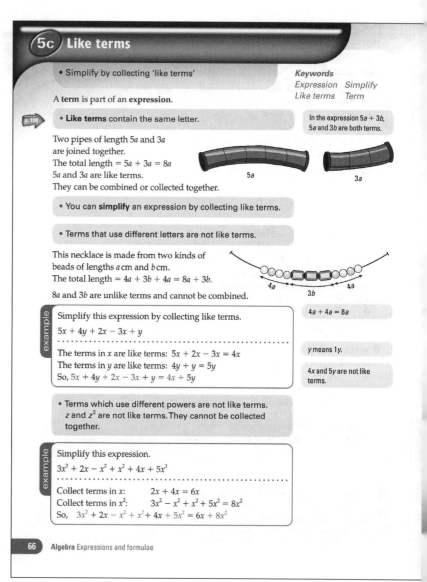

5c **Like terms**

- Simplify by collecting 'like terms'

Keywords
Expression Simplify
Like terms Term

A **term** is part of an **expression**.

p. 198

- **Like terms** contain the same letter.

In the expression $5a + 3b$, $5a$ and $3b$ are both terms.

Two pipes of length $5a$ and $3a$ are joined together.
The total length $= 5a + 3a = 8a$
$5a$ and $3a$ are like terms.
They can be combined or collected together.

$5a$

$3a$

- You can **simplify** an expression by collecting like terms.

- Terms that use different letters are not like terms.

This necklace is made from two kinds of beads of lengths a cm and b cm.
The total length $= 4a + 3b + 4a = 8a + 3b$.

$8a$ and $3b$ are unlike terms and cannot be combined.

$4a$ $3b$ $4a$

$4a + 4a = 8a$

example

Simplify this expression by collecting like terms.
$5x + 4y + 2x - 3x + y$

The terms in x are like terms: $5x + 2x - 3x = 4x$
The terms in y are like terms: $4y + y = 5y$
So, $5x + 4y + 2x - 3x + y = 4x + 5y$

y means $1y$.

$4x$ and $5y$ are not like terms.

- Terms which use different powers are not like terms. z and z^2 are not like terms. They cannot be collected together.

example

Simplify this expression.
$3x^2 + 2x - x^2 + x^2 + 4x + 5x^2$

Collect terms in x: $2x + 4x = 6x$
Collect terms in x^2: $3x^2 - x^2 + x^2 + 5x^2 = 8x^2$
So, $3x^2 + 2x - x^2 + x^2 + 4x + 5x^2 = 6x + 8x^2$

66   Algebra Expressions and formulae

Plenary

Invite students to challenge each other by writing a 'complicated' expression for their friend to simplify and to swap examples. They must be sure to work out the answer to their own example first, of course! (CT)

Simplification

Encourage students to rewrite the expressions with like terms placed together before collecting. Further examples that do not involve powers and follow the progression in the early questions may be helpful.

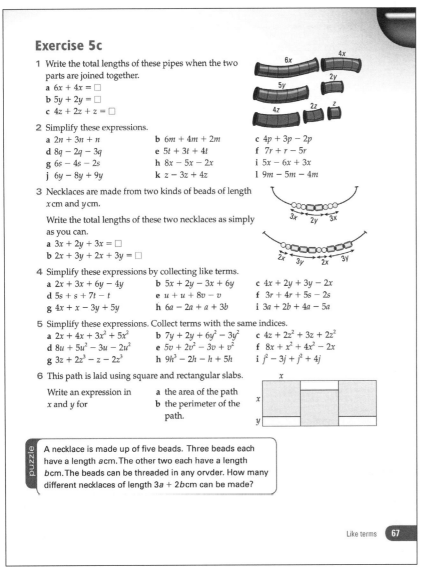

Exercise 5c

1 Write the total lengths of these pipes when the two parts are joined together.
   a $6x + 4x = \square$
   b $5y + 2y = \square$
   c $4z + 2z + z = \square$

2 Simplify these expressions.
   a $2n + 3n + n$ b $6m + 4m + 2m$ c $4p + 3p - 2p$
   d $8q - 2q - 3q$ e $5t + 3t + 4t$ f $7r + r - 5r$
   g $6s - 4s - 2s$ h $8x - 5x - 2x$ i $5x - 6x + 3x$
   j $6y - 8y + 9y$ k $z - 3z + 4z$ l $9m - 5m - 4m$

3 Necklaces are made from two kinds of beads of length x cm and y cm.

   Write the total lengths of these two necklaces as simply as you can.
   a $3x + 2y + 3x = \square$
   b $2x + 3y + 2x + 3y = \square$

4 Simplify these expressions by collecting like terms.
   a $2x + 3x + 6y - 4y$ b $5x + 2y - 3x + 6y$ c $4x + 2y + 3y - 2x$
   d $5s + s + 7t - t$ e $u + u + 8v - v$ f $3r + 4r + 5s - 2s$
   g $4x + x - 3y + 5y$ h $6a - 2a + a + 3b$ i $3a + 2b + 4a - 5a$

5 Simplify these expressions. Collect terms with the same indices.
   a $2x + 4x + 3x^2 + 5x^2$ b $7y + 2y + 6y^2 - 3y^2$ c $4z + 2z^2 + 3z + 2z^2$
   d $8u + 5u^2 - 3u - 2u^2$ e $5v + 2v^2 - 3v + v^2$ f $8x + x^2 + 4x^2 - 2x$
   g $3z + 2z^3 - z - 2z^3$ h $9h^3 - 2h - h + 5h$ i $j^2 - 3j + j^2 + 4j$

6 This path is laid using square and rectangular slabs.

   Write an expression in x and y for
   a the area of the path
   b the perimeter of the path.

---

puzzle
A necklace is made up of five beads. Three beads each have a length a cm. The other two each have a length b cm. The beads can be threaded in any orvder. How many different necklaces of length $3a + 2b$ cm can be made?

Like terms **67**

Extension

Encourage more able students to try to generate a puzzle of their own that uses their understanding of expressions, using the one at the end of the exercise as a model (IE).

Exercise 5c commentary

Question 1 – An introductory example: check that students understand what $6x$ and $4x$ mean.

Question 2 – Develops question **1** by also involving subtraction.

Question 3 – Develops question **1** by introducing two variables.

Question 4 – This develops question **3** by also involving subtraction.

Question 5 – Similar to the second example. The presence of different indices is likely to cause difficulty: suggest first re-writing the expression, collecting all the like terms together. Check that like terms are correctly identified and that signs are kept with individual terms (1.3).

Question 6 – Students may need reminding about area and perimeter and how they are calculated.

Puzzle – This could be done as a paired activity, or a practical activity for the visual or kinaesthetic learners.

Links

Beads have been found dating back 100 000 years and are the oldest form of jewellery ever found. There are pictures of ancient shell beads at http://www.nhm.ac.uk/about-us/news/2007/june/news_11808.html How long is a necklace made from 12 shells each 1cm long, 6 shells each 1.5 cm long and 2 shells each 2 cm long?

- Multiply a single term over a bracket   (L5)   *Useful resources*
- Use index notation for small positive integer powers (L5)

Starter – Expressions

$a = 2$, $b = 3$, $c = 5$

Ask students for expressions that have a value of 24, for example, $3b + 3c$.

Can be extended by changing the target number or values of a, b and c.

Teaching notes

The visual prompt of the rectangle provided here will be very helpful to visual learners and also to students who find this work more difficult as it is a useful reminder strategy to fall back on (1.2).

The abstract nature of the work often leads to difficulty. The emphasis on beginning with a word problem from which to generate an expression places this topic in a context which many will find helpful. The process skill of analysing through using appropriate mathematical procedures is developed where students expand expressions.

Where x^2 is involved, some students will need support to recall their previous work on powers. Many make careless errors where there are subtraction signs.

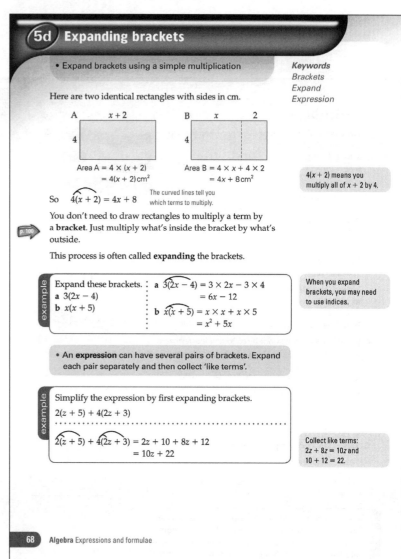

Plenary

Display three examples similar to those in question **6**. Ask students to choose one and to write an annotated answer that they think would explain this work to a student who had been absent from the lesson. Pair students who have answered different examples and ask them to read through each other's work and identify two things they thought were particularly helpful and one thing they could improve. This supports students in inviting feedback and dealing positively with praise, setbacks and criticism (RL).

Simplification

Offer further graduated examples similar to those in questions **1–3**. The word problems, questions **4** and **5**, can be completed as a paired activity with the starting point being to identify the key information.

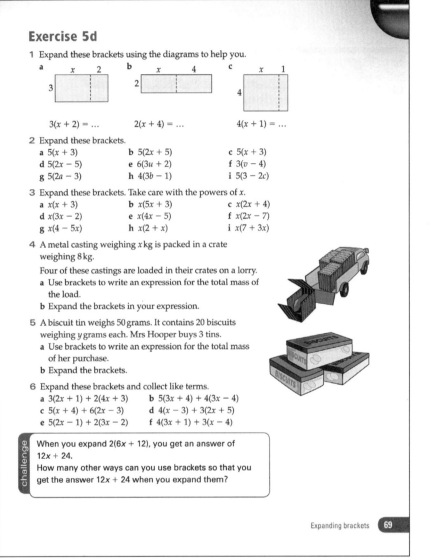

Exercise 5d

1 Expand these brackets using the diagrams to help you.

a x 2   b x 4   c x 1

   3   2   4

   $3(x + 2) = ...$ $2(x + 4) = ...$ $4(x + 1) = ...$

2 Expand these brackets.
   - **a** $5(x + 3)$ **b** $5(2x + 5)$ **c** $5(x + 3)$
   - **d** $5(2x - 5)$ **e** $6(3u + 2)$ **f** $3(v - 4)$
   - **g** $5(2a - 3)$ **h** $4(3b - 1)$ **i** $5(3 - 2c)$

3 Expand these brackets. Take care with the powers of x.
   - **a** $x(x + 3)$ **b** $x(5x + 3)$ **c** $x(2x + 4)$
   - **d** $x(3x - 2)$ **e** $x(4x - 5)$ **f** $x(2x - 7)$
   - **g** $x(4 - 5x)$ **h** $x(2 + x)$ **i** $x(7 + 3x)$

4 A metal casting weighing x kg is packed in a crate weighing 8 kg.

   Four of these castings are loaded in their crates on a lorry.
   - **a** Use brackets to write an expression for the total mass of the load.
   - **b** Expand the brackets in your expression.

5 A biscuit tin weighs 50 grams. It contains 20 biscuits weighing y grams each. Mrs Hooper buys 3 tins.
   - **a** Use brackets to write an expression for the total mass of her purchase.
   - **b** Expand the brackets.

6 Expand these brackets and collect like terms.
   - **a** $3(2x + 1) + 2(4x + 3)$ **b** $5(3x + 4) + 4(3x - 4)$
   - **c** $5(x + 4) + 6(2x - 3)$ **d** $4(x - 3) + 3(2x + 5)$
   - **e** $5(2x - 1) + 2(3x - 2)$ **f** $4(3x + 1) + 3(x - 4)$

> **challenge**
>
> When you expand $2(6x + 12)$, you get an answer of $12x + 24$.
> How many other ways can you use brackets so that you get the answer $12x + 24$ when you expand them?

Extension

More able students could produce annotated PowerPoint presentations showing how to solve one of the questions that was generally found to be most challenging. This could be shown to the class and/or could be made available as a student led tutorial on the school website. Parents would also find this helpful to support their children with homework (CT, SM).

Exercise 5d commentary

All questions (1.3)

Question 1 – Simple introductory examples.

Question 2 – Similar to the first example; remind students that negative signs occur.

Question 3 – Similar to the first example; that $x \times x = x^2$ can be linked to the previous spread.

Question 4 and **5** – Word problems: a whole class discussion may be necessary to model how to extract the relevant information (1.1).

Question 6 – Similar to the second example.

Challenge – Requires understanding of factors. This is best carried out as a paired activity to promote mathematical dialogue (IE).

Assessment Criteria – UAM, level 5: show understanding of situations by describing them mathematically using symbols, words and diagrams.

Links

Bring in some dictionaries for the class to use. The word *bracket* can have several meanings and can be used as a noun or a verb. What do the meanings have in common? (group or hold something together)

- Substitute integers into simple formulae   (L5)   **Useful resources**
- Use formulae from mathematics and other subjects   (L5)   *Mini whiteboards*

Starter – Priceless!

If A costs 1p, B costs 2p, C costs 3p etc how much is your name worth?

Are you more expensive than the person beside you?

Which of your school subjects is worth the most?

How much does your favourite hobby cost?

Teaching notes

This spread develops process skills in algebra. The capacity of students to represent problems is developed through constructing formulae. Their evaluation skills are developed through communicating and reflecting on links to related problems or to different problems with a similar structure.

Students may need a brief recap of substitution and the first two worked examples could be used as a whole class discussion activity.

The use of brackets in formulae and the occasions where two values will be substituted will need greater clarification for some and again offer an opportunity to make connections with prior learning.

There is an opportunity here to make links to formulae specifically relevant to the students, such as mobile phone tariffs.

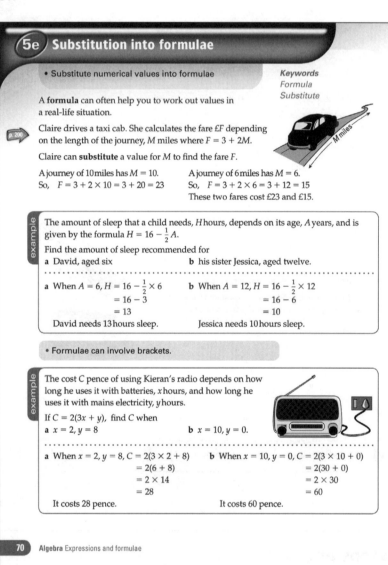

5e Substitution into formulae

- Substitute numerical values into formulae

Keywords
Formula
Substitute

A **formula** can often help you to work out values in a real-life situation.

Claire drives a taxi cab. She calculates the fare £F depending on the length of the journey, M miles where F = 3 + 2M.

Claire can **substitute** a value for M to find the fare F.

A journey of 10 miles has M = 10.   A journey of 6 miles has M = 6.
So, F = 3 + 2 × 10 = 3 + 20 = 23   So, F = 3 + 2 × 6 = 3 + 12 = 15
These two fares cost £23 and £15.

example

The amount of sleep that a child needs, H hours, depends on its age, A years, and is given by the formula $H = 16 - \frac{1}{2}A$.
Find the amount of sleep recommended for
a David, aged six   **b** his sister Jessica, aged twelve.

a When A = 6, $H = 16 - \frac{1}{2} \times 6$ **b** When A = 12, $H = 16 - \frac{1}{2} \times 12$
            = 16 − 3   = 16 − 6
            = 13   = 10
David needs 13 hours sleep.   Jessica needs 10 hours sleep.

- Formulae can involve brackets.

example

The cost C pence of using Kieran's radio depends on how long he uses it with batteries, x hours, and how long he uses it with mains electricity, y hours.
If C = 2(3x + y), find C when
a x = 2, y = 8   **b** x = 10, y = 0.

a When x = 2, y = 8, C = 2(3 × 2 + 8)   **b** When x = 10, y = 0, C = 2(3 × 10 + 0)
                     = 2(6 + 8)   = 2(30 + 0)
                     = 2 × 14   = 2 × 30
                     = 28   = 60
It costs 28 pence.   It costs 60 pence.

70   Algebra Expressions and formulae

Plenary

Ask students to identify as many formulae as possible that they know or use in other areas of the curriculum or in real life – for example in science or when working out cooking times, *etc.* (EP)

Simplification

Students struggling with this unit will need to consolidate simple substitution, questions **1** and **2**, and build confidence through working on more of the simple formulae before working with brackets or using two values, questions **3** onwards.

The emphasis on using formulae in real-life contexts supports the embedding of functional skills in mathematics.

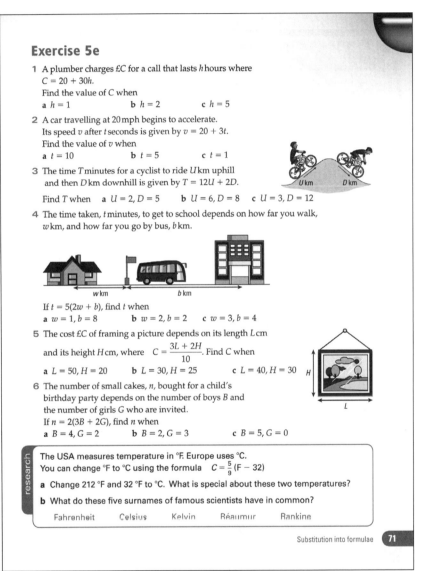

Exercise 5e

1 A plumber charges £C for a call that lasts h hours where
$C = 20 + 30h$.
Find the value of C when
   a $h = 1$ **b** $h = 2$ **c** $h = 5$

2 A car travelling at 20 mph begins to accelerate.
Its speed v after t seconds is given by $v = 20 + 3t$.
Find the value of v when
   a $t = 10$ **b** $t = 5$ **c** $t = 1$

3 The time T minutes for a cyclist to ride U km uphill
and then D km downhill is given by $T = 12U + 2D$.
Find T when   **a** $U = 2, D = 5$ **b** $U = 6, D = 8$ **c** $U = 3, D = 12$

4 The time taken, t minutes, to get to school depends on how far you walk, w km, and how far you go by bus, b km.

If $t = 5(2w + b)$, find t when
   a $w = 1, b = 8$ **b** $w = 2, b = 2$ **c** $w = 3, b = 4$

5 The cost £C of framing a picture depends on its length L cm
and its height H cm, where $C = \dfrac{3L + 2H}{10}$. Find C when
   a $L = 50, H = 20$ **b** $L = 30, H = 25$ **c** $L = 40, H = 30$

6 The number of small cakes, n, bought for a child's birthday party depends on the number of boys B and the number of girls G who are invited.
If $n = 2(3B + 2G)$, find n when
   a $B = 4, G = 2$ **b** $B = 2, G = 3$ **c** $B = 5, G = 0$

> **research**
>
> The USA measures temperature in °F. Europe uses °C.
> You can change °F to °C using the formula $C = \frac{5}{9}(F - 32)$
>
> **a** Change 212 °F and 32 °F to °C. What is special about these two temperatures?
>
> **b** What do these five surnames of famous scientists have in common?
>
> Fahrenheit   Celsius   Kelvin   Réaumur   Rankine

Substitution into formulae **71**

Extension

Ask students to investigate a number of mobile phone tariffs to make a recommendation as to which represents best value and justify their decision.

Exercise 5e commentary

Question 1 – This could be a whole class activity, using mini whiteboards, to allow diagnosis of misconceptions (1.3).

Question 2 – Similar to the first example (1.3).

Question 3 – Similar to the second example (1.3).

Question 4 – Discuss whether it is better to work out the value inside the brackets first and then multiply, or expand the brackets first. Ask students to try both methods and then justify which method they think is easiest (1.3).

Question 5 – Despite the apparent complexity, this is numerically straightforward (1.3).

Question 6 – Is it realistic that boys eat more cakes at parties than girls? (1.3)

Research – This science based activity lends itself to further investigation and findings could be presented back next lesson. Students are encouraged to plan and carry out research (IE, TW)

Links

Anders Celsius was a Swedish astronomer who proposed the Celsius temperature scale in 1742. He chose 0° as the boiling point of water and 100° as the freezing point and named the scale the centigrade scale (from the Latin for a hundred steps). Carl Linnaeus, another Swedish scientist, reversed the scale in 1745 and the scale is now used across the World. There is more information about Anders Celsius at http://en.wikipedia.org/wiki/Anders_Celsius

- Substitute integers into simple formulae, including examples that lead to an equation to solve   (L5)
- Derive simple formulae   (L5)

Useful resources

. . . .

Starter – T, S or F

Write the following on the board and ask students which are always **T**rue, **S**ometimes true, always **F**alse:

$x^2 = (-x^2)$ S, $x = 0$

$2(x - 3) = 2x - 3$ F

$x - y = y - x$ S, $x = y$

$x^2 = (-x)^2$ T

$3(2x + 4) = 2(3x + 6)$ T

Ask students to justify their answers using substitution.

Teaching notes

This unit fully supports the development of links within mathematics, using geometry and measures as a foundation upon which to build skills in algebra. There is also a major emphasis throughout on using formulae in real-life functional situations (1.2).

A stumbling block for many students is known to be weak literacy skills and the main thrust of the work here is around word problems. Emphasise the need to focus on identifying key information, pairing weaker students as necessary.

Maximum benefit to learning will be gained through encouraging discussion of each of the worked examples, wherever possible asking students to explain both the question and their solution in their own words. Encourage students to think creatively and make connections of their own through asking questions to extend their thinking (CT, EP).

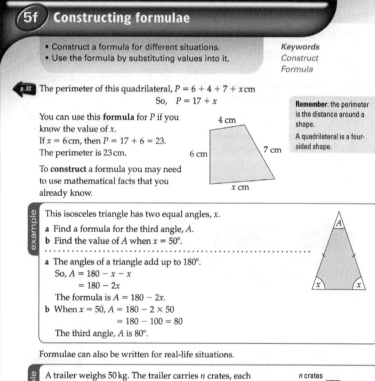

5f Constructing formulae

- Construct a formula for different situations.
- Use the formula by substituting values into it.

Keywords
Construct
Formula

p.22 The perimeter of this quadrilateral, $P = 6 + 4 + 7 + x$ cm
So, $P = 17 + x$

You can use this **formula** for P if you know the value of x.
If $x = 6$ cm, then $P = 17 + 6 = 23$.
The perimeter is 23 cm.

Remember: the perimeter is the distance around a shape.
A quadrilateral is a four-sided shape.

4 cm
6 cm
7 cm
x cm

To **construct** a formula you may need to use mathematical facts that you already know.

example

This isosceles triangle has two equal angles, x.
a Find a formula for the third angle, A.
b Find the value of A when $x = 50°$.
. . . .
a The angles of a triangle add up to 180°.
So, $A = 180 - x - x$
$= 180 - 2x$
The formula is $A = 180 - 2x$.
b When $x = 50$, $A = 180 - 2 \times 50$
$= 180 - 100 = 80$
The third angle, A is 80°.

A
x x

Formulae can also be written for real-life situations.

example

A trailer weighs 50 kg. The trailer carries n crates, each weighing 20 kg.
a Write a formula for the total mass of the loaded trailer.
b Find the total load if there are 10 crates.
. . . .
a The mass of all the crates is $20 \times n$. A formula for the total load is $L = 20n + 50$, where L is the mass of the load in kg.
b If $n = 10$, $L = 20 \times 10 + 50$
$= 200 + 50$
$= 250$.
So, with 10 crates, the total load is 250 kg.

n crates

You need to explain what any letters you invent actually mean.

72 Algebra Expressions and formulae

Plenary

Ask students to reflect on the work they have done on using and constructing formulae and on substitution and to list three key points that they think are the most important to remember. Take feedback and agree a whole class 'top three tips'.

Simplification

Where students struggle with this work, it will be necessary to identify whether it is the algebra work or the particular prior content knowledge that is the stumbling block to progress. Encourage paired work, as developing dialogue between students is supportive of overcoming difficulties.

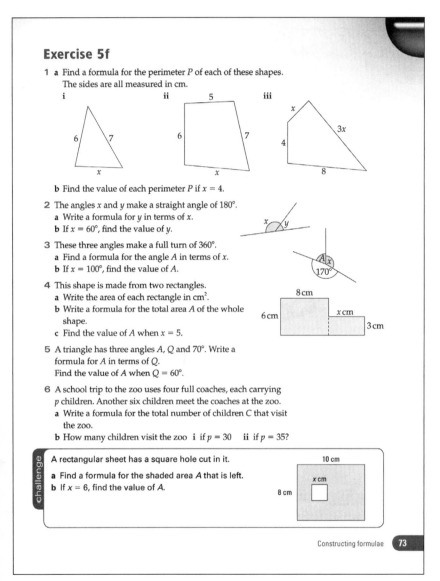

Exercise 5f

1 a Find a formula for the perimeter P of each of these shapes. The sides are all measured in cm.

  i   ii   5   iii

  6   7   6   7   4   3x

  x   x   8

  b Find the value of each perimeter P if x = 4.

2 The angles x and y make a straight angle of 180°.
  a Write a formula for y in terms of x.
  b If x = 60°, find the value of y.

3 These three angles make a full turn of 360°.
  a Find a formula for the angle A in terms of x.
  b If x = 100°, find the value of A.

4 This shape is made from two rectangles.
  a Write the area of each rectangle in cm².
  b Write a formula for the total area A of the whole shape.
  c Find the value of A when x = 5.

  8 cm
  6 cm   x cm   3 cm

5 A triangle has three angles A, Q and 70°. Write a formula for A in terms of Q.
  Find the value of A when Q = 60°.

6 A school trip to the zoo uses four full coaches, each carrying p children. Another six children meet the coaches at the zoo.
  a Write a formula for the total number of children C that visit the zoo.
  b How many children visit the zoo i if p = 30   ii if p = 35?

A rectangular sheet has a square hole cut in it.
  a Find a formula for the shaded area A that is left.
  b If x = 6, find the value of A.

  10 cm
  x cm
  8 cm

Constructing formulae   73

Extension

More able students can be challenged to draw further upon their prior knowledge in geometry and measures to create their own questions, using the examples as a model. This supports the consolidation of understanding of constructing formulae and substitution and also helps to embed functional application of skills.

Exercise 5f commentary

This exercise strengthens students' capacity to approach word problems (1.1).

Question 1 – Students could check answers to part **b** by replacing x by 4 in the diagrams and adding up the lengths directly (1.3).

Question 2 – The meaning of 'in terms of x' should be emphasised given its wide spread use.

Question 3 – Similar to the first example.

Question 4 – Highlight the strategy of breaking a problem down into smaller parts.

Question 5 – In principle, very similar to question **3** but its potential to cause confusion may require whole class discussion and agreement.

Question 6 – If the language proves troublesome, ask students to work in pairs to write a 'student friendly' version. Take feedback as a whole class.

Challenge – The diagram clarifies this problem; to develop fuller understanding, challenge students to describe the question in words, imagining that the diagram was not there.

Assessment Criteria – Algebra, level 5: construct, express in symbolic form and use simple formulae involving one or two operations.

Links

Albert Einstein's famous formula $E = mc^2$ says that mass can be converted into energy. The amount of energy contained in a piece of matter can be found by multiplying the mass m by the square of the speed of light, c. This means that, in theory, there is enough energy in a grain of sand to boil 10 million kettles.

5a

1 Write these in a simpler way.

a $x + x + x$ b $y + y + y + y$ c $2 \times 3 \times z$

2 There are x biscuits in a packet. You buy five packets.

a How many biscuits do you buy?

b You open one of the packets and eat six biscuits. How many biscuits do you now have altogether?

c If $x = 10$, how many biscuits are you left with?

3 If $p = 12$ and $q = 4$, find the values of

a $p + 2q$ b $2p - q$ c $\dfrac{p}{q}$ d $\dfrac{5q + 4}{p}$

5b

4 Find the values of

a $3 \times 3 \times 3$ b 2^4 c 10^2 d $5^2 \times 10^3$

5 Simplify each of these, using indices in your answers.

a $a \times a \times a \times b \times b \times c \times c \times c$ b $3^4 \times 3^2$ c $6^5 \times 6^3$

6 What values of n makes these statements true?

a $2^n = 16$ b $5^3 \times 5^n = 5^7$ c $3^6 \times 3^n \times 3^2 = 3^{10}$

5c

7 A necklace is made from two kinds of beads of length x cm and y cm.

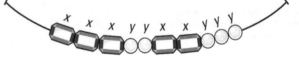

Write the total length of this necklace as simply as you can.

$3x + 2y + 2x + 3y = \square$

8 A patio with this pattern of paving slabs uses four identical hexagons and four identical triangles.

Write an expression for the perimeter of the shape in terms of x and y.

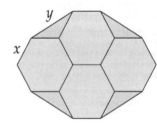

9 Simplify these expressions by collecting like terms.

a $3p + 2p + 5q - 2q$ b $4m + 2n + m - 3n$

c $3x^2 + 4x + 6x^2 - 5x$ d $z^2 + 5z + z^2 - 2z - 2z^2 - 3z$

10 a Find the total area of rectangles A and B together.
b Expand this bracket. $5(x + 2)$

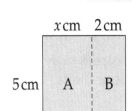

11 a Expand these brackets. **i** $4(x + 3)$ **ii** $3(5x + 4)$
b Expand these brackets and simplify your answers by collecting like terms.
i $2(y + 3) + 4(5y + 2)$ **ii** $3(2z + 4) + 2(z - 5)$

12 a A box weighing 20 grams contains 10 screws weighing x grams each. Write an expression for the total weight of the box and its contents.
b Mr Sturman buys five of these boxes. Write an expression (using brackets) for the total weight of these five boxes and their contents.
c Expand the brackets.

13 The time T hours to cook a turkey weighing W pounds is given by $T = \dfrac{W}{3} + 1$.

Find T when **a** $W = 12$ **b** $W = 18$ **c** $W = 20$

14 The charge £C for excess baggage when you fly depends on the weight W kg of your luggage where $C = 5(W - 20)$.
Find C when **a** $W = 32$ **b** $W = 65$ **c** $W = 20$

15 If $p = 4, q = 2$ and $r = 5$, find the values of
a $2p + q$ **b** $4r - 5p$ **c** $3(p + 2q - r)$

16 This trapezium has two sides of 8 cm and 5 cm and two equal unknown sides.
a Write a formula for its perimeter P.
b Find the value of P when $x = 4$ cm.

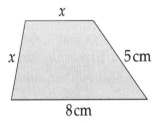

17 A triangle has three angles x, y and A.

a Write a formula for angle A in terms of x and y.
b Find the value of A when $x = 60°$ and $y = 45°$.

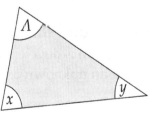

18 a Expand **i** $5(x + 2)$ **ii** $3(5x + 4)$
b Expand these and simplify by collecting like terms.

i $2(y + 3) + 4(5y + 2)$ **ii** $3(2z + 4) + 2(z - 5)$

5 Summary

Assessment criteria

- Show understanding of situations by describing them mathematically using symbols, words and diagrams   Level 5

- Construct, express in symbolic form and use simple formulae involving one or two operations.   Level 5

Question commentary

Example

The example illustrates a straightforward question about expanding brackets. Part **b** requires students to add together their answers from part **a** and simplify the resulting expression. Some students will add the expressions but not simplify the result. Ask questions such as "is the expression in its simplest form?" Emphasise thorough reading of the question and remind students to deal with letters and numbers separately.

a i $2x + 6$

  ii $3x - 6$

b $2x + 6 + 3x - 6 = 5x$

Past question

The question requires students to match several expressions to their algebraic form. Some students may find this difficult for the expressions involving multiplication and division. Ask questions such as "How do we write a division sign in algebra?" and "How do we know when two things have to be multiplied together?" Ask students to read out the algebraic forms of the expressions and explain what each one means before matching to the word forms.

Answer

Level 5

2 Add 2 to a $a + 2$

Subtract 2 from a $a - 2$

Multiply a by 2 $2a$

Divide a by 2 $\frac{a}{2}$

Multiply a by itself a^2

Development and links

The topic of algebra is developed further in Chapter 7 where students will solve equations with an unknown on both sides. Formulae are developed further in Chapter 13.

Students will encounter algebra in areas of the curriculum where formulae are used, especially in technology and in science where formulae are used extensively to describe the physical world. The language of algebra is used for computer programming in ICT.

6 Angles and 3-D shapes   Geometry

Objectives

- Use geometric properties of cuboids and shapes made from cuboids **5**
- Know that if two 2-D shapes are congruent, corresponding sides and angles are equal **5**
- Identify alternate angles and corresponding angles **6**
- Understand a proof that the angle sum of a triangle is 180° and of a quadrilateral is 360° **6**
- Understand a proof that the exterior angle of a triangle is equal to the sum of the two interior opposite angles **6**
- Solve geometrical problems using side and angle properties of equilateral, isosceles and right-angled triangles and special quadrilaterals, explaining reasoning with diagrams and text **6**
- Classify quadrilaterals by their geometric properties **6**
- Visualise 3-D shapes from their nets **6**
- Use simple plans and elevations **6**

Level

MPA

1.1	6a, f, g
1.2	6a, b, d, e, f, g
1.3	6a, b, c, d, d², f, g
1.4	6c, d
1.5	

PLTS

IE	6b, c, d, g
CT	6b, d, d², f, g
RL	6b, c, d
TW	6d
SM	6c, d, d², f, g
EP	6a, b, c, e

Introduction

This chapter consolidates and develops Year 7 knowledge of angle to include the sum of angles at a point, on a straight line and in a triangle. Students learn to recognise vertically opposite angles, alternate angles and corresponding angles which become important in later work when solving geometric problems. Students will investigate the properties of triangles and quadrilaterals and consider congruence. They begin to explore the properties of 3-D shapes using 2-D representations in the form of nets and plan and elevation views on isometric paper.

The student book discusses the Pyramid at the entrance to the Louvre Museum in Paris. The properties of shapes are very important in building, both for strength and for aesthetic appeal. Designers use plans and elevations to show how the completed building will look. Some examples of the use of shapes in various structures can be found at http://www.architecture. com/WhatsOn/Exhibitions/AtTheVictoriaAndAlbertMuseum/ ArchitectureGallery/Structures/Introduction.aspx

⚡Fast-track

All spreads

Extra Resources

6   Start of chapter presentation

6a   Worked solution: Q4b

6a   Consolidation sheet

6b   Animation: Angles in a triangle

6b   Consolidation sheet

6c   Starter: Missing angle bingo

6c   Worked solution: Q3a

6d   Animation: Quadrilaterals

6d   Consolidation sheet

6d² Starter: Angle-name matching

6d² Starter: 2D shape-name matching

6e   Animation: Congruent shapes

6e   Consolidation sheet

6f   Consolidation sheet

6g   Animation: Plans and elevations

6g   Consolidation sheet

Assessment: chapter 6

- Use correctly the vocabulary notation and labeling conventions for angles (L5)
- Know the sum of angles at a point and on a straight line (L5)
- Recognise vertically opposite angles (L5)

Useful resources

• •

Starter – Angle estimation

Draw a mixture of acute, obtuse and reflex angles on the board.

Ask students to estimate the size of each angle in degrees then measure the angles. Students score 6 points for an exact answer, 4 points for within 10 degrees and 2 points for within 15 degrees.

Teaching notes

Many students find it difficult to describe angles using three letters and it is important to emphasise the use of three letters when describing angles. It should also be realised that there are often two angles at angle ABC: the reflex and the acute or obtuse.

Draw a straight line and another line coming off it at a specified angle, say 45° or 120°, and ask students to work out the supplementary angle (135°, or 60°). Now extend the line and ask students if they can work out the size of the angle vertically opposite the original angle (45° or 120°) – again angles on a straight line. Draw out the general result and set students a problem similar to question **3b** and then a problem similar to the second example (1.2). Encourage students to tell you how to solve the problems and show them how to lay out their answers: giving reasons, 'angles on a straight line', *etc.*, and showing their workings. Emphasise the need to set out their answers as a series of logical steps.

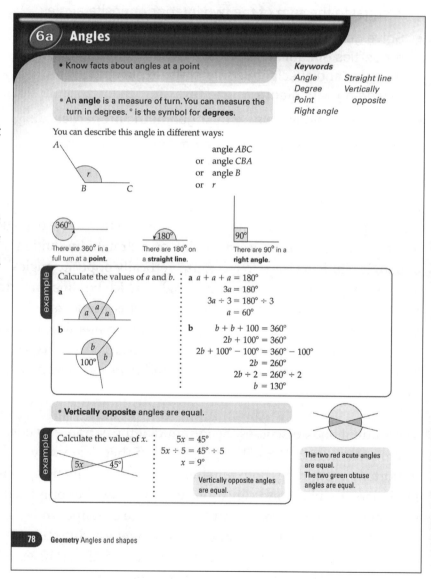

Plenary

Ask students to find the missing angles in this diagram. Ask them to tell you what arguments to use and how to apply them (EP).

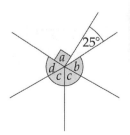

Simplification

Allow students to work in pairs, so that they can discuss which results they need to use and how they are applied to each problem. Offer more simple 'one step' problems, like question **3a**, for each question type.

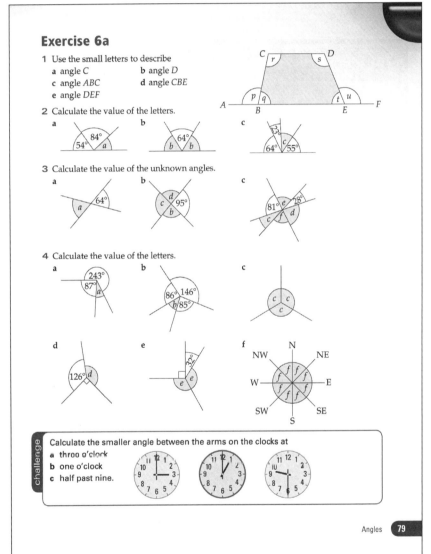

Exercise 6a

1 Use the small letters to describe
   a angle *C*   b angle *D*
   c angle *ABC*   d angle *CBE*
   e angle *DEF*

2 Calculate the value of the letters.
   a   b   c

3 Calculate the value of the unknown angles.
   a   b   c

4 Calculate the value of the letters.
   a   b   c

   d   e   f

challenge

Calculate the smaller angle between the arms on the clocks at
   a three o'clock
   b one o'clock
   c half past nine.

Angles **79**

Extension

Challenge students to find the angle between the hands of a clock showing 12:15 or 10:20, *etc*.

Exercise 6a commentary

Question 1 – Ask students to give a third way of describing each angle, for example, C, *r*, BCD or DCB (1.3).

Question 2 – Similar to part **a** of the first example; uses the angle on a straight line (1.1).

Question 3 – Similar to the second example. Parts **b** and **c** use the angle on a straight line (1.1).

Question 4 – Similar to part **b** of the first example. For part **f**, ask students to calculate the angles measured clockwise from North of the compass directions (1.1).

Challenge – It will help if students work out the angles around the clock: 0°, 30°, 60°, 90°, *etc*. In part **c**, remind students that the hour hand also moves.

Assessment Criteria – SSM, level 5: use language associated with angle and know and use the angle sum of a triangle and that of angles at a point.

Links

A sundial is a device that measures time. As the sun moves across the sky, a vertical pole or plate (the gnomon) casts a shadow which moves across a dial marked with hours like a clock. The largest sundial in the World is the Samrat Yantra (The Supreme Instrument) at Jantar Mantar in Jaipur in India. It is 27m tall and can be used to tell the time to an accuracy of about two seconds.

There are pictures of the Samrat Yantra at http://commons. wikimedia. org/wiki/ Image:Samrat_Yantra,_Jantar_ Mantar,_Delhi,_early_19th_century. jpg and at http://en.wikipedia.org/ wiki/Jantar_Mantar_(Jaipur)

- Solve geometric problems using side and angle properties of equilateral, isosceles and right-angled triangles and special quadrilaterals, explaining reasoning with diagrams and text (L5)
- Understand a proof that the exterior angle of a triangle is equal to the sum of the two interior opposite angles (L6)

Useful resources
Paper
Ruler
Scissors
Dictionaries

Starter – Straight line pairs

Write the following list of angles on the board:

73°, 156°, 116°, 13°, 107°, 104°, 35°, 64°, 145°, 49°, 89°, 151°, 131°, 55°, 24°, 76°, 29°, 167°, 125°, 91°.

Challenge students to match up the pairs (that add up to 180°) in the shortest possible time. Can be extended by using pairs that add up to 360°.

Teaching notes

The **activity** could be done as a whole class. This would then serve as motivation (but not proof) for stating that the sum of the angles in a triangle equals the angle on a straight line. If the alternative approach of tearing off the corners is used, make sure that the students first label the angles A, B, and C, so that they can easily identify which vertices to place together on a line.

In geometric arguments, it is important that students learn to state the results that they are using as well as writing down an equation, so that their logic is clear. Model this approach with a suitable example, taking the opportunity to remind students how to lay out the algebra.

Give students a problem to work on and then ask them to swap books with a partner. Go through the answer on the board assigning marks for stating the results used, layout of workings, checking the answers and 1 mark for the numerical answer. Ask students to assign a score and to write two sentences one saying what was good about the answer and one saying how it could be improved (EP, RL).

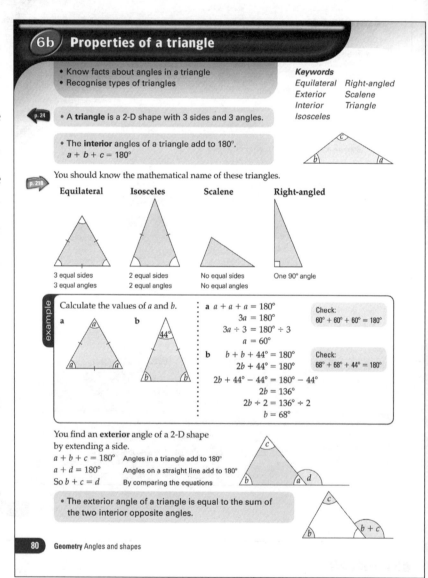

Plenary

Ask students to define acute, right, obtuse and reflex angles. Then challenge them to say what types of angles can be in a triangle (A-A-A or A-A-Ri, A-A-O). Why can't you have two obtuse angles? What are the allowed ranges of values of the angles in an isosceles triangle? (CT)

Simplification

Get students to work in pairs, so that they can discuss which results they need to use and how they are applied in each problem. Offer more simple 'one step' problems similar to question **1a**, for each question type.

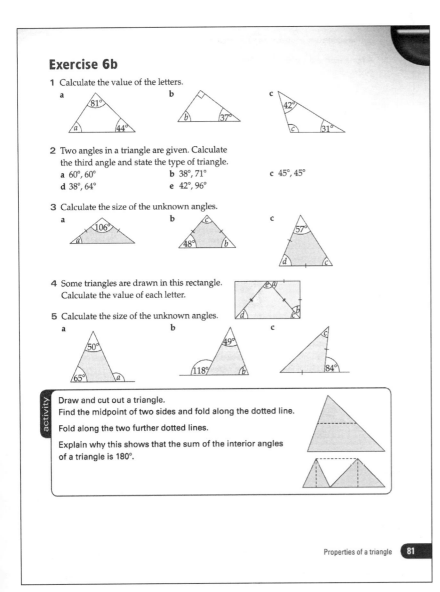

Exercise 6b

1 Calculate the value of the letters.

a   b   c

2 Two angles in a triangle are given. Calculate the third angle and state the type of triangle.
a 60°, 60°   **b** 38°, 71°   **c** 45°, 45°
d 38°, 64°   **e** 42°, 96°

3 Calculate the size of the unknown angles.

a   b   c

4 Some triangles are drawn in this rectangle. Calculate the value of each letter.

5 Calculate the size of the unknown angles.

a   b   c

activity
Draw and cut out a triangle.
Find the midpoint of two sides and fold along the dotted line.

Fold along the two further dotted lines.

Explain why this shows that the sum of the interior angles of a triangle is 180°.

Properties of a triangle   **81**

Extension

Challenge students to find the sum of the angles in a quadrilateral; hint at drawing in a diagonal and looking at the two triangles formed. Can they generalise this to other polygons (IE).

Exercise 6b commentary

Question 1 – Make sure that students state the result they are using and show their workings (1.3).

Question 2 – Encourage students to sketch triangles with the given angles to make the problem less abstract.

Question 3 – Similar to the example. In part **c**, suggest students rotate the page if the orientation causes problems.

Question 4 – To find a starting point, students may need reminding that they know the corner angles of the rectangle.

Question 5 – Students could approach these problems using the result that an exterior angle of a triangle is equal to the sum of the other two interior angles or using the results for the sum of the angles on a straight line and in a triangle. Both are correct (1.2).

Activity – Students will need to be fairly accurate when finding the midpoint and folding the paper: it may help to place the edge of a ruler along the fold line.

Assessment Criteria – SSM, level 5: use language associate d with angle and know and use the angle sum of a triangle and that of angles at a point.

Links

Bring in some dictionaries for the class to use. The word *isosceles* derives from the Greek *isos* meaning "equal", and *skelos* meaning "leg". Ask the class to find other words beginning with *iso* that are related to the word equal. For example, *isobar* – a line on a map linking points with the same atmospheric pressure, *isometric* – having equal dimensions.

- Identify parallel and perpendicular lines (L5)
- Identify alternate and corresponding angles (L6)
- Solve problems using properties of angles, of parallel and intersecting lines, and of triangles, justifying inferences and explaining reasoning with diagrams and text (L6)

Useful resources
Ruler
Protractor
Coloured pencils
Mini whiteboards

Starter – Triangle bingo

Ask students to draw a 3 × 3 grid and enter nine angles from the following list:

30°, 35°, 40°, 45°, 50°, 55°, 60°, 65°, 70°, 75°, 80°, 85°, 90°, 95°, 100°, 105°, 110°, 115°, 120°, 125°.

Give two angles of a triangle, for example, 83° and 42°.

If students have the third angle in their grid (55°) they cross it out. The winner is the first student to cross out all nine angles.

Teaching notes

It is important for all students to realise that, when there are parallel lines, many angles are the same and, where possible, this should be easy to identify by inspection. The introduction to the lesson should show alternate (**Z**) and corresponding (**F**) angles. Students should be reminded to write these reasons in brackets whenever they use them to solve a problem.

Measuring angles, question **1**, can often lead to problems and it is useful go through the procedure for doing this. First, estimate the angle (is it acute, obtuse or reflex?). Using a protractor place the cross at the vertex, the base line along one edge and read off from the scale starting at zero. Finally, check the measurement agrees with the initial estimate. If the diagram has short lines, place a ruler over the top of the protractor to see exactly where the line will cross the scale.

Question **5** offers the opportunity to develop one of the standard proofs that the sum of the angles in a triangle is 180°.

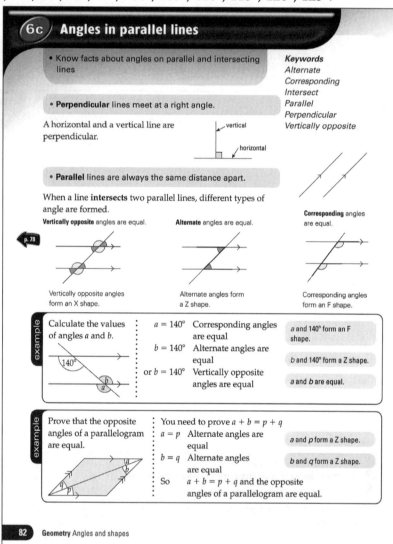

Plenary

Put up a multifaceted question, like **5**, and go round the class asking students how to solve it and what to write at each stage. It would be useful to introduce some deliberate mistakes or omissions to keep students focused. Emphasise the need to always consider whether the solution makes sense. It is likely that some students will want to solve the problem in a different sequence; it will be instructive to follow this though and demonstrate that the same answers are obtained (EP) (1.4).

Simplification

Students may find it difficult to identify alternate and corresponding angles given the many choices available, especially in a complex case such as question **5**. Suggest that students work in pairs to agree which angles go together. Ask them to copy the diagrams and, with a coloured pencil, mark on a Z or an F to help match angles.

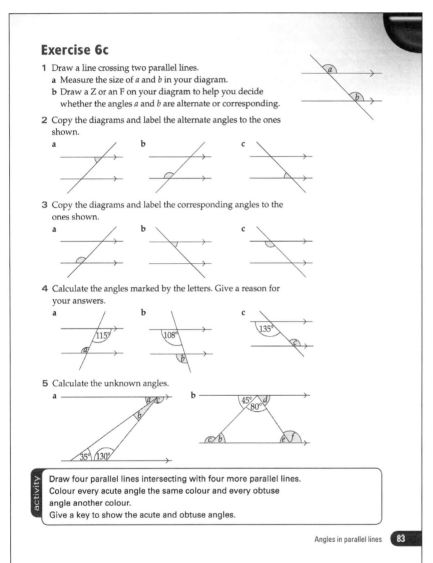

Exercise 6c

1 Draw a line crossing two parallel lines.
  a Measure the size of *a* and *b* in your diagram.
  b Draw a Z or an F on your diagram to help you decide whether the angles *a* and *b* are alternate or corresponding.

2 Copy the diagrams and label the alternate angles to the ones shown.
  a   b   c

3 Copy the diagrams and label the corresponding angles to the ones shown.
  a   b   c

4 Calculate the angles marked by the letters. Give a reason for your answers.
  a   b   c

5 Calculate the unknown angles.
  a   b

Angles in parallel lines **83**

Extension

Put students in pairs and ask them to make up their own question like those in question **5**, together with a mark scheme for their answer. Then get pairs to swap questions and then mark one another's answers giving one good feature of the answer and one area for improvement (RL).

Exercise 6c commentary

Question 1 – Ensure that students' diagrams are big enough to enable easy measurement of the angles.

Question 2 and **3** – These could be completed as a class using mini whiteboards. Check that parallel lines are correctly marked (1.3).

Question 4 – Similar to the first example.

Question 5 – This question brings together material from spreads **6a**–**c**. Let students work in pairs to decide which results to apply to find each angle before writing out carefully argued solutions (IE) (1.4).

Challenge – Ask students how many intersections, acute and obtuse angles there are in the diagram? How does these change as the number of parallel lines is changed; are there patterns? (SM)

Assessment Criteria – SSM, level 6: identify alternate and corresponding angles. SSM, level 6: solve geometric problems using properties of angles, of parallel and intersecting lines, and of triangles and other polygons.

Links

Parallel lines are used in road markings. Yellow lines laid parallel to the kerb indicate that vehicles must not park at certain times. Double yellow lines mean no waiting is permitted at any time. Red lines prevent all stopping, parking and loading. Double white lines down the centre of the road are used to prevent overtaking and reduce speeds. There is more information about road markings at http://www.direct.gov.uk/en/ TravelAndTransport/Highwaycode/ Signsandmarkings/index.htm? IdcService=GET_FILE&dID= 95931&Rendition=Web

6d Properties of a quadrilateral

- Classify quadrilaterals by their geometrical properties (L6)
- Understand a proof that the angle sum of a triangle is 180° and of a quadrilateral is 360° (L6)

Useful resources
Protractor
Ruler
Scissors
Glue

Starter – One hundred and eighty!

Ask students to write down the 180 times table.

Extend by asking students questions, for example,

How many 360s in 1080? (3)
What is the angle sum of 8 triangles? (1440°)
How many triangles will give an angle sum of 900°? (5)
How many 180s are there in 4500? (25).

Teaching notes

To introduce the topic, ask students to name and describe different quadrilaterals. These could be drawn by the students on the board and all characteristics marked in as a whole class activity. Alternatively students could work in pairs to draw the named quadrilateral marking in appropriate characteristics (TW).

Having shown the students how to establish the sum of the angles inside a quadrilateral, challenge them to repeat the proof for a concave quadrilateral (1.4).

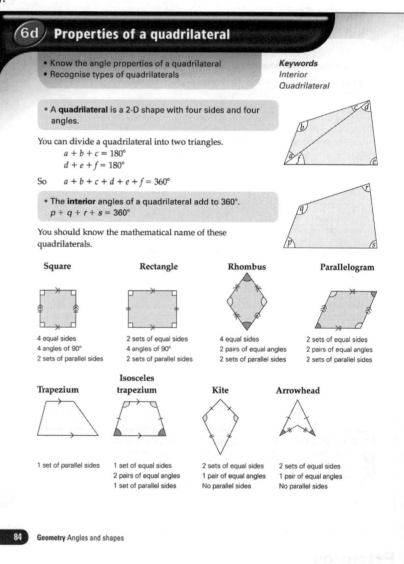

6d Properties of a quadrilateral

- Know the angle properties of a quadrilateral
- Recognise types of quadrilaterals

Keywords
Interior
Quadrilateral

- A **quadrilateral** is a 2-D shape with four sides and four angles.

You can divide a quadrilateral into two triangles.
$a + b + c = 180°$
$d + e + f = 180°$
So $a + b + c + d + e + f = 360°$

- The **interior** angles of a quadrilateral add to 360°.
$p + q + r + s = 360°$

You should know the mathematical name of these quadrilaterals.

Square
4 equal sides
4 angles of 90°
2 sets of parallel sides

Rectangle
2 sets of equal sides
4 angles of 90°
2 sets of parallel sides

Rhombus
4 equal sides
2 pairs of equal angles
2 sets of parallel sides

Parallelogram
2 sets of equal sides
2 pairs of equal angles
2 sets of parallel sides

Trapezium
1 set of parallel sides

Isosceles trapezium
1 set of equal sides
2 pairs of equal angles
1 set of parallel sides

Kite
2 sets of equal sides
1 pair of equal angles
No parallel sides

Arrowhead
2 sets of equal sides
1 pair of equal angles
No parallel sides

84 **Geometry** Angles and shapes

Plenary

Use some large quadrilaterals cut from paper or card – include a parallelogram, kite, square, rectangle, rhombus and trapezium. In turn, hold each shape behind a piece of card and gradually reveal it so that students can make educated guesses about the shape you are holding. For example, as you reveal the rectangle it may look like a square (or a kite) if you reveal one corner first (CT).

Simplification

For question **2**, provide students with a sheet showing the five quadrilaterals, without names. After identifying them students can draw in the diagonals and make measurements. Likewise have available pre-prepared sheets for question **3** and the **activity**.

Exercise 6d

1 Three angles in a quadrilateral are given. Calculate the fourth angle in each and state the type of quadrilateral.
There could be several answers for each question.
a 90°, 90°, 90°   **b** 60°, 120°, 120°   **c** 90°, 90°, 110°
d 30°, 90°, 210°   **e** 63°, 87°, 110°

2 Copy and complete the table to show the properties of the diagonals of these quadrilaterals. Use Yes or No for each answer.

	The diagonals		
	are equal in length	bisect each other	are perpendicular
Parallelogram			
Kite			
Rhombus			
Square			
Rectangle			

3 Name the different types of quadrilaterals in the regular pentagon on the right.

4 Give the names of quadrilaterals that have
  a 4 equal angles   **b** only one pair of equal angles
  c 4 equal sides.
  There may be more than one answer for each question.

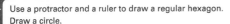

activity

Use a protractor and a ruler to draw a regular hexagon.
Draw a circle.
Use a protractor to mark off points at 60° intervals at 0°, 60°, 120°, 180°, 240°, 300° and 360° (same as 0°).
Join up the points with straight lines and cut out the 6 equilateral triangles.
a Rearrange 6 triangles to make a parallelogram.
   Show that the opposite angles of this parallelogram are equal.
b Rearrange 5 triangles to make an isosceles trapezium.
   Show that this isosceles trapezium has two pairs of equal angles.
c Rearrange 4 triangles to make an equilateral triangle.
   Explain why you know the triangle is equilateral.
d Rearrange 3 triangles to make an isosceles trapezium.
   Calculate the sum of the interior angles of a trapezium
e Rearrange 2 triangles to make a rhombus.
   Show that the opposite angles of this rhombus are equal.

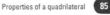

Properties of a quadrilateral **85**

Extension

In the final activity ask students to use two equilateral triangles, then three, then four, *etc.* How many different shapes they can make joining these along a common side?

Exercise 6d commentary

Question 1 – Allow students to work in pairs on finding the missing angles and naming the quadrilaterals. It will help to draw the shapes. Can any pairs find all the possible solutions? They may need to reorder the angles (RL, IE).

Question 2 – Ask students to make accurate drawings of the various quadrilaterals so that they can draw in the diagonals and make measurements in order to complete the table (SM).

Question 3 – Have multiple copies of the pentagonal diagrams available so that students can mark on the various quadrilaterals.

Question 4 – Encourage students to try to find all the possible answers (1.2).

Challenge – Have a sheet available showing a number of circles with six equal markings on the circumference to help students make regular hexagons. This should provide sufficient triangles to make all the target shapes which can be put on a small poster with space for students to write the required explanations (1.3).

Assessment Criteria – SSM, level 6, understand a proof that the sum of the angles of a triangle is 180° and of a quadrilateral is 360°.

Links

The Trapezium cluster is a bright cluster of stars in the constellation of Orion discovered by Galileo in 1617. The four brightest stars form the shape of a trapezium. There is more information about the Trapezium cluster at http:// en.wikipedia.org/wiki/Trapezium_ cluster and at http://www. astropix.com/HTML/B_WINTER/ TRAPEZ.HTM

- Explain why inscribed regular polygons can be
  constructed by equal divisions of a circle   (L6)

Useful resources
Squared paper
Protractor
Ruler
Scissors

Starter – SHP!

Write words on the board missing out the vowels, for example, shp.

Ask students for the original words: shape. Possible 'words':

NGL, PRLLL, RHMBS, LTRNT, RCTNGL, DGNL, SSCLS, KT, PRPNDCLR, QLTRL

(angle, parallel, rhombus, alternate, rectangle, diagonal, isosceles, kite, perpendicular, equilateral).

Can be extended by asking students to make up their own examples.

Teaching notes

It is appropriate for all students to have a copy of the table showing the names of polygons from three to ten sides. In pairs ask students to draw a blank table for number of sides and name of polygon (do not use the text book!), now give them two minutes to complete the table writing in the names of polygons that they know. Many will write square for quadrilateral, a discussion point, and most will need help with heptagon and nonagon.

If students have drawn a regular hexagon previously through the equal divisions of a circle method then they will easily be able to draw a regular pentagon as in the example, otherwise some guidance will be needed (1.3).

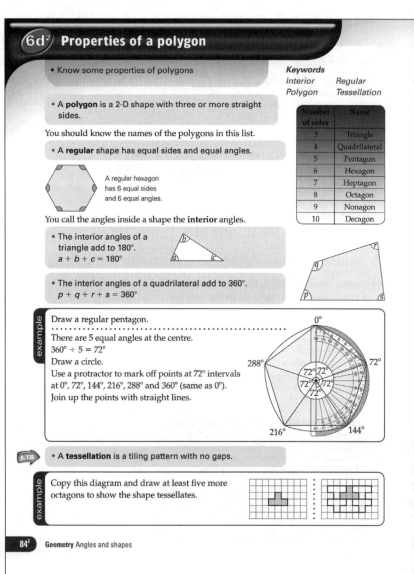

6d² Properties of a polygon

- Know some properties of polygons

Keywords
Interior   Regular
Polygon   Tessellation

- A **polygon** is a 2-D shape with three or more straight sides.

You should know the names of the polygons in this list.

- A **regular** shape has equal sides and equal angles.

A regular hexagon has 6 equal sides and 6 equal angles.

Number of sides	Name
3	Triangle
4	Quadrilateral
5	Pentagon
6	Hexagon
7	Heptagon
8	Octagon
9	Nonagon
10	Decagon

You call the angles inside a shape the **interior** angles.

- The interior angles of a triangle add to 180°.
  $a + b + c = 180°$

- The interior angles of a quadrilateral add to 360°.
  $p + q + r + s = 360°$

example
Draw a regular pentagon.
There are 5 equal angles at the centre.
$360° ÷ 5 = 72°$
Draw a circle.
Use a protractor to mark off points at 72° intervals at 0°, 72°, 144°, 216°, 288° and 360° (same as 0°).
Join up the points with straight lines.

- A **tessellation** is a tiling pattern with no gaps.

example
Copy this diagram and draw at least five more octagons to show the shape tessellates.

84²   Geometry Angles and shapes

Plenary

Consider tessellations of triangles, squares and hexagons: how many of each meet at a point to form a tessellation (6 at 60°, 4 at 90° and 3 at 120°). Why can't other polygons tessellate? (CT)

Simplification

Accurate drawing and construction can take a long time: for those students where support is needed allow extra time or ask them to complete only one tessellation pattern accurately. Do not complete the drawings for them.

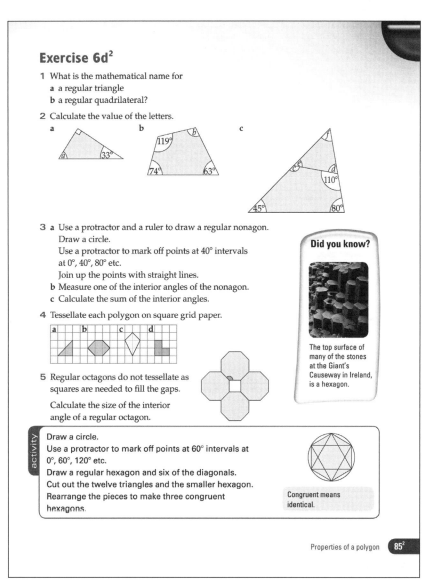

Exercise 6d²

1 What is the mathematical name for
   a a regular triangle
   b a regular quadrilateral?

2 Calculate the value of the letters.

3 a Use a protractor and a ruler to draw a regular nonagon.
     Draw a circle.
     Use a protractor to mark off points at 40° intervals at 0°, 40°, 80° etc.
     Join up the points with straight lines.
   b Measure one of the interior angles of the nonagon.
   c Calculate the sum of the interior angles.

4 Tessellate each polygon on square grid paper.

5 Regular octagons do not tessellate as squares are needed to fill the gaps.

   Calculate the size of the interior angle of a regular octagon.

Did you know?

The top surface of many of the stones at the Giant's Causeway in Ireland, is a hexagon.

activity

Draw a circle.
Use a protractor to mark off points at 60° intervals at 0°, 60°, 120° etc.
Draw a regular hexagon and six of the diagonals.
Cut out the twelve triangles and the smaller hexagon.
Rearrange the pieces to make three congruent hexagons.

Congruent means identical.

Properties of a polygon 85²

Extension

In question **3,** can students find an easy way of working out the interior angle of a nonagon? $(180 - 40)°$ How can they find the angle sum of other regular polygons?

Exercise 6d² commentary

In the constructions, emphasise the importance of accuracy.

Question 1 – It may help to draw the shapes.

Question 2 – Part c uses the sum of angles on a straight line.

Question 3 – Similar to the first example. It may help to draw a circle the same radius as the protractor. Ask students to check that their answers to parts **b** and **c** are consistent (1.3).

Question 4 – Similar to the second example. There are multiple ways to tessellate the shapes in parts **a** and **d**.

Question 5 – Ask students if they can see why a regular octagon alone cannot tessellate; question can be extended to other regular polygons.

Activity – If completed accurately, the students should have six identical isosceles and six identical equilateral triangles. Check that students do not confuse themselves by drawing in the diameters. The small central hexagon can be used as a base on which to build the other hexagons (SM). (1.3)

Assessment Criteria – SSM, level 5: use a wider range of properties of 2-D and 3-D shapes.

Links

The Giant's Causeway is a formation of thousands of columns of basalt which jut into the sea. It resulted from a volcanic eruption 60 million years ago. According to local legend, the Causeway was a bridge for two giants who wanted to cross the sea to do battle. There is more information about the Giant's Causeway at http://www.giantscausewayofficialguide.com/home.htm

- Know that if two 2-D shapes are congruent, corresponding sides and angles are equal   (L6)

Useful resources
Squared paper
Tracing paper
Scissors
Coloured pencils

Starter – Guess the polygons

I have exactly four lines of symmetry. (square)

Each of my angles is exactly 120°. (hexagon)

I have one pair of parallel lines and my angle sum is half of 720°. (trapezium)

I have one angle of 90°. All my other angles are less than this. (right-angled triangle)

My angle sum is 540°. (pentagon)

Can be extended with students' own clues.

Teaching notes

Explain to students that congruent means exactly the same size and shape. Challenge then to think of pairs of letters that can be congruent: b—d—p—q, W—M—E, i—l, H—I, u—n—c

Identifying corresponding sides and angles can be a little tricky. Display two congruent triangles, with sides and angles labelled, and ask the class how they should match up corresponding pairs. Agree on a correct way to do this and test understanding with another example, perhaps with triangles in a new set of orientations.

Work through a question similar to the second example to illustrate how congruence can be used to gain information.

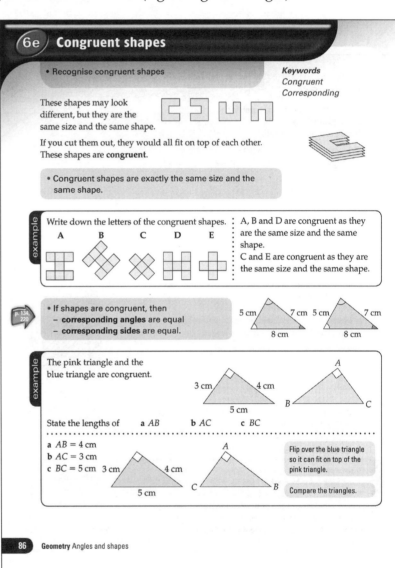

Plenary

Draw an isosceles triangle and join the mid-point of the base to the opposite vertex. Ask students what they can say about the two sub-triangles: agree there are three pairs of equal length sides. Explain that this enough to conclude that the triangles are congruent; what can they say about the angles? The two base angles must be equal. The two angles formed by the base and the line to the vertex must be equal, and, since they add up to a straight line, must both be right angles: the line is a perpendicular bisector.

Simplification

It may help students to trace a copy of one shape so that they can test whether it fits on top of another shape (is congruent) and to help in identifying the corresponding angles and sides.

Exercise 6e commentary

Question 1 – Similar to the first example (1.2).

Question 2 and **3** – Similar to the second example.

Activity – The diagram will be easier to copy on 1 cm² grid paper. It may help to identify congruent shapes if students cut up copies of the diagram. For part **c**, students will need to work systematically (SM).

Links

Patchwork is a form of needlework in which pieces of different fabric are cut into shapes and then joined together to form a larger design. Congruent shapes are often used and tessellated to form decorative quilts. There are examples of quilt patterns using congruent shapes at http://quilting.about.com/od/picturesofquilts/ig/Scrap-Quilts-Photo-Gallery/ (EP)

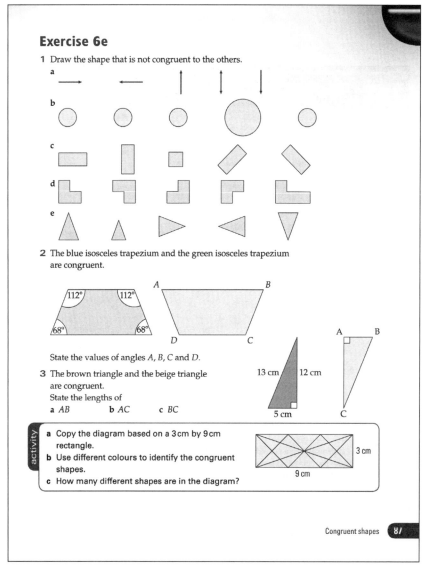

Exercise 6e

1 Draw the shape that is not congruent to the others.

a

b

c

d

e

2 The blue isosceles trapezium and the green isosceles trapezium are congruent.

State the values of angles *A*, *B*, *C* and *D*.

3 The brown triangle and the beige triangle are congruent.
State the lengths of
a *AB*   b *AC*   c *BC*

13 cm   12 cm

5 cm

activity
a Copy the diagram based on a 3 cm by 9 cm rectangle.
b Use different colours to identify the congruent shapes.
c How many different shapes are in the diagram?

3 cm

9 cm

Extension

Part **c** of the **activity** is already challenging. Students could also be asked to count how many of each shape can be found within the diagram or how many triangles there are in the diagram.

- Use 2-D representations to visualise 3-D shapes and deduce some of their properties   (L5)
- Visualise 3-D shapes from their nets   (L6)

Useful resources
Isometric paper
Squared paper
Scissors

Starter – How many angles?

Two lines meet exactly at a point. Excluding reflex angles, one angle is made.

Ask students how many non-reflex angles will be made if three lines meet at a point. (3)

What if four lines meet at a point? (6) Five lines? (10)

[*n* lines, *n*(*n* – 1)/2 angles]

Teaching notes

Nets seen in a book do not always give the impression of a plan that will fold into a solid shape. It is important that students are able to identify this relationship and the challenge as a teacher is to make this exercise a real constructional activity for all students. Where possible have a number of nets available that can be easily folded into a 3-D solid.

In addition to making students aware of nets it is also desirable to practice drawing solids on isometric paper. As a whole-class activity it may be helpful to complete the example on the introductory page; make sure that the isometric paper is the right way round!

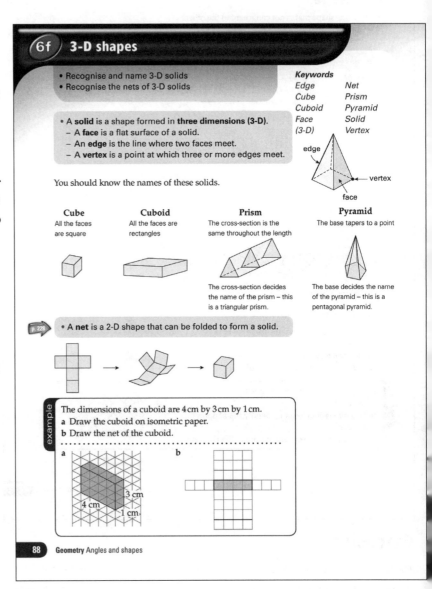

6f   3-D shapes

- Recognise and name 3-D solids
- Recognise the nets of 3-D solids

Keywords
Edge   Net
Cube   Prism
Cuboid   Pyramid
Face   Solid
(3-D)   Vertex

- A **solid** is a shape formed in **three dimensions (3-D)**.
  - A **face** is a flat surface of a solid.
  - An **edge** is the line where two faces meet.
  - A **vertex** is a point at which three or more edges meet.

You should know the names of these solids.

Cube
All the faces are square

Cuboid
All the faces are rectangles

Prism
The cross-section is the same throughout the length

The cross-section decides the name of the prism – this is a triangular prism.

Pyramid
The base tapers to a point

The base decides the name of the pyramid – this is a pentagonal pyramid.

- A **net** is a 2-D shape that can be folded to form a solid.

example

The dimensions of a cuboid are 4 cm by 3 cm by 1 cm.
a   Draw the cuboid on isometric paper.
b   Draw the net of the cuboid.

a   3 cm   4 cm   1 cm   b

88   Geometry Angles and shapes

Plenary

Using squared paper, challenge students to draw three nets for a cube: one should have four square faces in a row, one should have no more than three square faces in a row and one should have no more than two square faces in a row.

Simplification

Wherever possible have 3-D solids available for students to compare with the nets in the book. This will also help with counting faces, vertices and edges.

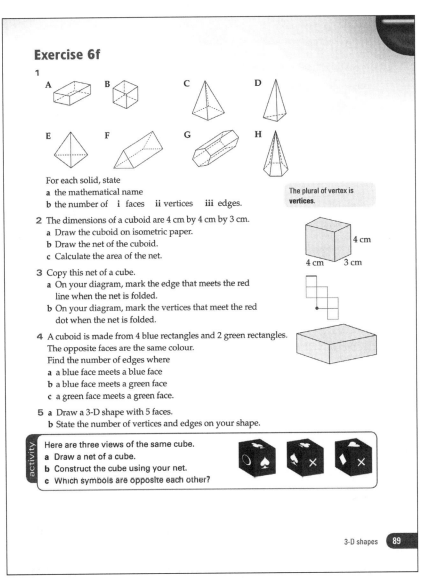

Exercise 6f

1

A B C D

E F G H

For each solid, state
a the mathematical name
b the number of i faces ii vertices iii edges.

> The plural of vertex is **vertices**.

2 The dimensions of a cuboid are 4 cm by 4 cm by 3 cm.
 a Draw the cuboid on isometric paper.
 b Draw the net of the cuboid.
 c Calculate the area of the net.

4 cm
4 cm 3 cm

3 Copy this net of a cube.
 a On your diagram, mark the edge that meets the red line when the net is folded.
 b On your diagram, mark the vertices that meet the red dot when the net is folded.

4 A cuboid is made from 4 blue rectangles and 2 green rectangles. The opposite faces are the same colour.
 Find the number of edges where
 a a blue face meets a blue face
 b a blue face meets a green face
 c a green face meets a green face.

5 a Draw a 3-D shape with 5 faces.
 b State the number of vertices and edges on your shape.

activity
Here are three views of the same cube.
 a Draw a net of a cube.
 b Construct the cube using your net.
 c Which symbols are opposite each other?

3-D shapes **89**

Extension

Challenge students to make the net of a solid with eight faces so that each face is a regular shape. What is the name of this solid? Can they make other solids using only regular shapes with different numbers of faces?

Exercise 6f commentary

Question 1 – Students should work in pairs. Ask them to put their results in a table and to see if they can spot a relationship: Euler's formula is F + V − E = 2 (CT) (1.2).

Question 2 – Similar to the example. Check that students orient the isometric paper so as to have vertical (not horizontal) lines. Having drawn one face it may help to draw the matching back face faintly to make it easier to join the parallel edges. Students may need reminding how to calculate the surface area (1.1, 1.3).

Question 3 – It will help to have available copies of the net so that students can see how it fits together (1.3).

Question 4 – Encourage students to draw a net, using the example as a model, and mark the coloured faces as an aid to visualisation (1.2).

Question 5 – Two solutions are possible: square pyramid (5 vertices, 8 edges), triangular prism (6 vertices and 9 edges). Challenge students to find both.

Activity – Encourage students to hold their cube in the same orientation as one of the diagrams and to copy over the three images, then move on to the next diagram. The orientation of the symbols should be correct (SM) (1.2).

Assessment Criteria – SSM, level 6: visualize and use 2-D representations of 3_d objects.

Links

There is a collection of nets for paper models of more complex solids at http://www. korthalsaltes. com/index.html

- Use geometric properties of cuboids and shapes made from cuboids (L6)
- Use simple plans and elevations (L6)

Useful resources
Isometric paper
Squared paper
Multilink cubes

Starter – Faceless

Ask students questions involving the numbers of faces, edges and vertices of 3-D shapes, for example,

> The sum of faces on a cuboid and vertices on a triangular prism. (12)
> The product of edges on a cube and faces on a square-based pyramid. (60)
> The difference between vertices on a cuboid and edges on a pentagonal pyramid. (2)

Teaching notes

The use of multi-link cubes in the classroom is challenging. Start with a short exercise to help introduce views and elevations. Show the class four cubes in an L shape: ask them (in pairs) to draw the front elevation, the side elevation and the plan view. This is similar to the example given. It will allow you to highlight the language and also the need for bold lines where the level of cubes changes.

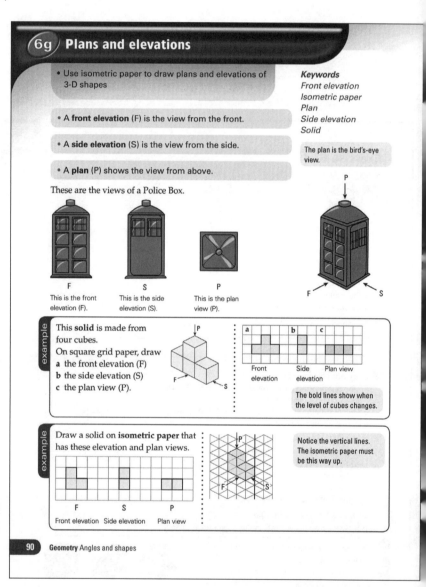

6g Plans and elevations

- Use isometric paper to draw plans and elevations of 3-D shapes

Keywords
Front elevation
Isometric paper
Plan
Side elevation
Solid

- A **front elevation** (F) is the view from the front.
- A **side elevation** (S) is the view from the side.
- A **plan** (P) shows the view from above.

The plan is the bird's-eye view.

These are the views of a Police Box.

F — This is the front elevation (F).
S — This is the side elevation (S).
P — This is the plan view (P).

example

This **solid** is made from four cubes.
On square grid paper, draw
a the front elevation (F)
b the side elevation (S)
c the plan view (P).

The bold lines show when the level of cubes changes.

example

Draw a solid on **isometric paper** that has these elevation and plan views.

F Front elevation
S Side elevation
P Plan view

Notice the vertical lines. The isometric paper must be this way up.

90 **Geometry** Angles and shapes

Plenary

Give students four or five multi-link cubes each. Projected or drawn on paper show them a front elevation, for example, two squares as a rectangle. What could the shape be? Show them the side elevation, for example, again two squares as a rectangle, what could the shape be? Show them the plan view, for example, this could be four squares as a 2 × 2 square. At each stage they should make the solid they think is being represented. Other shapes could be an L shape using three cubes, a T shape using five cubes, or be more adventurous! To plan this, make the shape yourself using cubes and then draw the views.

Simplification

Make available multilink cubes for all the questions. If possible, actual 3-D solids that can be handled will also help with visualisation.

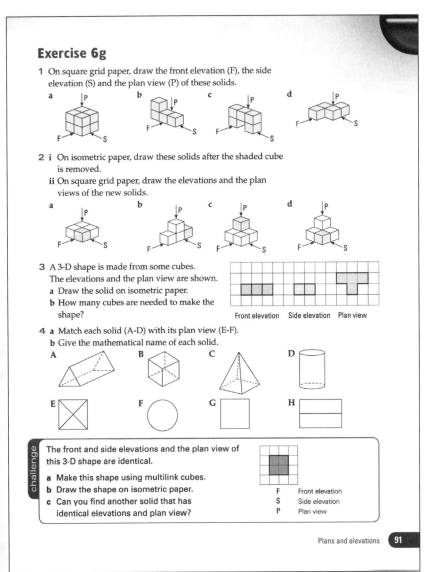

Extension

Give students the challenge of drawing three different solids on isometric paper that will have the same plan view, which they should also draw. How different can they make the solids? (SM)

Exercise 6g commentary

Question 1 – Similar to the first example. Check that students show 'bold lines' where appropriate (1.1, 1.2, 1.3).

Question 2 – Part **i** is similar to the second example. Check the orientation of the isometric paper. It may help to first draw the solid including the shaded cube and then to remove it. Part **ii** is similar to the first example (IE) (1.1, 1.2, 1.3).

Question 3 – Similar to the second example. Remind students to pay attention to the 'heavy' lines (1.1, 1.2, 1.3).

Question 4 – Insist on correct and full mathematical names: triangular prism, not just prism (or toblerone!) (1.1, 1.2).

Challenge – For part **c**, there is a whole family of various sized cubes but encourage other solutions such as '3-D crosses' (CT) (1.3).

Assessment Criteria – SSM, level 6: visualize and use 2-D representations of 3_d objects

Links

Engineers and architects use drawings showing plan and elevation views of parts, products and buildings. Traditionally drawings were produced by hand but computers have revolutionised the process and most drawings are now produced using computer-aided design (CAD). There is an example of an engineering drawing showing plan and elevation views at http://en.wikipedia.org/wiki/Image: Schneckengetriebe.png

6a

1 Calculate the value of the unknown angles.

a

b

c

2 Arrange these six angle values so they fit on a straight line and at a point.

14°   22°   24°   124°   142°   214°

6b

3 Two angles in a triangle are given. Calculate the third angle and state the type of triangle.

a 30°, 75°   b 43°, 47°   c 36°, 108°

d 35°, 64°   e 45°, 90°

4 One angle in an isosceles triangle is 70°.
What is the size of the other two angles?

> There are two possible answers to this question.

6c

5 Calculate the unknown angles.
Give a reason in each case.

a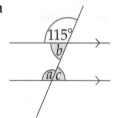

b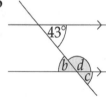

c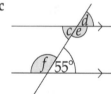

6d

6 Calculate the unknown angles in these quadrilaterals.

a

Kite

b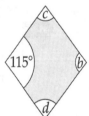

Rhombus

c

Arrowhead

7 State which shapes are the same in this regular hexagon and give the mathematical name of each shape.

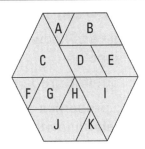

8 The diagram shows five regular pentagons and a rhombus.
One angle in the rhombus is 36°.

Calculate the values of *a* and *b*.

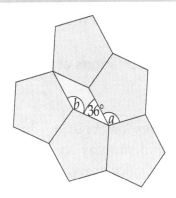

9 Identify the triangles that are congruent to the green triangle.

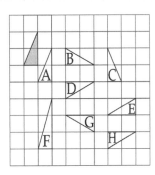

10 A prism is shown.
  a State the number of   **i** faces
                               ii vertices
                               iii edges.
  b Draw the solid on isometric paper.

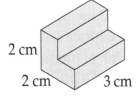

11 Sketch the front elevation (F), the side elevation (S) and the plan view (P) of this dice.

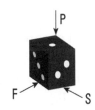

6 Summary

Assessment criteria

- Use a wider range of properties of 2-D and 3-D shapes   Level 5
- Use language associated with angle and know and use the angle sum of a triangle and that of angles at a point   Level 5
- Classify quadrilaterals by their geometric properties   Level 6
- Solve geometrical problems using properties of angles, of parallel and intersecting lines, and of triangles and other polygons   Level 6
- Identify alternate and corresponding angles   Level 6
- Understand a proof that the sum of the angles of a triangle is 180° and of a quadrilateral is 360°   Level 6
- Visualise and use 2-D representations of 3-D objects   Level 6

Question commentary

Example

The example illustrates a question about calculating angles. Some students may try to measure angles instead of calculating, or assume that angles that look similar must be equal. Emphasise the importance of checking the answers by adding. This is an opportunity to ask probing questions such as "How could you convince me that the sum of the angles in a triangle is 180" or "What other information do you know about triangles/ parallel lines?"	**a** $a = 43°$, alternate angles **b** $b = 180 - (43 + 72) = 65°$, angles on a straight-line/in a triangle **c** $c = 72°$, alternate angles

Past question

The question requires students to visualise a 3-D shape from a drawing on isometric paper. Some students will count only the faces showing. In part **b**, students are required to draw a shape with six faces. Encourage students to think of the shape first, before trying to draw it. Ask questions such as "What shape has six faces?"	**Answer** **Level 5** **2 a** 8 faces **b** Drawing of a cuboid or a cube or a pentagonal pyramid on isometric paper.

Development and links

Students will apply knowledge of congruence in work on transformations and enlargements in Chapter 9. The topic of 3-D shapes is developed in chapter 15 when students will also construct triangles and quadrilaterals using a ruler and compasses. Knowledge of the sum of angles at a point is important when constructing pie charts in Chapter 11.

The sum of angles at a point and in a triangle is important when working with bearings and maps and so links to the geography curriculum. Angle work is especially important in physics when calculating forces and in any subject where pie charts are used. The properties of shapes are used in art and design to create geometric patterns and tessellations.

7 Equations and graphs   Algebra

Objectives

- Use formulae from mathematics and other subjects 5
- Know the meanings of the words *formula* and *function* 5
- Generate points in all four quadrants 5
- Construct and solve linear equations with integer coefficients (unknown on either or both sides, without and with brackets) using appropriate methods (e.g. inverse operations, transforming both sides in same way) 6
- Express simple functions algebraically and represent them in mappings 6
- Plot the graphs of linear functions, where *y* is given explicitly in terms of *x*, on paper and using ICT 6

Introduction

In this chapter, students will review linear equations and extend techniques for solving them to include using inverse operations, solving equations with an unknown on both sides and equations containing brackets. The chapter develops to using mapping diagrams to represent functions, generating tables of values, plotting straight line graphs and beginning to explore their properties.

The student book discusses using equations to model aerodynamics. Equations are used in science and technology to describe and explain physical phenomena. Constructing and solving equations gives us the ability to understand and predict what is likely to happen in given circumstances. For example, a bridge has to be able to withstand the daily stresses and strains due to both passing traffic and extreme weather conditions. The Millennium Bridge in London is a famous example where the engineers didn't get it quite right first time. The bridge had to be closed shortly after opening as it started swaying violently when people walked over it. Engineers reconsidered the problem, drew up some more equations, the bridge was modified and the swaying eliminated. There is more information about the Millennium Bridge at http://www.arup.com/MillenniumBridge/index.html

🢒Fast-track

7a, b, e, f

Level

MPA

1.1	7a, b, c, d, e
1.2	7c, d, f
1.3	7a, b, c, e, f
1.4	7b, f
1.5	7c

PLTS

IE	7a, b, e, f
CT	7a, b, c, d, e
RL	7a, b, c, e
TW	7a, b
SM	7c, d, f
EP	7a, b, c, d, e

Extra Resources

7	Start of chapter presentation
7a	Animation: Solving equations
7a	Animation: Balancing equations
7a	Consolidation sheet
7b	Starter: Simplifying algebra time challenge
7b	Consolidation sheet
7c	Animation: Solving equations 2
7c	Animation: Solving equations with brackets
7c	Worked solution: Q2a
7c	Consolidation sheet
7e	Animation: Straight line equations
7e	Animation: Plotting coordinates
7e	Consolidation sheet
7f	Starter: Coordinates time challenge
7f	Worked solution: Q4a
	Assessment: chapter 7

- Solve simple linear equations with integer coefficients (unknown on one side) using an appropriate method (e.g. inverse operations) (L5)

Useful resources
Mini whiteboards
Scales and weights

Starter – Algebra scores 24

Each consonant scores 2 and each vowel scores 1. Multiply the total consonant score by the total vowel score to get the word score. Write down mathematical words and find their scores. Bonus points for scores that equal 24 (algebra, brackets, decimal, formulae, straight).
Can be differentiated by the score allocated to a consonant or vowel.

Teaching notes

Make links with students' prior knowledge of this topic through the use of the first questions in the exercise. This can be completed as a whole class mini whiteboard activity to ensure all students are involved and to identify any problems early on.

Exploration of the balancing method to explain inverse operations provides a solid foundation for many students to progress in this topic, as this approach helps students to develop a more concrete understanding. Using an actual pair of scales with mystery weights may consolidate understanding further. It is worth highlighting the meaning of inverse when discussing the example (1.1).

Use the later questions in the exercise as small group work to encourage student collaboration and develop dialogue early on, as many students lack confidence when solving equations. This approach also fosters the ability to reach agreements, managing discussions to achieve results (TW, EP).

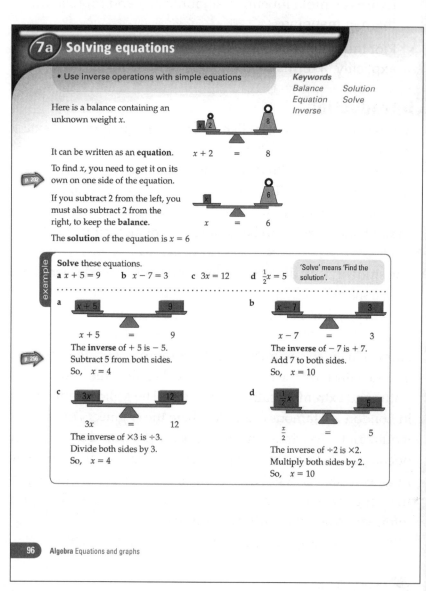

Plenary

Write three examples of the type of equation in the **challenge** involving more than one step – offer a range of difficulty levels – and invite students to choose one and write an annotated solution. They should then swap and compare, discussing any disagreements (RL).

Simplification

For those students who struggle here, small group work and the development of mathematical dialogue is crucial to build confidence and embed understanding. Ensure that students understand inverse operations and offer additional examples of single-step problems.

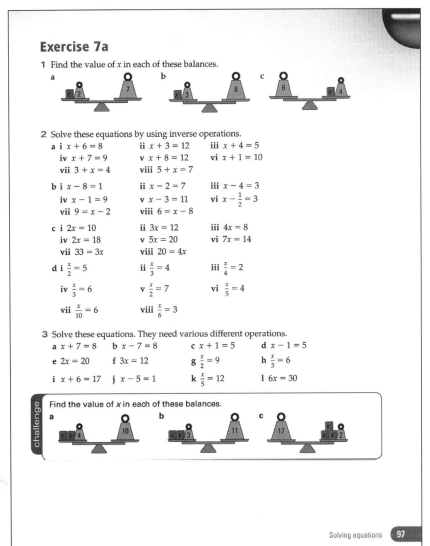

Exercise 7a

1 Find the value of x in each of these balances.

a   b   c

2 Solve these equations by using inverse operations.

a i $x + 6 = 8$ ii $x + 3 = 12$ iii $x + 4 = 5$
   iv $x + 7 = 9$ v $x + 8 = 12$ vi $x + 1 = 10$
   vii $3 + x = 4$ viii $5 + x = 7$

b i $x - 8 = 1$ ii $x - 2 = 7$ iii $x - 4 = 3$
   iv $x - 1 = 9$ v $x - 3 = 11$ vi $x - \frac{1}{2} = 3$
   vii $9 = x - 2$ viii $6 = x - 8$

c i $2x = 10$ ii $3x = 12$ iii $4x = 8$
   iv $2x = 18$ v $5x = 20$ vi $7x = 14$
   vii $33 = 3x$ viii $20 = 4x$

d i $\frac{x}{2} = 5$ ii $\frac{x}{3} = 4$ iii $\frac{x}{4} = 2$
   iv $\frac{x}{3} = 6$ v $\frac{x}{2} = 7$ vi $\frac{x}{5} = 4$
   vii $\frac{x}{10} = 6$ viii $\frac{x}{6} = 3$

3 Solve these equations. They need various different operations.

a $x + 7 = 8$ **b** $x - 7 = 8$ **c** $x + 1 = 5$ **d** $x - 1 = 5$
e $2x = 20$ **f** $3x = 12$ **g** $\frac{x}{2} = 9$ **h** $\frac{x}{3} = 6$
i $x + 6 = 17$ **j** $x - 5 = 1$ **k** $\frac{x}{5} = 12$ **l** $6x = 30$

challenge

Find the value of x in each of these balances.

a   b   c

Solving equations   97

Extension

In pairs, encourage students to challenge another with two step equations, as in the **challenge,** to which they know the solution (IE, CT).

Exercise 7a commentary

Question 1 – This should be relatively familiar and mini whiteboards could be used to assess students' prior learning.

Question 2 – Similar to the examples. Using mini whiteboards would ensure that all students have the opportunity to contribute and to help identify any difficulties early on. Ask students to also say what the inverse was that was needed.

Question 3 – Consider grouping students into threes and asking each group member to answer every third question and then to explain their examples to the other two. This will develop mathematical discussion and consolidate understanding (1.3).

Challenge – The balance approach of question **1** applied to two-step equations (CT).

Links

The equals sign was first used by the Welsh mathematician and physician Robert Recorde in 1557 in his book *The Whetstone of Witte*. He used two parallel lines in the symbol because 'noe 2, thynges, can be moare equalle' However, other symbols for 'is equal to' were still used until the 1700s including the Latin abbreviation *ae* or *oe* (for *aequalis* or 'equal') There is more information about Robert Recorde at http://en.wikipedia.org/wiki/Robert_Recorde

- Construct and solve linear equations with integer coefficients (unknown on either or both sides) using appropriate methods (e.g. inverse operations, transforming both sides in the same way) (L6)

Useful resources

Starter – Algebraic products

Draw a 4 × 4 table on the board. Label the columns a, b, a^2, 7 and the rows a, $2a$, b, c.

Fill in the table with the products, for example,

the top row in the table would read a^2, ab, a^3, $7a$.

Can be differentiated by the choice of terms.

Teaching notes

This spread emphasises the use of substitution to check answers. It merits particular attention as self-check is empowering for students, enabling them to become increasingly independent rather than waiting for teacher input. Also it supports them in using errors positively (RL).

A further focus in the worked examples is the issue of on which side of the equation the letters will be organised. Again this challenges a large number of students and misunderstandings are very common. Time spent discussing approaches here will benefit all abilities (1.4).

The work in this spread supports the process skill of analysing through using appropriate mathematical procedures, where students work towards solving equations and strengthens skills in interpreting and evaluating, where students consider different approaches, for example where another student has used a different method of finding a solution (TW) (1.3, 1.4).

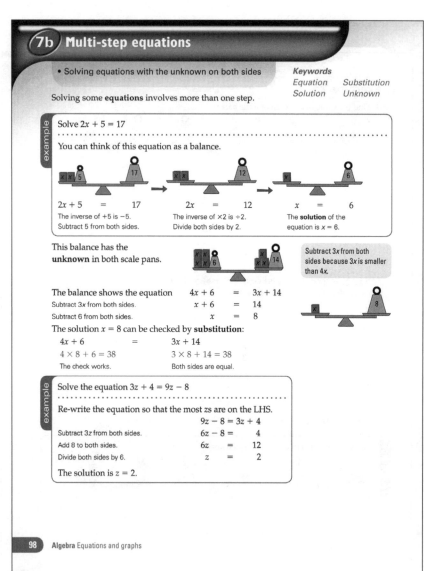

Plenary

Give students three examples of solutions to equations where two out of three are answered incorrectly. Ask them to identify the errors and to provide written feedback that they think would help the student who had made the errors (RL).

Simplification

Students who struggle with this work, may benefit from additional consolidation of solving equations with unknowns on one side only.

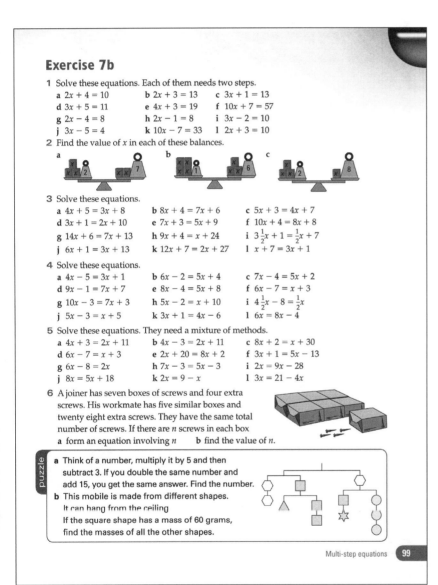

Exercise 7b

1 Solve these equations. Each of them needs two steps.

a $2x + 4 = 10$	**b** $2x + 3 = 13$	**c** $3x + 1 = 13$
d $3x + 5 = 11$	**e** $4x + 3 = 19$	**f** $10x + 7 = 57$
g $2x - 4 = 8$	**h** $2x - 1 = 8$	**i** $3x - 2 = 10$
j $3x - 5 = 4$	**k** $10x - 7 = 33$	**l** $2x + 3 = 10$

2 Find the value of x in each of these balances.

a   **b**   **c**

3 Solve these equations.

a $4x + 5 = 3x + 8$	**b** $8x + 4 = 7x + 6$	**c** $5x + 3 = 4x + 7$
d $3x + 1 = 2x + 10$	**e** $7x + 3 = 5x + 9$	**f** $10x + 4 = 8x + 8$
g $14x + 6 = 7x + 13$	**h** $9x + 4 = x + 24$	**i** $3\frac{1}{2}x + 1 = \frac{1}{2}x + 7$
j $6x + 1 = 3x + 13$	**k** $12x + 7 = 2x + 27$	**l** $x + 7 = 3x + 1$

4 Solve these equations.

a $4x - 5 = 3x + 1$	**b** $6x - 2 = 5x + 4$	**c** $7x - 4 = 5x + 2$
d $9x - 1 = 7x + 7$	**e** $8x - 4 = 5x + 8$	**f** $6x - 7 = x + 3$
g $10x - 3 = 7x + 3$	**h** $5x - 2 = x + 10$	**i** $4\frac{1}{2}x - 8 = \frac{1}{2}x$
j $5x - 3 = x + 5$	**k** $3x + 1 = 4x - 6$	**l** $6x = 8x - 4$

5 Solve these equations. They need a mixture of methods.

a $4x + 3 = 2x + 11$	**b** $4x - 3 = 2x + 11$	**c** $8x + 2 = x + 30$
d $6x - 7 = x + 3$	**e** $2x + 20 = 8x + 2$	**f** $3x + 1 = 5x - 13$
g $6x - 8 = 2x$	**h** $7x - 3 = 5x - 3$	**i** $2x = 9x - 28$
j $8x = 5x + 18$	**k** $2x = 9 - x$	**l** $3x = 21 - 4x$

6 A joiner has seven boxes of screws and four extra screws. His workmate has five similar boxes and twenty eight extra screws. They have the same total number of screws. If there are n screws in each box
   a form an equation involving n **b** find the value of n.

puzzle
a Think of a number, multiply it by 5 and then subtract 3. If you double the same number and add 15, you get the same answer. Find the number.
b This mobile is made from different shapes.
   It can hang from the ceiling
   If the square shape has a mass of 60 grams, find the masses of all the other shapes.

Multi-step equations   99

Extension

For more able students, offer additional examples of word problems that can be used to generate equations to be solved. This can be challenging for even the most able.

Students who grasp this work easily could try to design a puzzle of their own using question **6** as a model. This embeds awareness of the opportunity for skill application and strengthens skill transferability (CT).

Exercise 7b commentary

Question 1 – Similar to the first example.

Question 2 – An elaboration of question **1** with unknowns on both sides. Encourage students to check their solutions using back substitution as a way to promote independence and confidence (IE) (1.3).

Question 3 – Parts **i** (with fractions) and **l** (with more xs on the RHS) may cause difficulty.

Question 4 and **5** – Elaborations of question **3**, involving negative numbers (answers positive). (1.1)

Question 6 – Consider allowing time for students to work on this, in pairs is best, before taking feedback and working towards a whole class solution (EP).

Puzzle – Part **a** requires an equation to be constructed before being solved. Part **b**, uses the balancing method of solving equations in a real-life context that will engage students with the functional aspects of the topic (1.1)

Assessment Criteria – Algebra, level 6: construct and solve linear equations with integer coefficients, using an appropriate method.

Links

In Africa, India and the Far East, seeds were traditionally used as standard weights in balance pans to weigh small amounts. Carob seeds were often used because of their uniform size. A typical carob seed weighs 200 mg. The carat is the unit used to weigh gold and diamonds today and originates from the weight of a carob seed. The weight of one carat is precisely 200 mg, or 0.2 g. There is more information about the history of weighing at http://www.averyweigh-tronix.com/main.aspx?p=1.1.3.4

- Use formulae from mathematics and other subjects (L5)
- Construct and solve linear equations with integer coefficients (unknown on either or both sides, without and with brackets) using appropriate methods (e.g. inverse operations, transforming both sides in the same way) (L6)

Useful resources
Books, magazines

Starter – Budgies and hamsters

Luxmi had some budgies and hamsters. In total she counted 18 heads and 56 feet.

How many budgies and how many hamsters were there? (8 budgies, 10 hamsters)

Can be extended by asking students to make up their own bird and animal puzzles.

Teaching notes

The mathematics in this spread supports students' capacity to generate ideas and explore possibilities and to ask questions to extend their thinking (CT).

The spread brings together several aspects of prior learning: expanding brackets and collecting like terms, rearranging and solving equations using inverse operations and negative numbers. Be prepared to practice these skills separately before bringing them together in the examples. Carefully showing the necessary workings will provide students with a model for setting out their solutions and checking answers by back substitution (SM).

Many skills developed here involve making links with other learning and the activities promote the process skill of communicating and reflecting, particularly making links to related problems or problems with a similar structure (1.2, 1.5)

The initial discussion using scales will help the more visual learners grasp what the use of brackets in an equation represents. Where difficulties are experienced, encourage the use of this method of representation to solve problems.

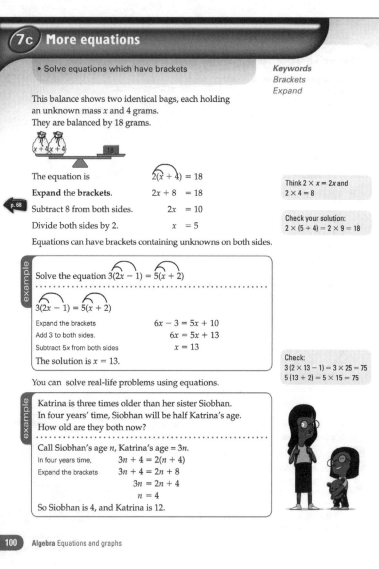

7c **More equations**

- Solve equations which have brackets

Keywords
Brackets
Expand

This balance shows two identical bags, each holding an unknown mass x and 4 grams.
They are balanced by 18 grams.

The equation is	$2(x + 4) = 18$
Expand the brackets.	$2x + 8 = 18$
Subtract 8 from both sides.	$2x = 10$
Divide both sides by 2.	$x = 5$

Think $2 \times x = 2x$ and $2 \times 4 = 8$

Check your solution:
$2 \times (5 + 4) = 2 \times 9 = 18$

Equations can have brackets containing unknowns on both sides.

example

Solve the equation $3(2x − 1) = 5(x + 2)$

$3(2x − 1) = 5(x + 2)$

Expand the brackets	$6x − 3 = 5x + 10$
Add 3 to both sides.	$6x = 5x + 13$
Subtract 5x from both sides	$x = 13$

The solution is $x = 13$.

Check:
$3(2 \times 13 − 1) = 3 \times 25 = 75$
$5(13 + 2) = 5 \times 15 = 75$

You can solve real-life problems using equations.

example

Katrina is three times older than her sister Siobhan. In four years' time, Siobhan will be half Katrina's age. How old are they both now?

Call Siobhan's age n, Katrina's age $= 3n$.

In four years time,	$3n + 4 = 2(n + 4)$
Expand the brackets	$3n + 4 = 2n + 8$
	$3n = 2n + 4$
	$n = 4$

So Siobhan is 4, and Katrina is 12.

100 Algebra Equations and graphs

Plenary

Ask students to vote on the example from the lesson that has generated most challenge and agree a whole class model answer on the board (RL). Indicate that a similar example may be used as a starter activity for next lesson.

Simplification

Encourage weaker students to refer to the visual representation using scales and to adapt this to support the solution of examples causing difficulty. Recap expanding brackets where this is a stumbling block.

Use this exercise to develop students' capacity to explain and justify their thinking (EP).

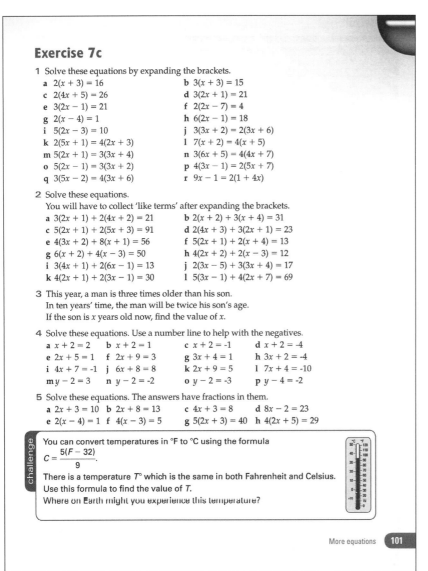

Exercise 7c

1 Solve these equations by expanding the brackets.

a $2(x + 3) = 16$ b $3(x + 3) = 15$
c $2(4x + 5) = 26$ d $3(2x + 1) = 21$
e $3(2x - 1) = 21$ f $2(2x - 7) = 4$
g $2(x - 4) = 1$ h $6(2x - 1) = 18$
i $5(2x - 3) = 10$ j $3(3x + 2) = 2(3x + 6)$
k $2(5x + 1) = 4(2x + 3)$ l $7(x + 2) = 4(x + 5)$
m $5(2x + 1) = 3(3x + 4)$ n $3(6x + 5) = 4(4x + 7)$
o $5(2x - 1) = 3(3x + 2)$ p $4(3x - 1) = 2(5x + 7)$
q $3(5x - 2) = 4(3x + 6)$ r $9x - 1 = 2(1 + 4x)$

2 Solve these equations.
You will have to collect 'like terms' after expanding the brackets.

a $3(2x + 1) + 2(4x + 2) = 21$ b $2(x + 2) + 3(x + 4) = 31$
c $5(2x + 1) + 2(5x + 3) = 91$ d $2(4x + 3) + 3(2x + 1) = 23$
e $4(3x + 2) + 8(x + 1) = 56$ f $5(2x + 1) + 2(x + 4) = 13$
g $6(x + 2) + 4(x - 3) = 50$ h $4(2x + 2) + 2(x - 3) = 12$
i $3(4x + 1) + 2(6x - 1) = 13$ j $2(3x - 5) + 3(3x + 4) = 17$
k $4(2x + 1) + 2(3x - 1) = 30$ l $5(3x - 1) + 4(2x + 7) = 69$

3 This year, a man is three times older than his son.
In ten years' time, the man will be twice his son's age.
If the son is x years old now, find the value of x.

4 Solve these equations. Use a number line to help with the negatives.

a $x + 2 = 2$ b $x + 2 = 1$ c $x + 2 = -1$ d $x + 2 = -4$
e $2x + 5 = 1$ f $2x + 9 = 3$ g $3x + 4 = 1$ h $3x + 2 = -4$
i $4x + 7 = -1$ j $6x + 8 = 8$ k $2x + 9 = 5$ l $7x + 4 = -10$
m $y - 2 = 3$ n $y - 2 = -2$ o $y - 2 = -3$ p $y - 4 = -2$

5 Solve these equations. The answers have fractions in them.

a $2x + 3 = 10$ b $2x + 8 = 13$ c $4x + 3 = 8$ d $8x - 2 = 23$
e $2(x - 4) = 1$ f $4(x - 3) = 5$ g $5(2x + 3) = 40$ h $4(2x + 5) = 29$

> **challenge**
> You can convert temperatures in °F to °C using the formula
> $$C = \frac{5(F - 32)}{9}.$$
> There is a temperature $T°$ which is the same in both Fahrenheit and Celsius.
> Use this formula to find the value of T.
> Where on Earth might you experience this temperature?

Extension

Students can be challenged with additional word problems from which they will need to generate an equation as in question **3**. Encourage paired working to develop student dialogue and to develop skills in justification. More involved equations leading to negative or fractional solutions may also be set (EP).

Exercise 7c commentary

Question 1 – Parts **a–i** require a single bracket to be expanded and parts **j–r**, which are similar to the first example, require expanding two brackets. Encourage self checking through substitution (1.3).

Question 2 – Requires brackets to be expanded and like terms to be collected. Students could be organised into small groups, sharing out the questions and explaining and justifying their solutions to each other (EP).

Question 3 – Similar to the second example. An opportunity to promote mathematical discussion by asking students how to tackle the problem. Check that any solution works both now and in ten years time (1.1).

Question 4 – Students may need reassurance that the negative solutions are valid.

Question 5 – Similar to question **4**, but now with fractions as answers.

Challenge – Here, students make links with previous work on formulae and generate an equation in a real-life context. There is also a link with work that will have been met in geography (1.2, 1.5).

Assessment Criteria – UAM, level 6: show understanding of situations by describing them mathematically using symbols, words and diagrams.

Links

Bring in some written text ,for example, books and magazines. Ask the students to find examples of the use of brackets. Round brackets (*parentheses*) are often used for explanations or to add to the information already given. They can also be used for translations and abbreviations. Why are the brackets used in the examples?

- Know the meanings of the words formula and function (L5)
- Express simple functions algebraically and represent them in mappings [or on a spreadsheet] (L5)

Useful resources

Starter – £20 more

A cd player and batteries together cost £23. The cd player cost £20 more than the batteries. How much do the batteries cost?

Hint: not £3 (£1.50)

Ask students to explain their methods.

Encourage the use of algebra.

See also the plenary to spread **7c**.

Teaching notes

The work on mappings and function machines is closely related to using position-to-term rules to generate sequences. A useful starting point would be to review work on sequences and use this to lead to the idea of a mapping diagram and a function (CT).

The worked examples here merit use as a whole-class discussion activity. To focus the discussion and ensure all are involved, invite students to generate some mappings and rules with which to challenge each other

Allow additional time to discuss how a table of values is constructed and emphasise the links with substitution. This is a skill that students will return to and require frequently. Questions **3** and **4** exercise this skill but it may be useful to complete another table as a whole-class feedback activity for additional support and consolidation (1.2).

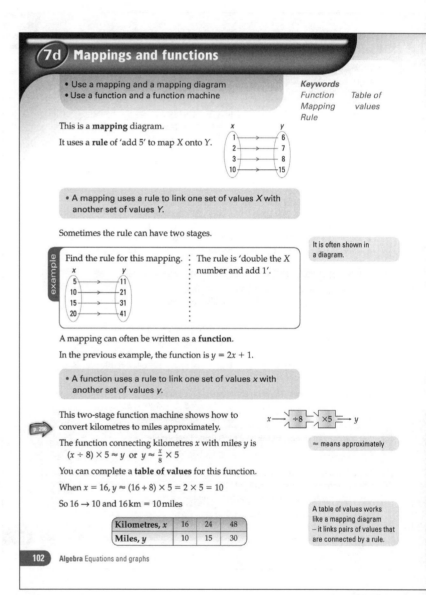

7d Mappings and functions

- Use a mapping and a mapping diagram
- Use a function and a function machine

Keywords
Function · Table of
Mapping · values
Rule

This is a **mapping** diagram.

It uses a **rule** of 'add 5' to map X onto Y.

- A mapping uses a rule to link one set of values X with another set of values Y.

Sometimes the rule can have two stages.

example

Find the rule for this mapping. : The rule is 'double the X number and add 1'.

It is often shown in a diagram.

A mapping can often be written as a **function**.

In the previous example, the function is $y = 2x + 1$.

- A function uses a rule to link one set of values x with another set of values y.

This two-stage function machine shows how to convert kilometres to miles approximately.

$x \rightarrow \boxed{\div 8} \rightarrow \boxed{\times 5} \rightarrow y$

The function connecting kilometres x with miles y is
$(x \div 8) \times 5 \approx y$ or $y \approx \frac{x}{8} \times 5$

≈ means approximately

You can complete a **table of values** for this function.

When $x = 16$, $y \approx (16 \div 8) \times 5 = 2 \times 5 = 10$

So $16 \rightarrow 10$ and $16\,km = 10\,miles$

A table of values works like a mapping diagram – it links pairs of values that are connected by a rule.

Kilometres, x	16	24	48
Miles, y	10	15	30

102 Algebra Equations and graphs

Plenary

Ask students to work in pairs, where each one writes a function to swap with their friend who then represents it in a table of values.

Simplification

Offer additional examples of the type in the first three questions of the exercise for consolidation and when completing tables of values, recap substitution. Encourage students to explain word problems to each other using their own words to help support weak literacy skills that may be causing a barrier to learning.

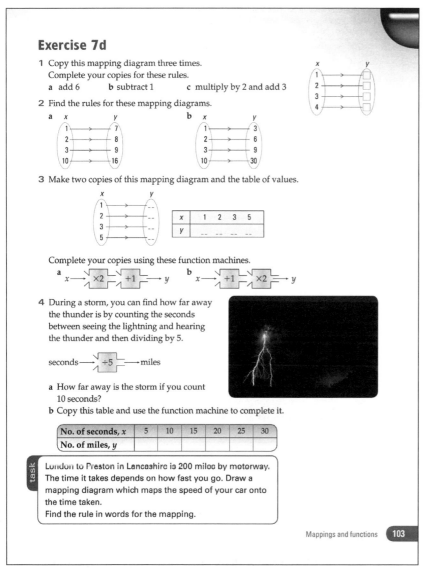

Exercise 7d

1 Copy this mapping diagram three times.
  Complete your copies for these rules.
  a add 6   **b** subtract 1   **c** multiply by 2 and add 3

2 Find the rules for these mapping diagrams.
  a
  b

3 Make two copies of this mapping diagram and the table of values.

x	1	2	3	5
y				

Complete your copies using these function machines.

  a $x \longrightarrow \boxed{\times 2} \longrightarrow \boxed{+1} \longrightarrow y$
  b $x \longrightarrow \boxed{+1} \longrightarrow \boxed{\times 2} \longrightarrow y$

4 During a storm, you can find how far away the thunder is by counting the seconds between seeing the lightning and hearing the thunder and then dividing by 5.

  seconds $\longrightarrow \boxed{\div 5} \longrightarrow$ miles

  a How far away is the storm if you count 10 seconds?
  b Copy this table and use the function machine to complete it.

No. of seconds, x	5	10	15	20	25	30
No. of miles, y						

task
London to Preston in Lancashire is 200 miles by motorway. The time it takes depends on how fast you go. Draw a mapping diagram which maps the speed of your car onto the time taken.
Find the rule in words for the mapping.

Mappings and functions **103**

Extension

Introduce more word-based problems as these prove challenging for even the more able. Again, rewriting a problem in a student's own words will help clarify the problems.

Exercise 7d commentary

Question 1 – An opportunity to practice drawing mapping diagrams.

Question 2 – Similar to the example (1.2).

Question 3 – Extends question **1** to include completing a table of values: this important skill may merit use as an example for whole class feedback for agreement.

Question 4 – Relates function machines and tables of values to a real-life context which should consolidate understanding (EP).

Task – Encourage refinement of the rule and steer discussion around to the magnitudes of speeds that are sensible to consider. Under what circumstances might someone be travelling at those speeds? (SM) (1.1)

Links

Lightning is a giant spark which can suddenly heat the surrounding air to around 30 000°C. The air expands and sends out a vibration or sound wave known as thunder. Sound travels more slowly than light through air, so the lightning is usually seen before the thunder is heard. Thunder rumbles because the sound wave reflects from hills, trees and other obstacles. There is more about thunderstorms at http://www.bbc.co.uk/weather/weatherwise/factfiles/extremes/lighting.shtml

• Generate points in all four quadrants and plot graphs
of linear functions, where y is given explicitly in terms
of x, on paper [and using ICT]   (L6)

Useful resources

Starter – ABC

If $A = 3$, $B = 5$, $C = 7$, ask students to form as many equations
as they can in four minutes.
Score 1 point for each different operation or pair of brackets
used, for example, $2(C - B) = 4$ uses multiplication, brackets
and subtraction and scores 3.

Teaching notes

The exercises focus on the first
quadrant but it will be useful revision
to ensure students are confident
plotting points in all four quadrants.
Ask students to supply ways to
remember which is the x- and y-axis:
'along the corridor and up the stairs',
'X a cross, and Y to the sky', *etc*

A physical way to cement
understanding of vertical and
horizontal lines is as follows.

Stand students in a position defined
as $x = 0$ then instructed them to move
left or right to represent, for example,
$x = 5$ or $x = -3$. Similarly they can
hold out their arms to represent $y = 0$
and then bend or stretch to represent,
for example, $y = 4$ or $y = -2$.

When drawing tables of values,
encourage students to look for a
pattern in the numbers or a lack of
one that would suggest an error. This
is likely to be particularly useful when
they start to use negative numbers.

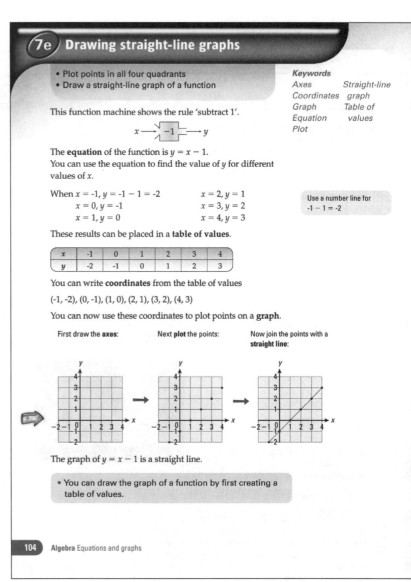

7e Drawing straight-line graphs

• Plot points in all four quadrants
• Draw a straight-line graph of a function

Keywords
Axes   Straight-line
Coordinates   graph
Graph   Table of
Equation   values
Plot

This function machine shows the rule 'subtract 1'.

$$x \longrightarrow \boxed{-1} \longrightarrow y$$

The **equation** of the function is $y = x - 1$.
You can use the equation to find the value of y for different
values of x.

When $x = -1$, $y = -1 - 1 = -2$ $x = 2$, $y = 1$
$x = 0$, $y = -1$ $x = 3$, $y = 2$
$x = 1$, $y = 0$ $x = 4$, $y = 3$

Use a number line for
$-1 - 1 = -2$

These results can be placed in a **table of values**.

x	-1	0	1	2	3	4
y	-2	-1	0	1	2	3

You can write **coordinates** from the table of values

$(-1, -2)$, $(0, -1)$, $(1, 0)$, $(2, 1)$, $(3, 2)$, $(4, 3)$

You can now use these coordinates to plot points on a **graph**.

First draw the **axes**:   Next **plot** the points: Now join the points with a
                                                                  straight line:

The graph of $y = x - 1$ is a straight line.

• You can draw the graph of a function by first creating a
table of values.

104   Algebra Equations and graphs

Plenary

Ask students to draw a flow diagram that shows a step by
step strategy for drawing a graph from a given equation.

Simplification

Suggest students begin by thinking of the equation as a position-to-term rule, to generate coordinate pairs in the first quadrant. Negative coordinates and negative gradients can then be introduced as confidence builds (CT).

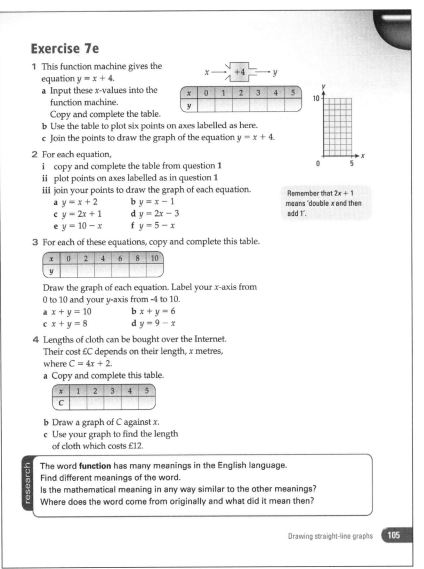

Exercise 7e

1 This function machine gives the equation $y = x + 4$.
  a Input these x-values into the function machine.
  Copy and complete the table.
  b Use the table to plot six points on axes labelled as here.
  c Join the points to draw the graph of the equation $y = x + 4$.

x	0	1	2	3	4	5
y						

2 For each equation,
  i copy and complete the table from question **1**
  ii plot points on axes labelled as in question **1**
  iii join your points to draw the graph of each equation.
    a $y = x + 2$ **b** $y = x - 1$
    c $y = 2x + 1$ **d** $y = 2x - 3$
    e $y = 10 - x$ **f** $y = 5 - x$

> Remember that $2x + 1$ means 'double x and then add 1'.

3 For each of these equations, copy and complete this table.

x	0	2	4	6	8	10
y						

Draw the graph of each equation. Label your x-axis from 0 to 10 and your y-axis from -4 to 10.
    a $x + y = 10$ **b** $x + y = 6$
    c $x + y = 8$ **d** $y = 9 - x$

4 Lengths of cloth can be bought over the Internet. Their cost £C depends on their length, x metres, where $C = 4x + 2$.
  a Copy and complete this table.

x	1	2	3	4	5
C					

  b Draw a graph of C against x.
  c Use your graph to find the length of cloth which costs £12.

> **research**
>
> The word **function** has many meanings in the English language.
> Find different meanings of the word.
> Is the mathematical meaning in any way similar to the other meanings?
> Where does the word come from originally and what did it mean then?

Extension

The application of straight line graphs in a real life setting and reading information from graphs as in the final question of the exercise can be developed to challenge more confident students.

Exercise 7e commentary

Question 1 – It may be helpful to discuss how to chose sensible scales for the axes. Check for careless errors: axes labelled '-2, -1, 1, 2' or points plotted (y, x) (RL).

Question 2 – Encourage students to look for the pattern in the numbers in their table so that they check for errors independently. Parts **e** and **f** involve negative gradients and may cause difficulty for some students.

Question 3 – Students may need reminding how to handle an implicit equation.

Question 4 – For students in difficulty here, suggest they call the cost y pounds to avoid confusion at this stage.

Research – This activity provides an opportunity to link terms used in mathematics with the use of language elsewhere which will be particularly valuable (IE, EP).

Assessment Criteria – Algebra, level 5: use and interpret coordinates in all four quadrants Algebra, level 6: plot graphs of linear functions, where y is given explicitly in terms of x.

Links

The word *axes* is a heteronym as it can be pronounced in two different ways, each with a different meaning (the plural of axis or the plural of axe). Ask the class to try to list some other heteronyms. Some examples are *minute, lead, wind, buffet, refuse, tear, wind, wound* and *sow*.

- Plot and interpret graphs of simple linear functions arising from real-life situations, [e.g. conversion graphs]   (L5)
- Recognise that equations of the form $y = mx + c$ correspond to straight-line graphs   (L6)

Useful resources
Mini whiteboards
cm graph paper

Starter – Coordinates

Ask students to give three sets of coordinates that lie on the following lines:

$y = 5x - 3$	for example (1,2) (2,7) (-1,-8)
$y = 5 - 3x$	(1,2) (2,-1) (-1,8)
$2y - x = 3$	(1,2) (2,2.5) (-1,1)

Encourage students to think of negative values.
Ask students what they notice when x 5 1: the point (1,2) lies on all three lines.

Teaching notes

Two points are sufficient to define a straight line but students should be encouraged to plot more as a check of their working and as an aide to appreciating the connection between the intercept and gradient terms in the equation and the pattern of coordinate pairs. An example of how to draw a line using a short table of values would help illustrate this.

Question **4** asks students to say whether a point lies on a given line. An example of how to approach this question and set out an answer is likely to prove useful.

Once students have gained some confidence in matching equations to straight lines, mini white boards (with grids) can be used to help develop further understanding. Ask students to show relative to the x- and y-axes where, for example, $y = 3x, y = x + 1, y = x - 1$ would lie and discuss how they know.

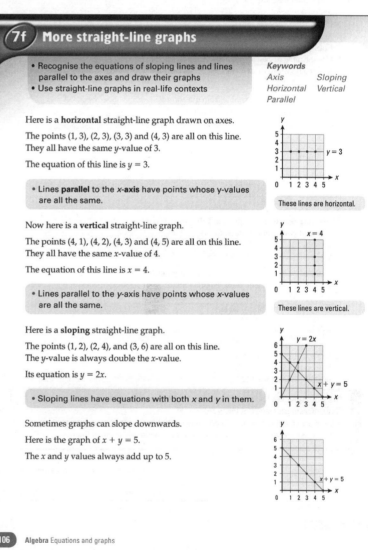

Plenary

Use a simplified version of the **investigation** activity in the exercise, asking students to draw the graphs of two equations where one is a horizontal line and locate the coordinates where the lines cross.

Simplification

Some students may need further time on locating horizontal and vertical lines.

Exercise 7f

1 Here are the equations of some straight lines.
Say whether each line is
i a line parallel to the x-axis   ii a line parallel to the y-axis
iii a sloping line.

  a $y = 3$ b $x = 5$ c $y = x - 4$ d $y = 2x + 3$
  e $x = 2$ f $x + y = 7$ g $y = 6$ h $y = 4x - 1$

2 Write the equations of these five straight lines A to E.

3 For each set of equations, draw axes, and label both axes from 0 to 10. Draw all three graphs on the same axes to make a triangle.

  a $y = x + 1$, $x = 8$, $y = 2$ b $y = 2x$, $x = 5$, $y = 4$
  c $y = 2x - 2$, $x = 1$, $y = 8$ d $y = 9 - x$, $x = 7$, $y = 8$

> You may need to make a table of values for the sloping lines.

4 Without drawing any diagrams, find whether each point lies on the given line.

  a $(1, 5)$ and $y = x + 4$ b $(5, 4)$ and $y = x - 1$
  c $(3, 7)$ and $y = 2x + 1$ d $(5, 13)$ and $y = 2x - 3$
  e $(0, 8)$ and $y = 6x + 2$ f $(3, 6)$ and $y = 9 - x$
  g $(5, -1)$ and $y = 4 - x$ h $(-1, 3)$ and $y = 4x + 1$

5 On axes labelled from 0 to 10, draw the lines
$x = 4, x = 6$, $y = 1$ and $y = 5$.
Find the area of the shape enclosed by these lines.

6 The average life-span of a man in the UK is eighty years. A man who is x years old now might expect to live for another y years, where $y = 80 - x$.

Copy and complete this table.

x	20	30	40	50	60
y					

Draw a graph of y against x on axes like these.

investigation
Learn how to use a computer software package to draw straight-line graphs.
Use a computer to draw the graphs of the two lines with these equations on the same axes. See if you can find the point where they cross.

$y = 7 - x$ $y = 2x + 1$

More straight-line graphs   107

Extension

Encourage more able students to think about what the c and the m values tell us in an equation of the form $y = mx + c$ and how they help us to draw the line (IE).

Exercise 7f commentary

Question 1 and **2** – involve classifying and defining horizontal and vertical lines: use whole class questioning with mini whiteboards to assess and ensure that all students have this basic skill (1.2).

Question 3 – It may help to ask students to start by drawing the horizontal and vertical lines. In part **b**, $y = 2x$ (no const.) and part **d**, $y = 9 - x$ (negative gradient) may cause difficulties (1.3).

Question 4 – This is a particularly useful activity to embed the use of substitution when identifying equations of lines and to ensure coordinate values are substituted in the correct way. A few examples may need to be worked through as a whole class (1.4)

Question 5 – This will be most straightforward if students use cm squared paper (1.3)

Question 6 – A straightforward real-life scenario (1.3).

Investigation – Should computer facilities not be available, this activity still has considerable value done by hand (SM) (1.3).

Assessment Criteria – Algebra, level 6: recognise that equations of the form $y = mx + c$ correspond to straight-line graphs.

Links

Question **6** discusses the life span of human beings. The oldest person ever to have lived was Jeanne Calment of France, who died in 1997 at the age of 122 years and 164 days. The oldest person ever to have lived in Britain was Charlotte Hughes who died in 1993 aged 115 years and 228 days. There is a list of the 100 longest lived people at http://en.wikipedia.org/wiki/List_of_the_verified_oldest_people

7a

1 Find the value of x in each of these balances.

a 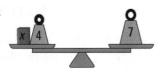 b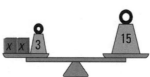

2 Solve these equations by using inverse operations.

a $x + 3 = 7$ b $x - 3 = 7$ c $2x = 8$ d $\frac{x}{2} = 5$

e $4 + x = 6$ f $x - 4 = 1$ g $3x = 18$ h $\frac{x}{4} = 2$

7b

3 Solve these equations. Each of them needs two steps.

a $2x + 4 = 14$ b $3x + 2 = 23$ c $2x - 1 = 11$

d $5x - 6 = 9$ e $\frac{x}{2} + 1 = 6$ f $\frac{x}{3} - 3 = 3$

4 Find the value of x in each of these balances.

a   b

5 Solve these equations. There are unknowns on both sides.

a $4x + 2 = 3x + 7$ b $6x + 1 = 5x + 13$ c $3x + 6 = x + 10$

d $7x + 2 = 4x + 8$ e $6x + 9 = x + 24$ f $7x = 3x + 20$

6 Solve these equations. Take care with the negative signs.

a $3x - 1 = 2x + 4$ b $7x - 2 = 5x + 6$ c $5x - 5 = 3x - 1$

d $8x - 11 = 5x - 2$ e $9x + 8 = 7x + 4$ f $6x + 14 = 3x + 5$

7 Sarah has six packets of Christmas cards and two loose cards. Her sister, Jane, has three similar packets of cards and seventeen loose cards. Each packet has x cards in it. When the sisters open all their boxes and count their cards, they find that they have the same total.
Write an equation and find the value of x.

7c

8 Solve these equations.

a $2(3x + 4) = 20$ b $3(2x - 1) = 21$ c $5(x - 2) = 20$

d $3(4x + 1) = 123$ e $4(2x + 1) = 6(x + 2)$ f $5(3x + 2) = 4(3x + 4)$

9 Solve these equations by expanding the brackets and collecting like terms.

a $3(2x + 4) + 2(3x + 1) = 38$ b $5(2x + 1) + 2(x + 3) = 35$

c $4(x + 3) + 6(x + 1) = 28$ d $2(2x + 3) + 4(2x + 1) = 18$

e $3(5x + 1) + 2(1 - 6x) = 9$ f $5(x + 5) + 2(2x - 3) = 31$

10 Copy this mapping diagram three times.

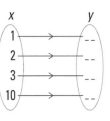

Complete your copies using these rules.

a add five **b** subtract two **c** double then add five

For each mapping diagram, copy and complete this table.

x	1	2	3	10
y				

11 Make two copies of this mapping diagram and the table of values.

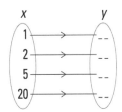

x	1	2	3	5	20
y	--	--	--	--	--

Complete your copies using these function machines.

a

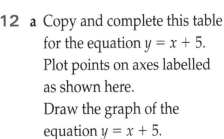

b

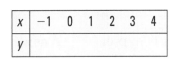

12 a Copy and complete this table
for the equation $y = x + 5$.
Plot points on axes labelled
as shown here.
Draw the graph of the
equation $y = x + 5$.

x	−1	0	1	2	3	4
y						

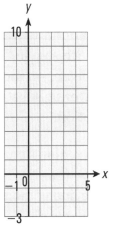

b Repeat for the equations

  i $y = x + 2$ **ii** $y = 2x - 1$ **iii** $x + y = 9$

13 Write the equations of these four straight lines A to D.

14 Here are the equations of some straight lines.
Which lines are
  i parallel to the x-axis
  ii parallel to the y-axis
  iii sloping lines?

  a $y = 2$ **b** $x = 4$ **c** $y = x + 3$
  d $y = 2x + 1$ **e** $x = 5$ **f** $y = -2$

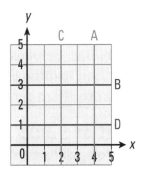

7 Summary

Assessment criteria

- Use and interpret coordinates in all four quadrants   Level 5
- Construct and solve linear equations with integer coefficients, using
  an appropriate method   Level 6
- Plot the graphs of linear functions, where y is given explicitly in terms of x Level 6
- Recognise that equations of the form $y = mx + c$ correspond to
  straight-line graphs   Level 6
- Show understanding of situations by describing them mathematically
  using symbols, words and diagrams   Level 6

Question commentary

Example	The example illustrates a problem where students are asked to solve an equation with the unknown on both sides. Ollie's answer uses inverse operations. Ask questions such as "How do you decide where to start" and "Does it matter if you subtract x from both sides before you subtract 4?" In case of difficulty, encourage students to sketch a balance to visualise each step in Ollie's answer. Emphasise the importance of checking the final answer by substituting into the original equation.	$5x + 4 = x + 12$ $5x - x = 12 - 4$ $\quad\quad 4x = 8$ $\quad\quad\ x = 2$
Past question	The question requires students to plot the graph of $x + y = 12$. Some students will draw the line without writing down coordinate pairs. To prevent error, encourage students to write down some coordinate pairs before drawing the line and to check that all the points generated lie on a straight line. Emphasise that a ruler must be used to draw the line and encourage use of a sharp pencil.	**Answer** **Level 6** 2 A straight line passing through (0, 12) and (12, 0)

Development and links

This topic is developed further in Chapter 13 where students will solve further equations including those containing directed numbers, consider the equation of a straight-line graph and interpret and draw real-life graphs.

The ability to solve equations is important in the sciences and technology where equations are used to model the behaviour of objects and phenomena in the real-world. Plotting and interpreting graphs is also important in science and geography and has links to handling and interpreting data.

8 Calculations

Objectives

	Level
• Strengthen and extend mental methods of calculation, working with decimals	4
• Solve problems mentally	4
• Use efficient written methods to add and subtract integers and decimals of any size, including numbers with differing numbers of decimal places	4
• Round positive numbers to any given power of 10	5
• Round decimals to the nearest whole number or to one or two decimal places	5
• Multiply and divide integers and decimals by 0.1, 0.01	5
• Use efficient written methods for multiplication and division of integers and decimals	5
• Understand where to position the decimal point by considering equivalent calculations	5
• Make and justify estimates and approximations of calculations	5
• Use the order of operations, including brackets, with more complex calculations	5
• Using a calculator, carry out more difficult calculations effectively and efficiently	5
• Use calculator brackets and the memory	5
• Using a calculator, enter numbers and interpret the display in different contexts (extend to negative numbers, fractions, time)	6

Introduction

The focus of this chapter is on reviewing and developing numerical methods. Students consider mental, standard written and calculator methods for addition, subtraction, multiplication and division, practise rounding whole numbers and decimals and use the order of operations including brackets and indices.

Calculations surround our daily lives. We calculate when shopping, measuring, calculating taxes, working out calorie intake and calculating CO_2 emissions. People have tried for hundreds of years to make calculating quicker and easier and the student book discusses Wilhelm Schickard who invented the "calculating clock" in 1623. There is more information about the history of computing including a picture of the Schickard's "calculating clock" at http://www.computersciencelab.com/ComputerHistory/History.htm(TW).

⚡Fast-track
All spreads

MPA

1.1	8a, b, c, e, f, g, h, i
1.2	8f
1.3	8a, b, c, d, e, f, g, h, i
1.4	8a, e, g, h, i
1.5	8c, e, g

PLTS

IE	8b, c, e, f, g, i
CT	8d, f, h, i
RL	8a, b, c, d, e, f, h
TW	8I, a, b, e, g, h
SM	8a, b, i
EP	8a, c, g, i

Extra Resources

8	Start of chapter presentation
8a	Animation: Rounding
8a	Consolidation sheet
8b	Starter: Addition with negatives matching
8c	Consolidation sheet
8d	Worked solution: Q1d
8d	Consolidation sheet
8e	Starter: Multiplying with negatives T or F
8e	Consolidation sheet
8f	Animation: Multiplication
8f	Animation: Multiplying decimals
8f	Consolidation sheet
8g	Worked solution: Q5a
8g	Consolidation sheet
8i	Starter: Rounded decimal matching

Assessment: chapter 8

- Round positive numbers to any given power of 10   (L4)
- Round decimals to the nearest whole number or to one or two decimal places   (L5)

Useful resources
Mini whiteboards

Starter – What is my number?

I am even but not square.

I am a multiple of 3.

I am greater than the number of days in November.

I am less than the product of 8×9.

I have 8 factors. (54)

Can be extended by asking students to make up their own puzzles.

Teaching notes

An emphasis in this unit is on rounding to a 'sensible degree of accuracy' and it is helpful to discuss this, thereby making explicit links to the use of mathematics in a functional context (EP).

Spend time talking through the procedure for rounding numbers and ensure that students understand that rounding down means keeping the last digit the same, not taking something away, as this is a common error (RL).

Draw attention to the visual representation of the reading on the weighing scale as this clearly shows how a value is closer to one unit than another.

The spread supports learning through the process skill of communicating and reflecting, focusing on the effects of rounding

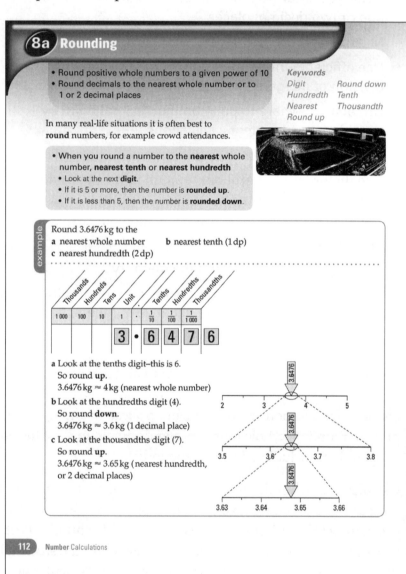

8a **Rounding**

- Round positive whole numbers to a given power of 10
- Round decimals to the nearest whole number or to 1 or 2 decimal places

Keywords
Digit   *Round down*
Hundredth   *Tenth*
Nearest   *Thousandth*
Round up

In many real-life situations it is often best to **round** numbers, for example crowd attendances.

- When you round a number to the **nearest** whole number, **nearest tenth** or **nearest hundredth**
- Look at the next **digit**.
- If it is 5 or more, then the number is **rounded up**.
- If it is less than 5, then the number is **rounded down**.

example

Round 3.6476 kg to the
a nearest whole number   **b** nearest tenth (1 dp)
c nearest hundredth (2 dp)

Thousands	Hundreds	Tens	Unit	.	Tenths	Hundredths	Thousandths
1 000	100	10	1	.	$\frac{1}{10}$	$\frac{1}{100}$	$\frac{1}{1\,000}$
			3	.	6	4	7	6

a Look at the tenths digit–this is 6.
So round **up**.
3.6476 kg ≈ 4 kg (nearest whole number)
b Look at the hundredths digit (4).
So round **down**.
3.6476 kg ≈ 3.6 kg (1 decimal place)
c Look at the thousandths digit (7).
So round **up**.
3.6476 kg ≈ 3.65 kg (nearest hundredth, or 2 decimal places)

112   **Number** Calculations

Plenary

Ask students to think of a real-life example (for example, measurement, money, *etc*.) where it is appropriate to round to the nearest, 1000, 10, whole number, two decimal places and take feedback. Produce a whole class 'best three examples' in each category (TW).

Simplification

For students who have difficulty with this activity, encourage them to use a diagram to help to make a decision as in the worked example. The visual support will often move students forward.

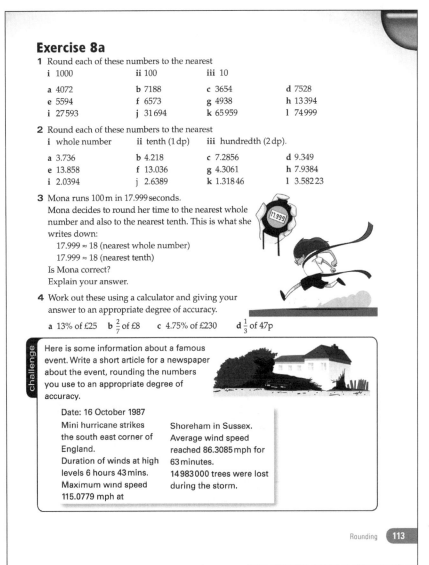

Exercise 8a

1 Round each of these numbers to the nearest

   i 1000   **ii** 100   **iii** 10

a 4072	**b** 7188	**c** 3654	**d** 7528
e 5594	**f** 6573	**g** 4938	**h** 13394
i 27593	**j** 31694	**k** 65959	**l** 74999

2 Round each of these numbers to the nearest

   i whole number   **ii** tenth (1 dp)   **iii** hundredth (2 dp).

a 3.736	**b** 4.218	**c** 7.2856	**d** 9.349
e 13.858	**f** 13.036	**g** 4.3061	**h** 7.9384
i 2.0394	**j** 2.6389	**k** 1.31846	**l** 3.58223

3 Mona runs 100 m in 17.999 seconds.
Mona decides to round her time to the nearest whole number and also to the nearest tenth. This is what she writes down:

   17.999 ≈ 18 (nearest whole number)
   17.999 ≈ 18 (nearest tenth)

Is Mona correct?
Explain your answer.

4 Work out these using a calculator and giving your answer to an appropriate degree of accuracy.

   a 13% of £25   **b** $\frac{2}{7}$ of £8   **c** 4.75% of £230   **d** $\frac{1}{3}$ of 47p

challenge

Here is some information about a famous event. Write a short article for a newspaper about the event, rounding the numbers you use to an appropriate degree of accuracy.

Date: 16 October 1987
Mini hurricane strikes the south east corner of England.
Duration of winds at high levels 6 hours 43 mins.
Maximum wind speed 115.0779 mph at

Shoreham in Sussex.
Average wind speed reached 86.3085 mph for 63 minutes.
14 983 000 trees were lost during the storm.

Rounding **113**

Extension

For more able students, consider discussing significant figures.

Exercise 8a commentary

There are a number of opportunities in this exercise to link to the strengthening of literacy skills.

Question 1 and **2** – Similar to the example. These questions could be used as an oral whole class activity or using mini whiteboards for assessment of understanding (1.3).

Question 3 – Insist on properly written explanations (1.3, 1.4).

Question 4 – As this question involves other skills it may benefit from being completed as a whole class exercise. Agree what is an appropriate degree of accuracy.

Challenge – This provides an excellent opportunity to link mathematics with improving students' skills in writing (SM) (1.3, 1.4).

Assessment Criteria – NNS, level 5: round decimals to the nearest decimal places. UAM, level 6: give solutions to an appropriate degree of accuracy.

Links

In parliamentary elections in Germany in 1992, a rounding error caused the wrong results to be announced. Under German law, a party can not have any seats in Parliament unless it has 5.0% or more of the vote. The Green Party appeared to have exactly 5.0%, until it was discovered that the computer that printed out the results only used one place after the decimal point and had rounded the vote up to 5.0%. The Green Party only had 4.97% of the vote and the results had to be changed.

- Strengthen and extend mental methods of calculation, working with decimals (L5)
- Solve problems mentally (L5)

Useful resources

Starter – NMBR!

Write words on the board missing out the vowels, for example, nmbr.
Ask students for the original words (number) Possible 'words':
chnc dcml hndrdth prbblty lvn frctn mlln tnth prcntg nt
(chance decimal hundredth probability eleven fraction million tenth percentage unit).
Can be extended by asking students to make up their own examples.

Teaching notes

The process skill of communicating and reflecting is central to the work in this spread as students are involved throughout in discussing and reflecting on different approaches to solving a numerical problem and comparing the efficiency of calculation procedures (TW).

When working through the examples, stress the point that there is no one correct method of working and that even if a favoured method is less efficient than another, that does not mean it is wrong to use it. The aim here should be to highlight a full range of strategies to ensure that students understand them and again to suggest that a different method could always be used as a mechanism for checking.

The worked examples offer two key points that need to be highlighted for students. First, the importance of approximating first as a checking mechanism and second, to practise the skill of analysing text to identify key information (1.1, 1.3).

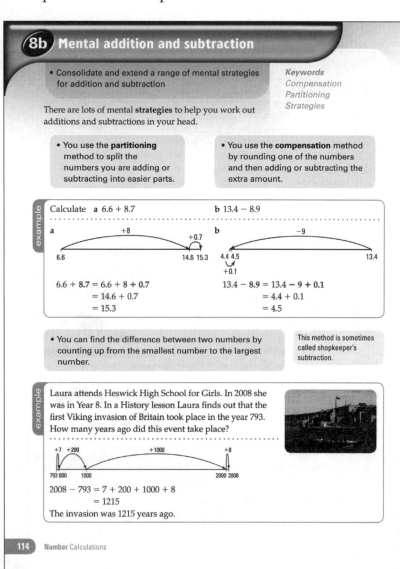

Plenary

Ask students to work in small groups to design a poster explaining a variety of mental methods that might be used to solve a word problem along similar lines to the worked example.

Simplification

To build confidence, encourage the use of a number line and show how to use this to break down the calculation as in the unit illustrations.

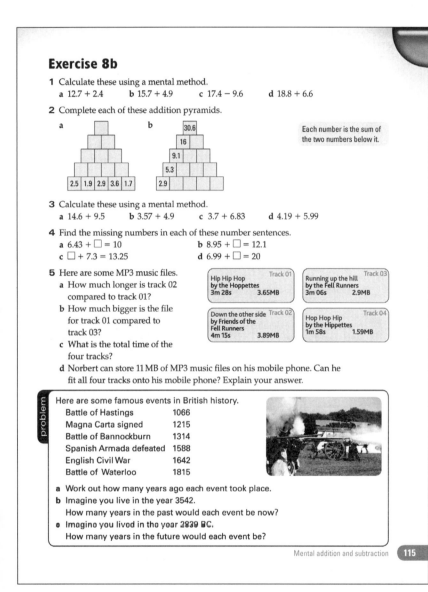

Exercise 8b

1 Calculate these using a mental method.
  a 12.7 + 2.4   **b** 15.7 + 4.9   **c** 17.4 − 9.6   **d** 18.8 + 6.6

2 Complete each of these addition pyramids.

  a   **b**

Each number is the sum of the two numbers below it.

| 30.6 |
| 16 |
| 9.1 |
| 5.3 |
| 2.9 |

| 2.5 | 1.9 | 2.9 | 3.6 | 1.7 |

3 Calculate these using a mental method.
  a 14.6 + 9.5   **b** 3.57 + 4.9   **c** 3.7 + 6.83   **d** 4.19 + 5.99

4 Find the missing numbers in each of these number sentences.
  a 6.43 + □ = 10   **b** 8.95 + □ = 12.1
  c □ + 7.3 = 13.25   **d** 6.99 + □ = 20

5 Here are some MP3 music files.
  a How much longer is track 02 compared to track 01?
  b How much bigger is the file for track 01 compared to track 03?
  c What is the total time of the four tracks?
  d Norbert can store 11 MB of MP3 music files on his mobile phone. Can he fit all four tracks onto his mobile phone? Explain your answer.

| Track 01 |
| Hip Hip Hop by the Hoppettes |
| 3m 28s   3.65MB |

| Track 03 |
| Running up the hill by the Fell Runners |
| 3m 06s   2.9MB |

| Track 02 |
| Down the other side by Friends of the Fell Runners |
| 4m 15s   3.89MB |

| Track 04 |
| Hop Hop Hip by the Hippettes |
| 1m 58s   1.59MB |

problem

Here are some famous events in British history.

Battle of Hastings	1066
Magna Carta signed	1215
Battle of Bannockburn	1314
Spanish Armada defeated	1588
English Civil War	1642
Battle of Waterloo	1815

  a Work out how many years ago each event took place.
  b Imagine you live in the year 3542.
    How many years in the past would each event be now?
  c Imagine you lived in the year 2839 BC.
    How many years in the future would each event be?

Mental addition and subtraction **115**

Extension

More able students will be familiar and confident with most of the strategies explored here and will gain most from discussion about the relative efficiency of various methods and when their use is most likely to be appropriate.

Exercise 8b commentary

All questions (1.3)

Question 1 and **2** – Ask students to discuss and compare the methods they used.

Question 3 – Ask students why using a written approach here would be inefficient?

Question 4 – This question could be used for whole class feedback to consider how many different options for calculation students can suggest and to discuss their suitability and efficiency.

Question 5 – Ask students to identify the operation and method they will use to solve each part. Beware of treating times as decimals, not minutes and seconds (RL).

Problem – For part **c**, remind students that there is no year 0, so students will need to add on one (RL). Possibly link this to some historical research and ask students to find out how many soldiers were involved in each battle, round the numbers to a sensible degree of accuracy (IE, SM).

Assessment Criteria – Calculating, Level 4: use a range of mental methods of computation with all operations..

Links

There are pictures and descriptions of many Viking artefacts at http://www.britishmuseum.org/explore/highlights/highlights_search_results.aspx?RelatedId=1672

There is an educational game about Viking Raiders at http://www.bbc.co.uk/history/ancient/vikings/launch_gms_viking_quest.shtml.

- Use efficient written methods to add and subtract integers and decimals of any size, including numbers with differing numbers of decimal places   (L5)

Useful resources

• •

Starter – Miss the multiples

Ask students to add up numbers you read out that are **not** multiples of 7.
For example, 16, 21, 3, 7, 28, 30, 15, 8, 35, 11, 5, 14, 12, 49, 17 (117).
Can be differentiated by choice of multiple and numbers.

Teaching notes

The worked examples in the unit stress the importance of correctly lining up the digits in written calculation as this causes many students to make errors. Draw particular attention to the extra zeros that have been added where necessary.

It is again worth emphasising that students should use an *appropriate* method of calculation, there is no *right* way and allow time for discussion of which method they will use and ask them to justify that decision (EP).

Students should be advised to estimate a solution first as a strategy for self-checking both to build confidence and for support in taking increased responsibility for their own learning (IE).

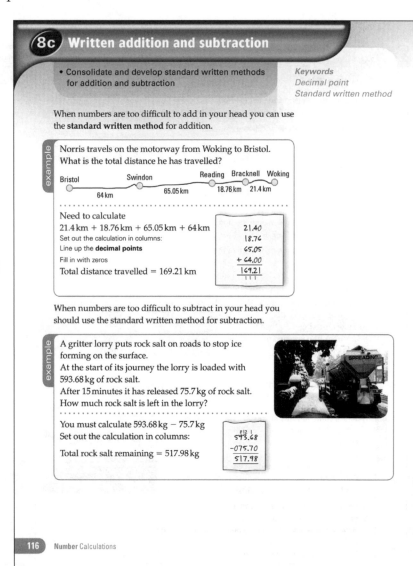

Plenary

Refer students back to the exercise and ask them to look at the questions where they were asked to choose an *appropriate* method. Ask them to justify the method they chose. Preferably work in pairs here, if possible pairing students who chose different methods (EP).

Simplification

Common difficulties here will be misplacing digits. Some students persist in setting out a calculation horizontally, which works perfectly well if they are using a mental method but often leads to disaster for a written one. Discuss with weaker students which method they plan to use and emphasise the value of approximating first (RL).

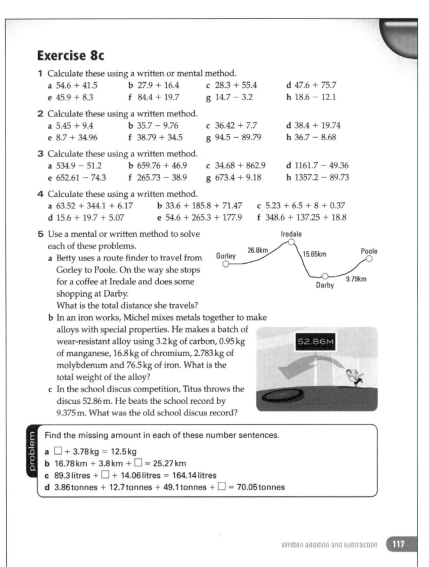

Exercise 8c

1 Calculate these using a written or mental method.
  a 54.6 + 41.5   **b** 27.9 + 16.4   **c** 28.3 + 55.4   **d** 47.6 + 75.7
  e 45.9 + 8.3   **f** 84.4 + 19.7   **g** 14.7 − 3.2   **h** 18.6 − 12.1

2 Calculate these using a written method.
  a 5.45 + 9.4   **b** 35.7 − 9.76   **c** 36.42 + 7.7   **d** 38.4 + 19.74
  e 8.7 + 34.96   **f** 38.79 + 34.5   **g** 94.5 − 89.79   **h** 36.7 − 8.68

3 Calculate these using a written method.
  a 534.9 − 51.2   **b** 659.76 + 46.9   **c** 34.68 + 862.9   **d** 1161.7 − 49.36
  e 652.61 − 74.3   **f** 265.73 − 38.9   **g** 673.4 + 9.18   **h** 1357.2 − 89.73

4 Calculate these using a written method.
  a 63.52 + 344.1 + 6.17   **b** 33.6 + 185.8 + 71.47   **c** 5.23 + 6.5 + 8 + 0.37
  d 15.6 + 19.7 + 5.07   **e** 54.6 + 265.3 + 177.9   **f** 348.6 + 137.25 + 18.8

5 Use a mental or written method to solve each of these problems.
  a Betty uses a route finder to travel from Gorley to Poole. On the way she stops for a coffee at Iredale and does some shopping at Darby.
What is the total distance she travels?
  b In an iron works, Michel mixes metals together to make alloys with special properties. He makes a batch of wear-resistant alloy using 3.2 kg of carbon, 0.95 kg of manganese, 16.8 kg of chromium, 2.783 kg of molybdenum and 76.5 kg of iron. What is the total weight of the alloy?
  c In the school discus competition, Titus throws the discus 52.86 m. He beats the school record by 9.375 m. What was the old school discus record?

problem

Find the missing amount in each of these number sentences.
  a ☐ + 3.78 kg = 12.5 kg
  b 16.78 km + 3.8 km + ☐ = 25.27 km
  c 89.3 litres + ☐ + 14.06 litres = 164.14 litres
  d 3.86 tonnes + 12.7 tonnes + 49.1 tonnes + ☐ = 70.05 tonnes

Written addition and subtraction   **117**

Extension

More able students will quickly achieve the correct solutions to the number problems and will need to focus more on word problems.

Exercise 8c commentary

Question 1 – Emphasise using the most appropriate method and that there is no 'right' method (1.3).

Question 2 to 4 – Similar to the two examples. Ensure digits are lined up and encourage first making an estimate (1.3).

Question 5 – Ask students to explain why they decide that their method of calculation was most appropriate (EP) (1.1, 1.3, 1.5).

Problem solving – Students will need to revisit inverse operations (1.3).

Assessment Criteria – Calculating, Level 4: use efficient written methods of addition and subtraction.

Links

Discus throwing is an ancient sport and was one of the five exercises in the original Greek pentathlon. The men's World record for the discus is held by Jürgen Schult (G.D.R.) with a throw of 74.08 m. The women's record is 76.80 m and is held by Gabriele Reinsch (G.D.R.). How much further are these distances than the school record in Question **5**?

- Multiply and divide integers and decimals by
  10, 100, 1000, and explain the effect **(L5)**
- Multiply and divide integers and decimals by 0·1
  and 0·01 **(L7)**

Useful resources

••

Starter – Pocket money

Anwar received £10 pocket money each week.

Bryony's pocket money started at £5 and increased by 50p each week:
£5 in week 1, £5.50 in week 2, *etc*.

Charlie said he would be happy if his pocket money started with 1p and
it doubled each week: 1p, 2p, 4p, *etc*.

Ask students who would have accumulated the most money after 10 weeks. (Anwar)
How about 20 weeks? (Charlie)

Teaching notes

Make links with students' prior
learning here, possibly using some of
the earlier questions in the exercise as
a whole class mini whiteboard activity
to facilitate assessment. Use the
response to these questions to decide
where to focus discussion when
looking at the worked examples.

The most likely area requiring
consolidation will be multiplication
and division by tenths and
hundredths. Support learning through
the process skill of analysing – use
mathematical reasoning about
connections between number
operations, for example, dividing by a
number is the same as multiplying by
its reciprocal (CT).

One way to help students see the
connection between dividing by, say
0.01, and multiplying by 100 is to pose
questions involving units. 'How many
centimetres are there in 3 m?' versus
'How many 0.01 m long items will fit
into 3 m?' compared with, 'What is
3 ÷ 0.01?' (1.2)

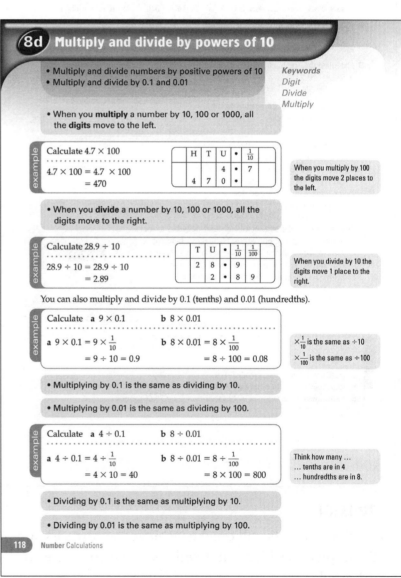

8d Multiply and divide by powers of 10

- Multiply and divide numbers by positive powers of 10
- Multiply and divide by 0.1 and 0.01

Keywords
Digit
Divide
Multiply

- When you **multiply** a number by 10, 100 or 1000, all
  the **digits** move to the left.

example
Calculate 4.7×100

$4.7 \times 100 = 4.7 \times 100$
$= 470$

H	T	U	•	$\frac{1}{10}$
		4	•	7
4	7	0	•	

When you multiply by 100
the digits move 2 places to
the left.

- When you **divide** a number by 10, 100 or 1000, all the
  digits move to the right.

example
Calculate $28.9 \div 10$

$28.9 \div 10 = 28.9 \div 10$
$= 2.89$

T	U	•	$\frac{1}{10}$	$\frac{1}{100}$
2	8	•	9	
	2	•	8	9

When you divide by 10 the
digits move 1 place to the
right.

You can also multiply and divide by 0.1 (tenths) and 0.01 (hundredths).

example
Calculate **a** 9×0.1 **b** 8×0.01

a $9 \times 0.1 = 9 \times \frac{1}{10}$
$\qquad = 9 \div 10 = 0.9$

b $8 \times 0.01 = 8 \times \frac{1}{100}$
$\qquad = 8 \div 100 = 0.08$

$\times \frac{1}{10}$ is the same as $\div 10$
$\times \frac{1}{100}$ is the same as $\div 100$

- Multiplying by 0.1 is the same as dividing by 10.
- Multiplying by 0.01 is the same as dividing by 100.

example
Calculate **a** $4 \div 0.1$ **b** $8 \div 0.01$

a $4 \div 0.1 = 4 \div \frac{1}{10}$
$\qquad = 4 \times 10 = 40$

b $8 \div 0.01 = 8 \div \frac{1}{100}$
$\qquad = 8 \times 100 = 800$

Think how many …
… tenths are in 4
… hundredths are in 8.

- Dividing by 0.1 is the same as multiplying by 10.
- Dividing by 0.01 is the same as multiplying by 100.

118   **Number** Calculations

Plenary

Extend the activity in question **7** of the exercise with
additional examples asking students to identify which
number will make the calculation correct. Convert this into
a mini whiteboard activity to identify whether any
problems in understanding persist here.

Simplification

If students struggle, discuss with them for any given calculation whether they expect the answer to be bigger or smaller after the operation. Discussion of this type exposes misconceptions about multiplying and dividing by numbers less than 1 (RL).

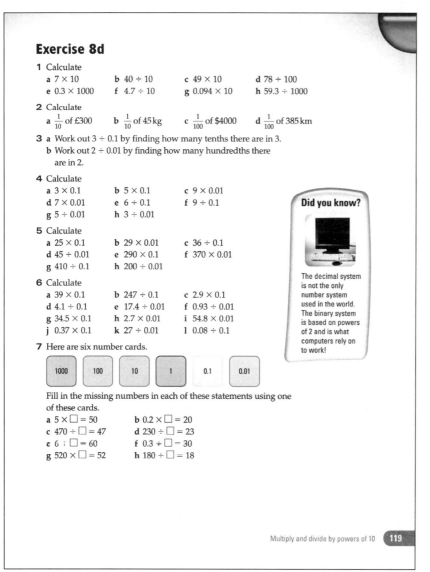

Exercise 8d

1 Calculate

  a 7×10 **b** $40 \div 10$ **c** 49×10 **d** $78 \div 100$

  e 0.3×1000 **f** $4.7 \div 10$ **g** 0.094×10 **h** $59.3 \div 1000$

2 Calculate

  a $\frac{1}{10}$ of £300   **b** $\frac{1}{10}$ of 45 kg   **c** $\frac{1}{100}$ of \$4000   **d** $\frac{1}{100}$ of 385 km

3 a Work out $3 \div 0.1$ by finding how many tenths there are in 3.

  b Work out $2 \div 0.01$ by finding how many hundredths there are in 2.

4 Calculate

  a 3×0.1 **b** 5×0.1 **c** 9×0.01

  d 7×0.01 **e** $6 \div 0.1$ **f** $9 \div 0.1$

  g $5 \div 0.01$ **h** $3 \div 0.01$

5 Calculate

  a 25×0.1 **b** 29×0.01 **c** $36 \div 0.1$

  d $45 \div 0.01$ **e** 290×0.1 **f** 370×0.01

  g $410 \div 0.1$ **h** $200 \div 0.01$

6 Calculate

  a 39×0.1 **b** $247 \div 0.1$ **c** 2.9×0.1

  d $4.1 \div 0.1$ **e** $17.4 \div 0.01$ **f** $0.93 \div 0.01$

  g 34.5×0.1 **h** 2.7×0.01 **i** 54.8×0.01

  j 0.37×0.1 **k** $27 \div 0.01$ **l** $0.08 \div 0.1$

7 Here are six number cards.

  | 1000 | | 100 | | 10 | | 1 | | 0.1 | | 0.01 |

Fill in the missing numbers in each of these statements using one of these cards.

  a $5 \times \square = 50$ **b** $0.2 \times \square = 20$

  c $470 \div \square = 47$ **d** $230 \div \square = 23$

  e $6 \div \square = 60$ **f** $0.3 \div \square = 30$

  g $520 \times \square = 52$ **h** $180 \div \square = 18$

> **Did you know?**
>
> The decimal system is not the only number system used in the world. The binary system is based on powers of 2 and is what computers rely on to work!

Extension

Provide work for more able students to 'mark' using examples such as those in the exercise that have been answered with some errors. This will offer consolidation in a more engaging format (RL).

Exercise 8d commentary

All questions (1.3)

Question 1 – Similar to the first two examples.

Question 2 – Make links here with fractions and division.

Question 3 – This question helps to re-emphasise that division by 0·1 is the same as multiplication by 10, *etc*.

Question 4 – Similar to the last two examples.

Question 5 – Develops question **4** to use larger numbers.

Question 6 – Develops question **4** to use decimals.

Question 7 – Reviews the material from a slightly different perspective.

Assessment Criteria – NNS, level 5: use understanding of place value to multiply and divide whole numbers and decimals by 10, 100 and 1000 and explain the effect. Calculating, level 7: understand the effects of multiplying and dividing by numbers between 0 and 1.

Links

Number 10 Downing Street in London is the official residence of the First Lord of the Treasury, who is usually also the prime minister of Great Britain. The first prime minister to live at Number 10 was Robert Walpole in the early 1700s and most prime ministers have lived there since. The house is mainly comprised of offices and the Prime Minister's apartment is at the top of the building. Originally the apartment rooms were used by servants. There is more information about Number 10 at http://www.number10.gov.uk/history-and-tour.

- Strengthen and extend mental methods of calculation, working with decimals (L5)

Useful resources

Starter – 105

Ask students to make 105 by

the product of 2 odd numbers ($3 \times 35, 5 \times 21, 7 \times 15$)

the product of 3 odd numbers ($3 \times 5 \times 7$)

the product of 4 odd numbers ($1 \times 3 \times 5 \times 7$)

the sum of a square number and a prime number ($4 + 101, 16 + 89, 64 + 41, 100 + 5$)

the difference between 2 square numbers ($361(19^2) - 256(16^2), 169 - 64, 121 - 4$)

Can any be done in more than one way?

Teaching notes

Whilst all the methods employed in this spread are mental methods it greatly helps understanding to see examples that are supported by written workings. These examples will also serve as model solutions.

As a preliminary to using the method based on factors, check that students can explain what they are and are able to quickly identify useful factors.

Forn the method of partitioning and compensation, it may help to remind students how to multiply out a single term over a bracket.

In all cases, students should be encouraged to estimate the answer first using an approximate sum before calculating the exact result.

When discussing the examples, where appropriate, invite students to say whether the strategy illustrated would be the one they would have chosen. This not only emphasises the validity of alternative approaches but helps to build student confidence in taking responsibility for their own learning and in making and justifying decisions themselves (IE).

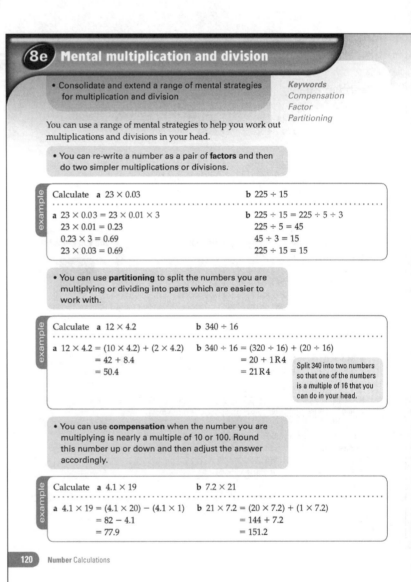

8e Mental multiplication and division

- Consolidate and extend a range of mental strategies for multiplication and division

Keywords
Compensation
Factor
Partitioning

You can use a range of mental strategies to help you work out multiplications and divisions in your head.

- You can re-write a number as a pair of **factors** and then do two simpler multiplications or divisions.

example

Calculate **a** 23×0.03 **b** $225 \div 15$

a $23 \times 0.03 = 23 \times 0.01 \times 3$
$23 \times 0.01 = 0.23$
$0.23 \times 3 = 0.69$
$23 \times 0.03 = 0.69$

b $225 \div 15 = 225 \div 5 \div 3$
$225 \div 5 = 45$
$45 \div 3 = 15$
$225 \div 15 = 15$

- You can use **partitioning** to split the numbers you are multiplying or dividing into parts which are easier to work with.

example

Calculate **a** 12×4.2 **b** $340 \div 16$

a $12 \times 4.2 = (10 \times 4.2) + (2 \times 4.2)$
$= 42 + 8.4$
$= 50.4$

b $340 \div 16 = (320 \div 16) + (20 \div 16)$
$= 20 + 1R4$
$= 21R4$

Split 340 into two numbers so that one of the numbers is a multiple of 16 that you can do in your head.

- You can use **compensation** when the number you are multiplying is nearly a multiple of 10 or 100. Round this number up or down and then adjust the answer accordingly.

example

Calculate **a** 4.1×19 **b** 7.2×21

a $4.1 \times 19 = (4.1 \times 20) - (4.1 \times 1)$
$= 82 - 4.1$
$= 77.9$

b $21 \times 7.2 = (20 \times 7.2) + (1 \times 7.2)$
$= 144 + 7.2$
$= 151.2$

120 **Number** Calculations

Plenary

Ask students to work in pairs, give three example questions and ask them to explain how they would calculate the answer mentally in two different ways for each question.

Simplification

Where students experience difficulty, allow them to decide which mental method they are most comfortable with and support them with this rather than placing too much emphasis on other strategies. Confidence building will be beneficial.

Exercise 8e

1 Calculate these mentally using factors.

a 8×30	**b** 5×40	**c** 6×300	**d** 7×500
e 25×20	**f** 17×30	**g** 12×50	**h** 33×200
i 24×0.2	**j** 5×0.03	**k** 18×0.5	**l** 15×0.04
m 32×0.3	**n** 51×0.06	**o** 44×0.4	**p** 26×0.02
q $156 \div 6$	**r** $140 \div 4$	**s** $264 \div 12$	**t** $360 \div 15$
u $216 \div 8$	**v** $126 \div 18$	**w** $135 \div 15$	**x** $108 \div 12$

2 Calculate these using the method of partitioning.

a 3.4×12	**b** 2.1×15	**c** 13×1.4	**d** 16×3.5
e 2.8×11	**f** 4.3×15	**g** 3.9×12	**h** 14×6.2
i $150 \div 7$	**j** $190 \div 8$	**k** $220 \div 12$	**l** $280 \div 13$
m $310 \div 14$	**n** $385 \div 15$	**o** $410 \div 16$	**p** $480 \div 22$

3 Calculate these using the method of compensation.

a 21×2.9	**b** 19×4.4	**c** 8.1×21	**d** 3.6×31
e 3.5×19	**f** 4.2×29	**g** 19×1.9	**h** 21×5.9
i 3.8×19	**j** 2.4×22	**k** 18×1.6	**l** 49×2.5

4 Use an appropriate mental method to solve each of these problems.

a A chocolate bar contains 6.3 g of fat. In a month, Ciaron eats fifteen of the chocolate bars. How much fat does he consume each month by eating the chocolate bars?

b Kiefer drinks 2.3 litres of water a day. How much water does he drink in the month of January?

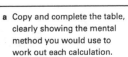

problem

a Copy and complete the table, clearly showing the mental method you would use to work out each calculation.

b Compare your answers with a partner. Discuss the questions where you used different methods.

Question	Method I would use
8×19	Work out 8×20 subtract 8
7×21	
13×15	
$156 \div 4$	
16×19	
12×34	
31×6	
$90 \div 7$	
9×30	
11×15	

Extension

More able students could consolidate their understanding by putting together a PowerPoint presentation illustrating how to use a variety of mental methods. This could be shown to the class and made available on the school system for additional 'tutorial style' support for weaker students.

Exercise 8e commentary

Question 1 – Similar to the first example (1.3).

Question 2 – Similar to the second example. Ask students if partitioning would have been their first choice (1.3).

Question 3 – Similar to the third example (1.3).

Question 4 – Ask students to identify and justify their chosen strategy (RL) (1.1, 1.3).

Problem – A good way to illustrate that there is no one right or best method (TW) (1.3, 1.4, 1.5).

Assessment Criteria – Calculating, level 4, use a range of mental methods of calculation with all operations.

Links

Question **4** refers to chocolate bars. Chocolate is made from the beans of the tropical cacao tree and was prized as a drink by the Aztecs. The first eating chocolate was produced by Joseph Fry in Bristol in 1848. On average, each person in the United Kingdom eats 10 kg of chocolate each year. The population of The United Kingdom is approximately 60 million. How many tonnes (= 1000 kg) of chocolate are eaten in the UK each year? (600 000 tonnes).

- Use efficient written methods for multiplication [and division] of integers and decimals (L5)
- Understand where to position the decimal point by considering equivalent calculations (L5)
- Make and justify estimates and approximations of calculations (L5)

Useful resources

Starter – Factor bingo

Ask students to draw a 3 × 3 grid and enter three factors of 24, three factors of 36 and three factors of 75.

Give possible answers (1, 2, 3, 4, 5, 6, 8, 9, 12, 15, 18, 24, 25, 36, 75).

Winner is the first student to cross out all their factors.

Can be differentiated by the choice of numbers.

Teaching notes

The worked examples in this unit emphasise the value of using approximation in the first instance to get an idea of what the answer to a calculation might be. This needs to be stressed to students, not only because of its usefulness as a checking mechanism here but also as a recap of rounding and the value of making sensible estimates in general.

There is scope to discuss how to choose the numbers to round to.

The use of using whole number calculations also helps to improve understanding of multiplication and division by 10 and 100.

The second example is especially useful in supporting links with literacy. Discuss as a whole class how the question should be analysed and identify what is the key information to be extracted (CT).

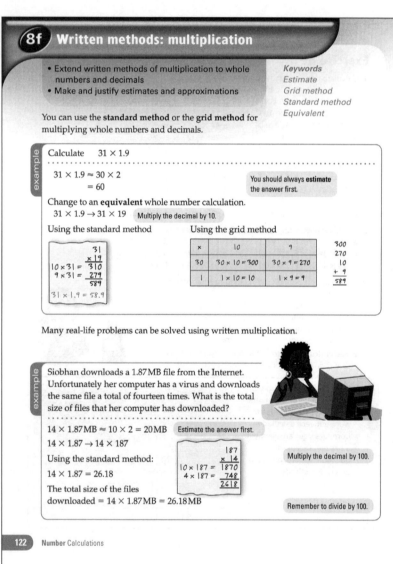

8f **Written methods: multiplication**

- Extend written methods of multiplication to whole numbers and decimals
- Make and justify estimates and approximations

Keywords
Estimate
Grid method
Standard method
Equivalent

You can use the **standard method** or the **grid method** for multiplying whole numbers and decimals.

Calculate 31×1.9

$31 \times 1.9 \approx 30 \times 2$
$= 60$

You should always estimate the answer first.

Change to an **equivalent** whole number calculation.
$31 \times 1.9 \rightarrow 31 \times 19$ *Multiply the decimal by 10.*

Using the standard method

$$
\begin{array}{r}
31 \\
\times 19 \\
\hline
10 \times 31 = 310 \\
9 \times 31 = 279 \\
\hline
589
\end{array}
$$

$31 \times 1.9 = 58.9$

Using the grid method

×	10	9
30	30 × 10 = 300	30 × 9 = 270
1	1 × 10 = 10	1 × 9 = 9

$$
\begin{array}{r}
300 \\
270 \\
10 \\
+ 9 \\
\hline
589
\end{array}
$$

Many real-life problems can be solved using written multiplication.

Siobhan downloads a 1.87 MB file from the Internet. Unfortunately her computer has a virus and downloads the same file a total of fourteen times. What is the total size of files that her computer has downloaded?

$14 \times 1.87\,\text{MB} \approx 10 \times 2 = 20\,\text{MB}$ *Estimate the answer first.*
$14 \times 1.87 \rightarrow 14 \times 187$
Using the standard method:
$14 \times 1.87 = 26.18$
The total size of the files
downloaded $= 14 \times 1.87\,\text{MB} = 26.18\,\text{MB}$

Multiply the decimal by 100.

$$
\begin{array}{r}
187 \\
\times 14 \\
\hline
10 \times 187 = 1870 \\
4 \times 187 = 748 \\
\hline
2618
\end{array}
$$

Remember to divide by 100.

122   Number Calculations

Plenary

Refer students back to questions **6** and **7**. Ask them to choose one part of one of these questions and to explain how they did it to a friend.

Simplification

Where students struggle, spend time allowing them to work in pairs on an activity of the kind illustrated in the investigation. This is useful as it develops many of the skills within the unit but the links between the different calculations is especially apparent in this format.

Allow students to use the grid method before switching to the standard method.

All questions (1.3)

Question 1 – Integers only. The mental approximations could all be made first as a whole class activity.

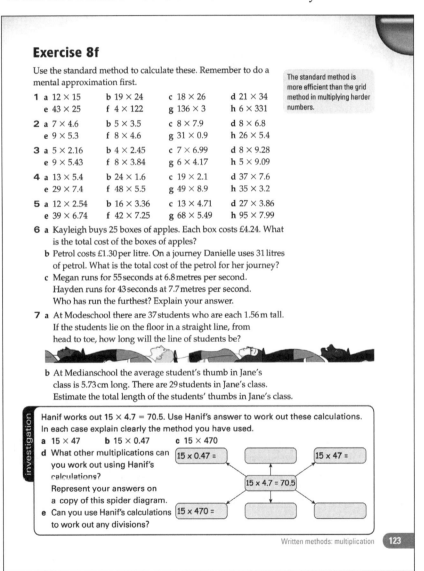

Exercise 8f

Use the standard method to calculate these. Remember to do a mental approximation first.

The standard method is more efficient than the grid method in multiplying harder numbers.

1 a 12 × 15   b 19 × 24   c 18 × 26   d 21 × 34
   e 43 × 25   f 4 × 122   g 136 × 3   h 6 × 331

2 a 7 × 4.6   b 5 × 3.5   c 8 × 7.9   d 8 × 6.8
   e 9 × 5.3   f 8 × 4.6   g 31 × 0.9   h 26 × 5.4

3 a 5 × 2.16   b 4 × 2.45   c 7 × 6.99   d 8 × 9.28
   e 9 × 5.43   f 8 × 3.84   g 6 × 4.17   h 5 × 9.09

4 a 13 × 5.4   b 24 × 1.6   c 19 × 2.1   d 37 × 7.6
   e 29 × 7.4   f 48 × 5.5   g 49 × 8.9   h 35 × 3.2

5 a 12 × 2.54   b 16 × 3.36   c 13 × 4.71   d 27 × 3.86
   e 39 × 6.74   f 42 × 7.25   g 68 × 5.49   h 95 × 7.99

6 a Kayleigh buys 25 boxes of apples. Each box costs £4.24. What is the total cost of the boxes of apples?
   b Petrol costs £1.30 per litre. On a journey Danielle uses 31 litres of petrol. What is the total cost of the petrol for her journey?
   c Megan runs for 55 seconds at 6.8 metres per second. Hayden runs for 43 seconds at 7.7 metres per second. Who has run the furthest? Explain your answer.

7 a At Modeschool there are 37 students who are each 1.56 m tall. If the students lie on the floor in a straight line, from head to toe, how long will the line of students be?

   b At Medianschool the average student's thumb in Jane's class is 5.73 cm long. There are 29 students in Jane's class. Estimate the total length of the students' thumbs in Jane's class.

investigation

Hanif works out 15 × 4.7 = 70.5. Use Hanif's answer to work out these calculations. In each case explain clearly the method you have used.
   a 15 × 47   b 15 × 0.47   c 15 × 470
   d What other multiplications can you work out using Hanif's calculations? Represent your answers on a copy of this spider diagram.
   e Can you use Hanif's calculations to work out any divisions?

15 x 0.47 =

15 x 47 =

15 x 4.7 = 70.5

15 x 470 =

Written methods: multiplication **123**

Extension

Encourage more able students to create their own spider diagram of the kind illustrated in the investigation to ensure that they have clarity of understanding of how calculations link together.

Exercise 8f commentary

Questions 2 to **5** – Similar to the first example.

Question 3 – Insist students first approximate and highlight how this helps to avoid careless errors.

Question 5 – Students can mark their work most effectively by initially comparing answers with their approximation and then with another student (RL).

Question 6 – Although the mathematics is no more challenging, some students will need support to analyse the text (1.1).

Question 7 – Part **b**, ask students why the answer is an estimate. This will link effectively to students' work on averages (IE) (1.1, 1.2).

Investigation – Ask students to explain their methods to each other. Talking through can clarify both understanding and any errors in a positive fashion (RL) (1.4).

Assessment Criteria – Calculating, level 5: understand and use an appropriate non-calculator method for solving problems that involve multiplying any three digit number by any two digit number. Calculating, level 5: apply inverse operations and approximate to check answers to problems are of the correct magnitude.

Links

A computer virus is an often malevolent program that can copy itself into other programs without the permission of the user. If a virus-infected email is sent to 10 people who each send it to 10 people who each send it to 10 more people, how many computers could become infected? (1000). There is more information at http://en.wikipedia.org/wiki/Computer_virus.

- Use efficient written methods for [multiplication and] division of integers and decimals **(L6)**
- Make and justify estimates and approximations of calculations **(L5)**

Useful resources

••

Starter – Double products

Draw a 4 × 4 table on the board.

Label the columns 9, 4, 18, 11 and the rows 3, 7, 14, 5.

Ask students to fill in the table with the products (no calculators).

For example, the top row in the table would read 27, 12, 54, 33.

Hint: 18 is double 9 and 14 is double 7.

Can be differentiated by the choice of numbers.

Teaching notes

This spread, together with the preceding few all develop the key processes in number: in representing, where students need to identify the type of problem and the operations needed to solve it; in analysing, where it is important to visualise images, such as a number line to support mental methods; and in communicating and reflecting where students compare the efficiency of calculation procedures and discuss and reflect on different approaches to solving a numerical problem.

The worked examples are very helpful for clarifying methods and strategies Refer students back to these, when they get stuck, in order to encourage them to take responsibility for their learning. The greatest emphasis however needs to be on the development of the students' capacity to apply their skills in a functional setting. The word problems provided here strongly support this (IE).

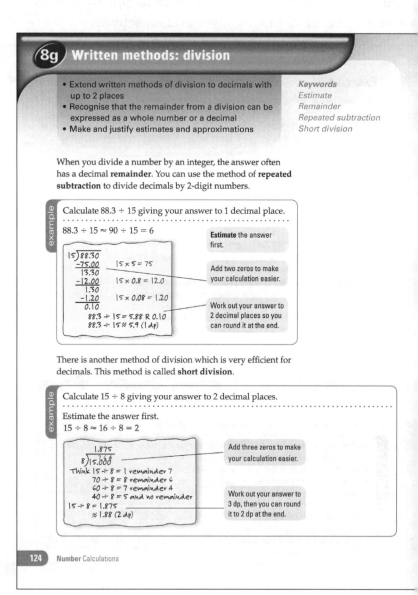

Plenary

Provide one word problem involving multiplication and one involving division and ask students to work in pairs to answer one each and to explain their working to each other. Encourage students to ask questions of each other for clarification. In addition to being a consolidation activity, this also offers the opportunity for students to provide constructive support and feedback to others (TW).

Simplification

For students who may struggle with some of the strategies discussed in this unit, re-emphasise the importance of estimating an answer in the first instance. Offer word problems with a decrease in technical demand and complexity and an increase in familiarity.

All questions (1.3)

Question 1 – Appropriate methods are short division or chunking; students should approximate first, whichever method they choose.

Question 2 – Discuss with students how they found their answers to parts **b** and **c**.

Exercise 8g

1 Calculate these using an appropriate method. Give your answer with a remainder where appropriate.
  a $161 \div 7$ **b** $156 \div 6$ **c** $279 \div 9$ **d** $272 \div 8$
  e $195 \div 13$ **f** $225 \div 14$ **g** $360 \div 15$ **h** $360 \div 17$

2 a What is $729 \div 27$?
  b What is the remainder when 750 is divided by 27?
  c What is the remainder when 720 is divided by 27?

3 Calculate these using an appropriate method.
  a $45.6 \div 6$ **b** $63.2 \div 8$ **c** $64.8 \div 9$ **d** $39.2 \div 7$
  e $64.8 \div 12$ **f** $81.9 \div 13$ **g** $99.4 \div 14$ **h** $99 \div 15$

4 Calculate these using an appropriate method. Give your answer as a decimal rounded to 1 decimal place where appropriate.
  a $55 \div 8$ **b** $20 \div 7$ **c** $35 \div 6$ **d** $46 \div 9$
  e $120 \div 11$ **f** $137 \div 12$ **g** $150 \div 16$ **h** $223 \div 18$

5 Calculate these using an appropriate method. Give your answer as a decimal rounded to 1 decimal place where appropriate.
  a $25.6 \div 8$ **b** $32.5 \div 7$ **c** $14.5 \div 6$ **d** $24.6 \div 9$
  e $34.8 \div 14$ **f** $37.5 \div 15$ **g** $46.3 \div 18$ **h** $55.4 \div 21$

6 In each calculation, decide whether to give your answer either as a remainder or as a decimal to 1 decimal place.
  a Devvon's car uses 27 litres of petrol to travel 400 km. Alec's car uses 25 litres of petrol to travel 365 km. Which car travels further for each litre of petrol? Explain your answer.
  b Farah has been alive for 765 hours. Is she more or less than a month old? Explain and justify your answer.
  c Xan downloads a file at 28 kB per second. The file is 95.2 kB in size. How many seconds does it take to download the file?

Did you know?

The world's most fuel-efficient car as of 2008 is the Volkswagen 285 MPG. See if you can work out why it's called 285 MPG!

puzzle
'Purfect' cat biscuits come in three different sizes.
  a Which size of biscuits is the best value for money?
  b Explain and justify your answer.

Extension

More able students will need to increase their skills at a functional level by considering (word) problems with decreased familiarity and an increase in complexity and the level of independence required.

Exercise 8g commentary

Question 3 – Encourage students to share their ideas about an 'appropriate' method. The decimals will be a stumbling block for some (EP) (1.5).

Question 4 – Remind students to work out 2 decimal places and then to round to 1 decimal place.

Question 5 – Extends question **4** to include decimals.

Question 6 – The functional setting will require analysis, and perhaps support, to tease out the mathematics to use (IE) (1.1, 1.4).

Puzzle – Use this activity to strengthen the development of functional skills, emphasising the element of explanation and justification (EP) (1.1).

Assessment Criteria – Calculating, level 5: understand and use an appropriate non-calculator method for solving problems that involve dividing any three digit number by any two digit number. Calculating, level 5: apply inverse operations and approximate to check answers to problems are of the correct magnitude.

Links

As a car moves it has to push air out of the way. This produces a force called air resistance, or drag, which slows down the car. The greater the drag, the more fuel is required to make the car drive the same distance. Sleek aerodynamic design reduces the drag and so improves the fuel economy. Other factors that affect fuel economy are weight and speed. There is more information about the Volkswagen 285 mpg at http://gas2.org/2008/ 03/12/the-worlds-most-fuel-efficient-car-285-mpg-not-a-hybrid/comment-page-4/

- Use the order of operations, including brackets, with more complex calculations (L5)
- Carry out more difficult calculations effectively and efficiently using the function key for powers (L5)
- Use brackets (L5)

Useful resources

Starter – Crazy clocks

One clock chimes every six minutes and a second every seven minutes.
The clocks chime together. How long before the clocks chime together again? (42 min)
What if a third clock chimes every half hour? (210 min)
Can be extended by asking how many times the clocks will chime together in 24 hours.
(2 clocks 34 times, 3 clocks 6 times).

Teaching notes

The order of operations is a matter of convention, so students may differ on how to evaluate an expression. Use question **1** to identify different ways of interpreting expressions and to agree on an order of precedence. Introduce the BIDMAS acronym as a useful mnemonic and, for visual learners, the 'traffic light' diagram. Encourage students to refer back to these when applying the rules. Use examples of increasing complexity to reinforce the correct interpretation.

The efficient use of calculators relies on correctly using rules of precedence to obtain a unique/correct answer. Encourage students to check how a calculator works by doing simple 'toy calculations' for which they already know the answer. This may vary between calculators, as may the appearance of some of the keys.

The process skill of analyzing, using appropriate mathematical procedures, is developed as students learn efficient use of the calculator.

Encourage students to work collaboratively in this spread, so that they can develop confidence in themselves and their contribution (TW).

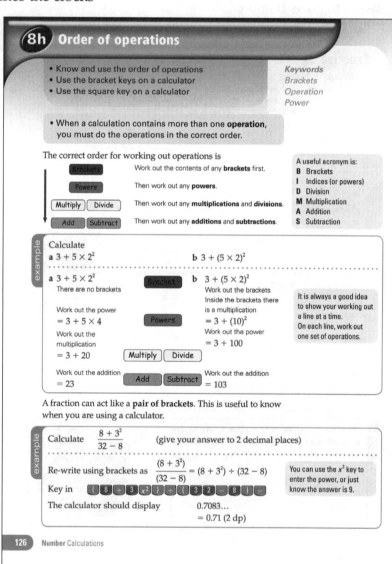

Plenary

Provide three calculations, of varying difficulty, which involve the correct order of operations. Ask students to choose one of these and write an annotated solution showing a step-by-step approach that would be useful for a classmate who had been absent and missed the work on this topic. Ask them to swap with someone who has answered a different question to check each other's work.

Simplification

Provide three questions that have been answered with some errors included. Ask the students to mark the work and include feedback on the errors. Use this as a paired activity where errors are analysed without exposure of students' own weaknesses (RL).

Exercise 8h

1 Work out these calculations using the order of operations.

a $5 + 4 \times 6$ b $15 - 3 \times 2$ c $12 + 2 \times 5$

d $24 - 16 \div 4$ e $5 \times 3 - 2 \times 3$ f $4 \times 7 + 6 \times 3$

g $18 \div 9 + 15 \div 5$ h $7 + 5 \times 3 - 2$

2 Match each question with the correct answer.

Question	Answer X	Answer Y
a $3 + 4^2 \times 5$	83	245
b $20 - 3^2 \times 2$	578	2
c $(19 - 4^2) \times 2$	250	6
d $5 \times (3 + 4)^2$	245	31
e $25 - 3 \times (2^2 - 1)$	87	16
f $3 \times 5^2 + 12 \div 3$	79	29
g $6 \times 7 - 3^2 \times 4$	132	6

Explain your method and reasoning for each answer.

3 Calculate these giving your answer to 1 decimal place where appropriate. You may with to use a calculator.

a $\dfrac{8+4}{5-2}$ b $\dfrac{5^2-1}{3^2+4}$ c $\dfrac{(28-3)}{10-3^2}$ d $\dfrac{(4+3)^2}{4^2-1}$

4 Copy and complete each of these by putting brackets in the correct place in the expression.

a $8 + 5 \times 4 - 3 = 49$ b $8 + 5 \times 4 - 3 = 13$

c $8 + 5 \times 4 - 3 = 25$

5 Use a calculator to work out these calculations. Give your answers to 2 decimal places where appropriate.

a $(5 + 2.8) \div 7$ b $34 - 1.7^2 \times 4$ c $6 \times (3.5 - 1.6)^2$

d $(4 + 2.5) \times 7$ e $\dfrac{7 + 3^2}{19 - 4^2}$ f $\dfrac{13^2 + 6^2}{13^2 - 6^2}$

g $\dfrac{48}{6 \times 8}$ h $\dfrac{5 + 2^2}{(18 - 9)}$

investigation

John and Vernon are working out this calculation $\dfrac{60}{3 \times 4}$

John works out $3 \times 4 = 12$, and then $60 \div 12 = 5$

Vernon works out $60 \div 3 = 20$, and then $20 \div 4 = 5$

Explain how and why both methods work.

Extension

Ask students to generate a small set of questions involving the order of operations with true/false answers. A selection of these could be compiled into a starter activity for a future lesson.

Exercise 8h commentary

Question 1 – Similar to the first example (1.3).

Question 2 – Ask students to collaborate, taking turns to explain the right answer or account for the wrong answer in order to build confidence and encourage participation (1.3).

Question 3 – If necessary, remind students how to round (1.3, 1.4).

Question 4 – Invite students to compare answers here as a checking strategy (RL) (1.3).

Question 5 – Similar to the second example. Ensure that weaker students are confident in using brackets on a calculator and, if necessary, discuss rounding to 2dp (1.3, 1.4).

Investigation – This is a very useful activity to strengthen understanding of number. Ask students to work in pairs initially and then take whole class feedback (CT) (1.1, 1.4).

Assessment Criteria – Calculating, Level 5:use known facts, place value, knowledge of operations and brackets to calculate including using all four operations with decimals to two places.

Links

The order of operations can be described using the mnemonic BIDMAS. (Brackets, Indices or Powers, Multiplication and Division, Addition and Subtraction). A mnemonic is a memory aid which uses words and letters to jog the memory. Some well-known mnemonics are *Richard of York Gave Battle In Vain* (colours of the rainbow), and the rhyme beginning *Thirty Days hath September* (number of days in the months of the year) How many other mnemonics can the class remember?

- Enter numbers and interpret the display in different contexts (extend to [negative numbers,] fractions, time)   (L6)

Useful resources

Starter – 9999

Ask students to make as many numbers between 1 and 20 inclusive using 4 nines and any operation(s), for example, $1 = (9 + 9) \div (9 + 9)$. All are possible.

Hint: The square root of 9 is 3.

Teaching notes

When looking at the worked examples here it is useful to emphasise that using a calculator to convert is not the method to use in every case and that mental methods may be more efficient at times. This will emphasise the key processes in number, communicating and reflecting, comparing the efficiency of calculation procedures.

Writing remainders in the context of the problem needs to be a key focus as this awareness is central to becoming functional in mathematics. Encourage students to discuss their reasoning as often as possible to develop mathematical talk around this process (IE).

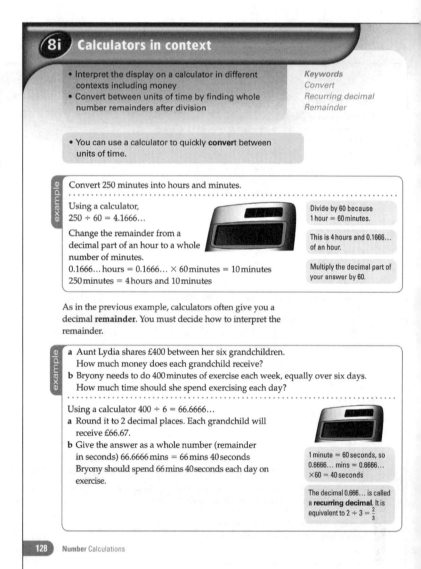

8i Calculators in context

- Interpret the display on a calculator in different contexts including money
- Convert between units of time by finding whole number remainders after division

Keywords
Convert
Recurring decimal
Remainder

- You can use a calculator to quickly **convert** between units of time.

example

Convert 250 minutes into hours and minutes.

Using a calculator,
$250 \div 60 = 4.1666...$

Change the remainder from a decimal part of an hour to a whole number of minutes.
$0.1666...\text{hours} = 0.1666... \times 60\,\text{minutes} = 10\,\text{minutes}$
$250\,\text{minutes} = 4\,\text{hours and } 10\,\text{minutes}$

Divide by 60 because 1 hour = 60 minutes.

This is 4 hours and 0.1666... of an hour.

Multiply the decimal part of your answer by 60.

As in the previous example, calculators often give you a decimal **remainder**. You must decide how to interpret the remainder.

example

a Aunt Lydia shares £400 between her six grandchildren. How much money does each grandchild receive?
b Bryony needs to do 400 minutes of exercise each week, equally over six days. How much time should she spend exercising each day?

Using a calculator $400 \div 6 = 66.6666...$
a Round it to 2 decimal places. Each grandchild will receive £66.67.
b Give the answer as a whole number (remainder in seconds) 66.6666 mins = 66 mins 40 seconds Bryony should spend 66 mins 40 seconds each day on exercise.

1 minute = 60 seconds, so 0.6666... mins = 0.6666... ×60 = 40 seconds

The decimal 0.666... is called a **recurring decimal**. It is equivalent to $2 \div 3 = \frac{2}{3}$

Plenary

Invite feedback on students' own problems that they have generated in part **c** of the investigation.

Simplification

Some students may have difficulty with the word problems. If this is the case, encourage them to rewrite the question in calculation form for themselves. Students could work together for support.

Exercise 8i

1 Calculate these divisions using your calculator, leaving the remainder part of the answer in the form specified.
   a £200 ÷ 9   (a decimal to 2 dp)
   b 50 cakes ÷ 6   (a whole number remainder)
   c 11 hours ÷ 4   (a fraction)
   d 45 kg ÷ 7   (a decimal to 2 dp)

2 Convert these measurements of time into the units indicated in brackets.
   a 200 mins   (into hours and mins)
   b 800 hours   (into days and hours)
   c 6450 seconds   (into minutes and seconds)
   d 6220 mins   (into hours and mins)

3 Solve each of these problems. Give each of your answers in a form appropriate to the question.
   a Samina sells free range eggs. She has 164 eggs. She packs them into boxes of 6. How many egg boxes does she need to pack the eggs?
   b Avril wins £1 000 000. She decides to share the money equally between the 13 people in her family (including herself). How much money will each of the relatives receive?
   c 76 pizzas are shared between 40 people. How much pizza does each person receive?
   d Cecil is an electrician. He cuts a 25 m length of cable into 7 identical pieces. How long is each piece of cable?

investigation

Lake Underdale contains 540 000 000 m³ of water.
There are 1000 litres in 1 m³ of water.
An average person should drink 2.5 litres of water per day.
There are 34 652 people living in the town of Underwater.
   a How long would it take one person to drink all the water in Lake Underdale?
   b How long would it take all the people of Underwater to drink all the water in Lake Underdale?
   c Investigate further problems of your own using these additional facts about water consumption.
   One person uses 89.6 litres per day in flushing the toilet.
   A washing machine uses 70.4 litres per wash.
   A shower uses 70.6 litres per use.
   A tap uses 48 litres in a typical day.

Calculators in context **129**

Extension

Encourage students to investigate additional water consumption facts (perhaps about water usage in a major hotel) to generate interesting problems. They could use these to create a 'question-posing poster' for display to engage other students (SM).

Exercise 8i commentary

Question 1 – Ask why the specified form of the remainder is appropriate.

Question 2 – Similar to the first example. Ask which parts could be most efficiently answered without a calculator (1.3).

Question 3 – Similar to the second example. Ask students to explain why the form they have chosen for their answer is appropriate (EP) (1.1, 1.3, 1.4).

Investigation – Encourage 'thinking time' where students decide on their approach before finding a solution. Using this as a paired activity is more effective (CT) (1.1, 1.3).

Assessment Criteria – UAM, Level 6: solve problems and carry through substantial tasks by breaking them into smaller, more manageable tasks, using a range of efficient techniques, methods and resources, including ICT.

Links

The deepest lake in the World is Lake Baikal in Siberia. It has a surface area of 31 500 km² and contains 23 000 km³ of water which is approximately 20% of all the fresh water on Earth. The largest lake in the UK is Lough Neagh in Northern Ireland which has a surface area of 385 km² and contains 3.45 km³ of fresh water. How much more water is in Lake Baikal than in Lough Neagh?

8a

1 Round each of these numbers to the nearest
  i 1000   ii 100   iii 10
  a 3182   b 6273   c 4765   d 8632   e 6713   f 7682
  g 5049   h 24 505   i 38 604   j 42 783   k 76 060   l 39 494

2 Round each of these numbers to the nearest
  i whole number   ii tenth (1 dp)   iii hundredth (2 dp).
  a 4.847   b 3.107   c 8.3967   d 8.238   e 24.969   f 22.623
  g 3.4172   h 8.0495   i 3.1405   j 3.0078   k 2.429 57   l 4.545 45

8b

3 Calculate these using a mental method.
   a 492 − 187   b 799 − 203   c 2615 − 616   d 3639 − 1009
   e 2215 − 1797   f 3011 − 1688   g 4383 − 3985   h 7473 − 4749

4 Calculate these using a mental method.
   a 11.8 + 7.4   b 2.68 + 8.9   c 4.8 + 5.92   d 3.07 + 2.98
   e 13.7 − 8.88   f 6.99 − 3.49   g 8.71 − 4.8   h 9.67 − 3.85

8c

5 Calculate these using a written method.
   a 257.3 − 67.9   b 540.87 + 55.9   c 45.79 + 753.4
   d 1252.6 − 38.79   e 763.72 − 85.4   f 376.62 − 49.8
   g 582.5 + 10.36   h 2476.2 − 78.67   i 1468.4 − 72.56
   j 816.3 − 95.9   k 923.28 + 359   l 43.5 + 2186.39

8d

6 Calculate
   a 3×10 b 6×100 c $90 \div 10$ d $400 \div 100$
   e 38×10 f 1.7×10 g $67 \div 100$ h $497 \div 10$
   i 0.075×10 j $6.1 \div 100$ k $48.2 \div 1000$ l 0.0032×100

7 Calculate
   a 4×0.1 b 6×0.1 c 7×0.01 d 2×0.01
   e 34×0.1 f 65×0.1 g 30×0.01 h 58×0.01
   i 85.4×0.01 j 0.73×0.1 k $68 \div 0.01$ l $0.03 \div 0.1$

8e

8 Calculate these using a mental method.
   a 7.3×11 b 6.4×9 c 14×5.2 d 13×31
   e 4.7×21 f $406 \div 7$ g 3.4×13 h $300 \div 9$
   i $235 \div 4$ j 16×1.9 k $576 \div 8$ l 3.7×15

9 Calculate these using the standard method.

    a 15×6.5 **b** 35×2.7 **c** 16×4.8 **d** 43×8.5

    e 39×9.2 **f** 57×6.7 **g** 38×7.6 **h** 88×7.7

10 Calculate these using the standard method.

    a 19×3.68 **b** 27×4.18 **c** 46×5.53 **d** 62×7.26

    e 49×5.69 **f** 74×8.57 **g** 79×8.37 **h** 99×9.99

11 Calculate these using an appropriate method. Give your answer as a decimal rounded to 1 decimal place where appropriate.

    a $76 \div 8$ **b** $40 \div 9$ **c** $85 \div 6$ **d** $99 \div 9$

    e $252 \div 18$ **f** $314 \div 19$ **g** $388 \div 21$ **h** $404 \div 25$

12 Calculate these using an appropriate method. Give your answer as a decimal rounded to 1 decimal place where appropriate.

    a $36.7 \div 8$ **b** $43.6 \div 7$ **c** $25.6 \div 6$ **d** $35.7 \div 9$

    e $50.4 \div 24$ **f** $52.7 \div 39$ **g** $91.6 \div 24$ **h** $41.8 \div 17$

13 Calculate

    a $2 \times 8 \div 4$ **b** $40 \div 2 \div 2$ **c** $3 \times 16 - (7 - 3)$

    d $216 \div 18 - (3^2 + 1)$ **e** $9 \times 19 - (14 - 5)$ **f** $90 \div 50 - (5^2 - 4^2)$

14 Use a calculator to work out these calculations. Give your answers to 2 decimal places where appropriate.

    a $(7 + 3.9) \div 5$ **b** $48 - 2.3^2 \times 5$ **c** $9 \times (7.2 - 1.9)^2$ **d** $(3 + 6.7) \times 4$

    e $\dfrac{8 + 5^2}{39 - 5^2}$ **f** $12^2 - 8^2$ **g** $\dfrac{256}{2^2 \times 2^3}$ **h** $\dfrac{11 + 5^2}{(7^2 - 13)}$

15 **a** Giovanni sells hot cross buns.

      He has 183 hot cross buns.

      He puts them into packs of 4.

      How many packs does he need?

    b Alison earns £32 000 per year.

      She shares $\frac{1}{4}$ of her money between her 3 children.

      How much does each child receive?

8 Summary

Assessment criteria

- Use a range of mental methods of computation with all operations   Level 4
- Use efficient written methods of addition and subtraction   Level 4
- Use understanding of place value to multiply and divide whole numbers and decimals by 10, 100 and 1000 and explain the effect   Level 5
- Round decimals to the nearest decimal place   Level 5
- Use known facts, place value, knowledge of operations and brackets to calculate including using all four operations with decimals to two places   Level 5
- Apply inverse operations and approximate to check answers to problems are of the correct magnitude   Level 5
- Understand and use an appropriate non-calculator method for solving problems that involve multiplying any three digit number by any two digit number   Level 5
- Give solutions to an appropriate degree of accuracy   Level 6
- Solve problems and carry through substantial tasks by breaking them into smaller, more manageable tasks, using a range of efficient techniques, methods and resources, including ICT.   Level 6
- Understand the effects of multiplying and dividing by numbers between 0 and 1   Level 7

Question commentary

Example — The example illustrates a straightforward question about multiplying and dividing by a decimal. Some students may think that the multiplication in part **a** should produce a larger result than the division in part **b**. Emphasise the importance of estimating the answer before calculating.	**a** $42 \times 6 \div 10 = 25.2$ **b** $42 \div 6 \times 10 = 70$
Past question — The question is a calculator question using the order of operations. Some students will forget to enter the hidden brackets in the divisor in part **b**. Encourage students to consider the order of operations before entering any numbers into the calculator. Ask questions such as "How would you calculate the question without a calculator" and "Which part of the calculation would you perform first?" Emphasise the importance of estimating the answer before performing the calculation.	**Answer** **Level 5** **2 a** 4410 **b** 2.5

Development and links

The methods of calculation discussed in this chapter are consolidated and practised in Chapter 16 and applied to problem solving in Chapter 17. Multiplication and division are important in the topic of ratio and proportion in Chapter 12.

Students will have opportunities to perform calculations involving all four operations across the curriculum, but particularly in science, technology and geography.

9 Transformations

Objectives
Level

- Transform 2-D shapes by rotation, reflection and translation **5**
- Try out mathematical representations of simple combinations of rotations, reflections and translations **5**
- Identify all the symmetries of 2-D shapes **5**
- Make scale drawings **5**
- Understand and use the language and notation associated with enlargement **6**
- Enlarge 2-D shapes, given a centre of enlargement and a positive integer scale factor **6**

MPA

1.1	9c, CS
1.2	9a, b, c, d, e, CS
1.3	9a, b, c, d, e, f, CS
1.4	9b, CS
1.5	9CS

PLTS

IE	9b, f, CS
CT	9a, b, c, d, e, CS
RL	9a, c, d, CS
TW	9a, b, d, CS
SM	9d, CS
EP	9a, b, e, f, CS

Introduction

This chapter builds on knowledge of transformations and symmetry gained in Year 7. Students identify line and rotational symmetry of 2-D shapes; reflect, rotate and translate 2-D shapes; and use combinations of transformations to produce tessellations. The topic is extended to include the enlargement of a shape using a centre of enlargement and a positive whole number scale factor and using scale drawings to represent real-life objects. Students begin to consider the congruence and similarity of the transformed shapes.

The student book discusses the design of the floor at the Church of St Giorgio Maggiore in Venice, Italy. Students encounter pattern design in art and in design and technology. Geometric tessellations may be used to decorate a surface after an object is created, or the pattern may be a fundamental part of the object design, for example in quilting and tiling. There is a variety of information about tessellations as a graphic art form at http://tessellations.org/ and examples of optical illusions involving tessellations at http://gwydir.demon.co.uk/jo/tess/optical.htm

Extra Resources

9   Start of chapter presentation
9a Animation: Transformations
9a Consolidation sheet
9b Animation: Transformations 2
9c Animation: Symmetry
9c Animation: Symmetry 2
9c Consolidation sheet
9d Animation: Enlargements
9d Consolidation sheet
9e Animation: Scale factors
9e Worked solution: Q3
9e Consolidation sheet
9f Consolidation sheet
    Maths life: Celtic knots

    Assessment: chapter 9

⫸Fast-track
9b, c, d, e, f

- Transform 2-D shapes by rotation, reflection and translation, on paper [and using ICT] (L5)

Useful resources
Large squared paper
Tracing paper
Isometric paper
Mirrors

Starter – Palindromic dates

1st November 2010 is palindromic if written using two digits for the day, month and year: 01.11.10
Ask students to find other palindromic dates occurring within the next 50 years. (11.11.11, 21.11.12, 02.11.20, 12.11.21, *etc.*)

Teaching notes

Students should have previous knowledge of basic translations; this can be reviewed in a class discussion.

Taking reflections, rotations and translations in turn, ask students to supply the facts needed to define the transformation and also to explain how to carry out the transformation. The emphasis can be placed on their practical know how and on their tips for avoiding any mistakes. Working together through examples on the board should consolidate the ideas (EP, RL).

If a large sheet of tracing paper is available, fixing a centre of rotation with your finger should allow you to demonstrate on the board how to use it to help rotate a shape.

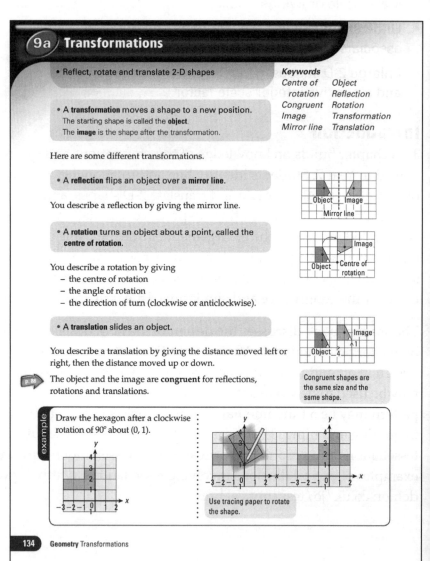

Plenary

As a literacy challenge, ask students to work in pairs and give them 4–5 minutes to write definitions (in less than 10 words) for each of the following: regular shape; translation; rhombus; centre of rotation; congruent. Ask pairs to share their answers and agree the best definitions (TW).

Simplification

Copying the diagrams accurately will need most support in this exercise. It will be helpful to provide a prepared grid for question **3**. Ask students to plot shape A first and then to translate that shape before plotting shape B.

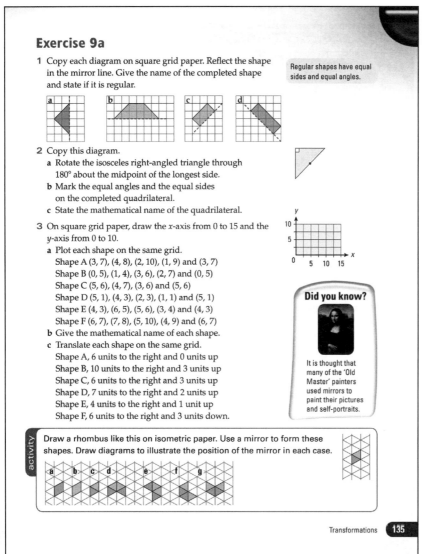

Extension

In question **3,** ask students to use the mathematical short hand to describe the translation, for example, 6 units to the right and 0 units up becomes $\begin{pmatrix} 6 \\ 0 \end{pmatrix}$. Students can then be asked to design their own translation problem using this short-hand (CT).

Exercise 9a commentary

If you usually use small squared paper (0.5mm or 0.7mm), it is advisable to use larger squares for this work. All questions (1.2, 1.3)

Question 1 – Emphasise the need to draw all shapes accurately and ensure that distances from the mirror line are equal.

Question 2 – Have available small pieces of tracing paper.

Question 3 – Students may need to be reminded of the order in which to plot coordinates. To avoid confusion, join up one set of points before moving on to the next. After drawing each translation, students should name the new shape A' or B', *etc.*

Activity – It may help to complete the first few parts collectively. The mirror line should be dashed.

Assessment Criteria – SSM, level 4: reflect simple shapes in a mirror line, translate shapes horizontally or vertically and begin to rotate a simple shape or object about its center or a vertex.

Links

The painting shown in the Did you know? is the Mona Lisa, also known as La Gioconda. It was painted by Leonardo Da Vinci in the 16th century and now hangs in The Louvre in Paris. There is an optical illusion based on a transformation of part of the Mona Lisa's face at http://www.exploratorium.edu/exhibits/mona/mona.html

- Try out mathematical representations of simple combinations of these transformations

(L5)

Useful resources
Squared paper
Tracing paper
Scissors

Starter – Moving triangles

Ask students to plot (or imagine) a triangle with vertices at (1, 1) (2, 1) (1, 5).
Then ask students to imagine the *x*-coordinates are multiplied by -1, that is, (-1, 1) (-2, 1) (-1, 5).
What will the triangle look like? What transformation has taken place? (reflection in *y*-axis)
What if the *y*-coordinates are multiplied by -1? (reflection in *x*-axis)
What if the *x* and *y* coordinates are reversed, that is, (1, 1) (1, 2) (5, 1)? (reflection in line *y = x*)

Teaching notes

Students will have seen tessellations before and will be familiar with the idea of a repeating shape filling the plane. They will be less practised at knowing which transformations to use to achieve tessellations. A good way into the topic would be to show examples of tessellations – paving, M.C. Escher, Islamic art, *etc.* – and invite students to explain how to create the pattern starting with a single shape. Move descriptions on from 'turn it upside down and move it over' to specifying rotations (centre-angle-direction) *etc.* with increasing mathematical precision (EP).

A collective activity is to ask groups of students to produce sets of identical shapes – accurate drawing will be very important here – and put them together to create tessellations. This may form the basis of posters for the classroom, which can be annotated with details of the transformations involved (TW).

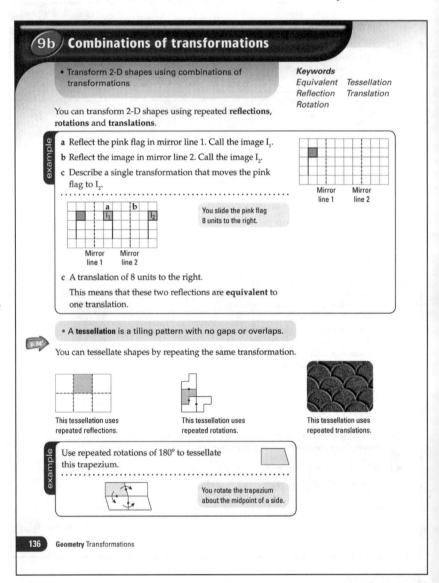

9b Combinations of transformations

- Transform 2-D shapes using combinations of transformations

Keywords
Equivalent Tessellation
Reflection Translation
Rotation

You can transform 2-D shapes using repeated **reflections**, **rotations** and **translations**.

example

a Reflect the pink flag in mirror line 1. Call the image I_1.
b Reflect the image in mirror line 2. Call the image I_2.
c Describe a single transformation that moves the pink flag to I_2.

You slide the pink flag 8 units to the right.

c A translation of 8 units to the right.
This means that these two reflections are **equivalent** to one translation.

- A **tessellation** is a tiling pattern with no gaps or overlaps.

You can tessellate shapes by repeating the same transformation.

This tessellation uses repeated reflections.

This tessellation uses repeated rotations.

This tessellation uses repeated translations.

example

Use repeated rotations of 180° to tessellate this trapezium.

You rotate the trapezium about the midpoint of a side.

136   Geometry Transformations

Plenary

Ask students to look at their answer to question **3** – it should be 1 unit right and 4 units down. Can they explain how they reached the answer one right and four down? The target is to show this as an arrow (vector) diagram (CT).

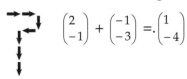

$$\begin{pmatrix} 2 \\ -1 \end{pmatrix} + \begin{pmatrix} -1 \\ -3 \end{pmatrix} = \begin{pmatrix} 1 \\ -4 \end{pmatrix}$$

Simplification

To simplify tessellation problems, let students cut out a template of the basic shape, transform it and then draw around the template.

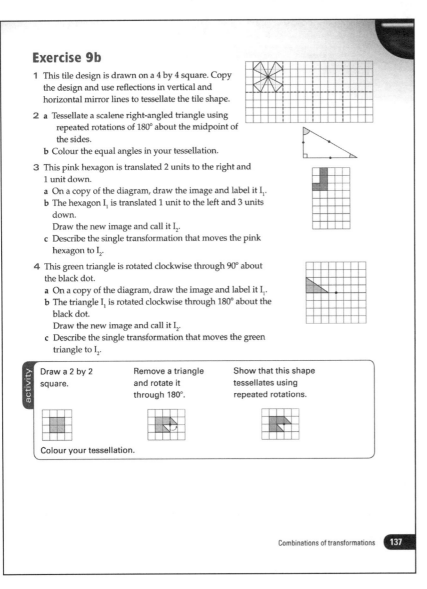

Exercise 9b

1 This tile design is drawn on a 4 by 4 square. Copy the design and use reflections in vertical and horizontal mirror lines to tessellate the tile shape.

2 a Tessellate a scalene right-angled triangle using repeated rotations of 180° about the midpoint of the sides.
  b Colour the equal angles in your tessellation.

3 This pink hexagon is translated 2 units to the right and 1 unit down.
  a On a copy of the diagram, draw the image and label it I_1.
  b The hexagon I_1 is translated 1 unit to the left and 3 units down.
    Draw the new image and call it I_2.
  c Describe the single transformation that moves the pink hexagon to I_2.

4 This green triangle is rotated clockwise through 90° about the black dot.
  a On a copy of the diagram, draw the image and label it I_1.
  b The triangle I_1 is rotated clockwise through 180° about the black dot.
    Draw the new image and call it I_2.
  c Describe the single transformation that moves the green triangle to I_2.

activity

| Draw a 2 by 2 square. | Remove a triangle and rotate it through 180°. | Show that this shape tessellates using repeated rotations. |

Colour your tessellation.

Combinations of transformations **137**

Extension

Any quadrilateral will tessellate – or will it? Challenge students to construct any quadrilateral and try to develop a tessellation. Remember angles at a point add up to 360° and all four angles of a quadrilateral add to 360° so there will need to be one of each of the quadrilateral angles at the meeting point! It can be done! (IE) (1.4).

Exercise 9b commentary

This exercise requires neat and accurate sketching, a major challenge for all students. Encourage students to be proud of their efforts. All questions (1.2, 1.3)

Question 1 – The same tessellation could be achieved with translations.

Question 2 – To help students get started, demonstrate one of the rotations about a mid-point.

Question 3 – Ensure that students do not apply both translations to the original hexagon.

Question 4 – Tracing paper may be needed for this question.

Activity – The dots show the centers of rotation. It may help to cut out a template to manipulate/draw around.

Assessment Criteria – SSM, level 5: reason about position and movement and transform shapes.

Links

Islamic art does not use images of living things, but instead uses geometric patterns and tessellations. The Alhambra palace in Granada, Spain is richly decorated with Islamic art. For more information about the Palace see http://en.wikipedia.org/wiki/Alhambra
There are examples of the patterns found at Alhambra at http://www2.spsu.edu/math/tile/grammar/moor.htm (TW).

- Identify all the symmetries of 2-D shapes   (L5)

Useful resources
Squared paper
Tracing paper
Mirrors
Cardboard shapes

Starter – Hexominoes

Ask students to draw hexominoes that will fold up to form a cube.
How many can they find? (11 possible nets).
Which ones are symmetrical?

Teaching notes

Symmetry is all around us and students have a fair knowledge of reflection symmetry. As an introduction to the topic ask pairs of students to think about the school and surrounding environment. Can they draw four objects with reflection or rotation symmetry or both? Remind students to show mirror lines as dashed lines. If these are drawn neatly on paper, a class poster could be made (CT).

Two common misconceptions should be addressed. First, explain that an object with rotation symmetry of order one is defined to have no rotation symmetry. Second, stress that a parallelogram has no lines of symmetry. This can be convincingly demonstrated by folding a paper cut out (RL).

Supply pairs of students with sets of small cardboard shapes (a square, rectangle, rhombus, parallelogram, isosceles trapezium, *etc.*). Ask them to identify each shape and to explain its defining features. Ask questions such as, which shape(s) have two lines of symmetry? (rhombus, rectangle); which shapes have rotational symmetry of order two? (rectangle, rhombus, parallelogram); which shape has no lines of symmetry? (parallelogram) *etc.* (TW)

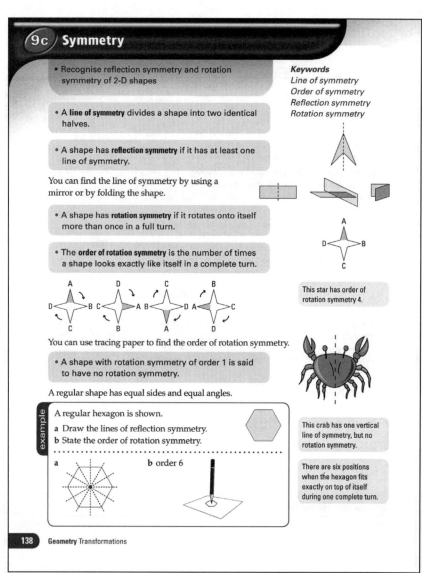

Plenary

Challenge students to draw a shape which satisfies various conditions: two lines of symmetry and order of rotation symmetry 2 (rectangle); a shape with order of rotation symmetry three (equilateral triangle) – does the shape have to have lines of symmetry? (no) *etc.* (CT)

Simplification

The use of mirrors to highlight lines of symmetry and the ability to fold shapes to show lines of symmetry (or not!) is always helpful to overcome areas of doubt.

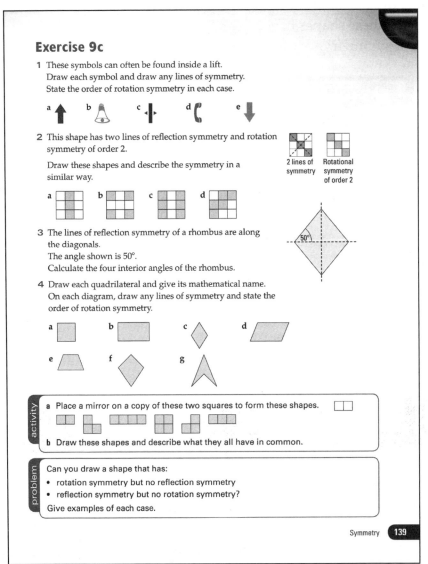

Exercise 9c

1 These symbols can often be found inside a lift.
Draw each symbol and draw any lines of symmetry.
State the order of rotation symmetry in each case.

a   b   c   d   e

2 This shape has two lines of reflection symmetry and rotation symmetry of order 2.

Draw these shapes and describe the symmetry in a similar way.

a   b   c   d

2 lines of symmetry   Rotational symmetry of order 2

3 The lines of reflection symmetry of a rhombus are along the diagonals.
The angle shown is 50°.
Calculate the four interior angles of the rhombus.

50°

4 Draw each quadrilateral and give its mathematical name.
On each diagram, draw any lines of symmetry and state the order of rotation symmetry.

a   b   c   d

e   f   g

activity

a Place a mirror on a copy of these two squares to form these shapes.

b Draw these shapes and describe what they all have in common.

problem

Can you draw a shape that has:
• rotation symmetry but no reflection symmetry
• reflection symmetry but no rotation symmetry?
Give examples of each case.

Symmetry   **139**

Extension

Give students three squares as a rectangle. By placing a mirror what different shapes can be made? Develop this so that the three squares form a small L shape and set the same problem.

Exercise 9c commentary

Require that student diagrams are accurate, so that any symmetries are clear (1.3).

Question 1 – Similar to the example (1.1, 1.2).

Question 2 – The marginal note provides an example of a model answer (1.2).

Question 3 – Students will need to use symmetry properties and the sum of the angles in a triangle.

Question 4 – Spread **6d** covers identifying and naming quadrilaterals. It may help students to identify symmetries if they physically rotate the page (1.2).

Activity – Several shapes have multiple answers. Pairs of students could compare answers (1.2).

Problem – Encourage students to find more than one solution and try to identify what they have in common (1.2).

Assessment Criteria – SSM, level 5: identify all the symmetries of 2-D shapes.

Links

https://www.cia.gov/library/
publications/the-world-factbook/
docs/flagsoftheworld.html
illustrates the flags of all the countries in the CIA World Factbook. The flag of the United Kingdom has rotation symmetry of order 2. Ask students to identify other flags with rotation symmetry.

- Understand and use the language and notation associated with enlargement (L6)

Useful resources
Squared paper
Protractor and ruler

Starter – Order 4

Ask students to draw
  shapes that have rotation symmetry of order four but no reflection symmetry
  shapes that have rotation symmetry of order four and do have reflection symmetry.
Ask students how many lines of symmetry these shapes have.
Can be extended using different orders of symmetry.

Teaching notes

Draw an object and its enlarged image. Ask students to describe what they have in common: corresponding angles are equal. What is different and how? Encourage students to check that all corresponding lengths (not just edges) are increased by the same 'scale factor'. Ask if the shapes are congruent and introduce the term 'similar'.

Demonstrate and explain the enlargement of a right-angled triangle. Emphasise the need to enlarge every side, not just some. Also clarify that, scale factor 2 means multiply by 2, not add 2: avoid using a shape with side 2 and scale factor 2 as 2 × 2 = 2 + 2 (RL). Once the diagram is complete, demonstrate the use of a protractor to measure all the angles. Students should aim for an accuracy of ±1° and ±1mm – the use of squared paper will help greatly with the accuracy of the drawings.

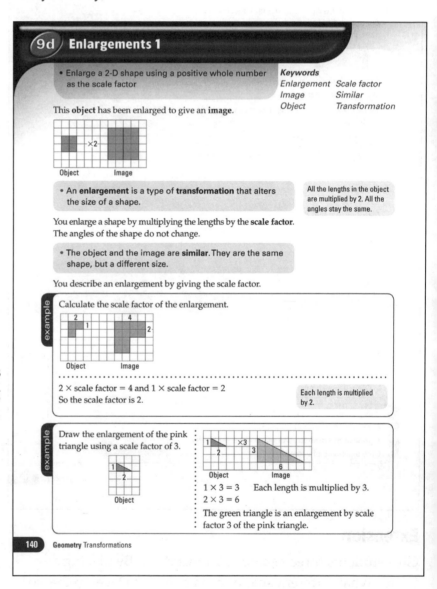

Plenary

Consider any rectangle, with clearly marked dimensions, under a scale factor two enlargement. What is the factor of enlargement for the area? Allow students to consider five or six different rectangles under the same enlargement. Can they see a pattern in the scale factor for the enlargement of the area? What if it was a scale factor 3 enlargement? (CT)

Simplification

As a preliminary to question **3**, give students three or four pairs of diagrams where there is a mistake in each enlargement. Ask them to identify this mistake and then correct the enlargement drawing (RL).

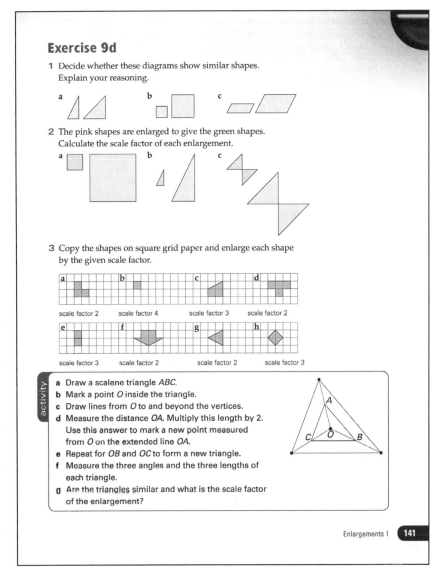

Exercise 9d

1 Decide whether these diagrams show similar shapes. Explain your reasoning.

a b c

2 The pink shapes are enlarged to give the green shapes. Calculate the scale factor of each enlargement.

a b c

3 Copy the shapes on square grid paper and enlarge each shape by the given scale factor.

a b c d

scale factor 2   scale factor 4   scale factor 3   scale factor 2

e f g h

scale factor 3   scale factor 2   scale factor 2   scale factor 3

activity

a Draw a scalene triangle *ABC*.
b Mark a point *O* inside the triangle.
c Draw lines from *O* to and beyond the vertices.
d Measure the distance *OA*. Multiply this length by 2. Use this answer to mark a new point measured from *O* on the extended line *OA*.
e Repeat for *OB* and *OC* to form a new triangle.
f Measure the three angles and the three lengths of each triangle.
g Are the triangles similar and what is the scale factor of the enlargement?

Enlargements 1   **141**

Extension

Scale factor enlargement can give a smaller shape? Introduce a scale factor half enlargement for question **2** parts **a**, **b**, **d**, **f** and **g**. Can students draw these?

Exercise 9d commentary

Question 1 – Students need to check that corresponding angles are equal and that corresponding sides are in the same ratio. Ask students to write down their reasons and then compare them with a partner. Each pair should agree a best explanation and then compare this with another pair's best explanation. Which is best now? (TW) (1.2)

Question 2 – Similar to the first example.

Question 3 – Similar to the second example. Encourage students to mark one another's work, giving neatly worded explanations when a question is marked incorrect (RL) (1.2, 1.3).

Activity – An introduction to using a centre of enlargement. It may help to follow the instructions if students work in pairs (but on individual diagrams), or by demonstrating the procedure for, say, a rectangle. Students may need a reminder of how to use a protractor (SM) (1.3).

Links

Magnifying glasses and microscopes are used to make an object appear larger. The magnification value is the scale factor. There are microscope images at different magnification values at http://micro.magnet.fsu. edu/primer/java/scienceopticsu/ virtual/magnifying/index.html

- Enlarge 2-D shapes, given a centre of enlargement and a positive integer scale factor (L6)

Useful resources
Squared paper

• •

Starter – Jumbled up

Write a list of anagrams on the board and ask students to unscramble them.
Possible anagrams are
ATTORNIO, SATTINNAROL, AGEMI, GRONNTUCE,
INFLECETOR, JOTBEC, TRYSMYME, DERRO
(rotation, translation, image, congruent, reflection, object, symmetry, order)
Can be extended by asking students to make a transformation word search.

Teaching notes

This develops the previous spread by introducing a centre of enlargement. Completing a worked example together will demonstrate the basic method. Discuss how a centre of enlargement fixes both the position and orientation of the image, unlike in the previous spread. Also emphasise that all the lines joining corresponding points on object and image pass through the centre of enlargement, as this is needed for question **1**.

The examples emphasise that distances should be measured from the centre of enlargement. Students should also confirm that the ratio of corresponding lengths between object and image are in the ratio 1 : scale factor.

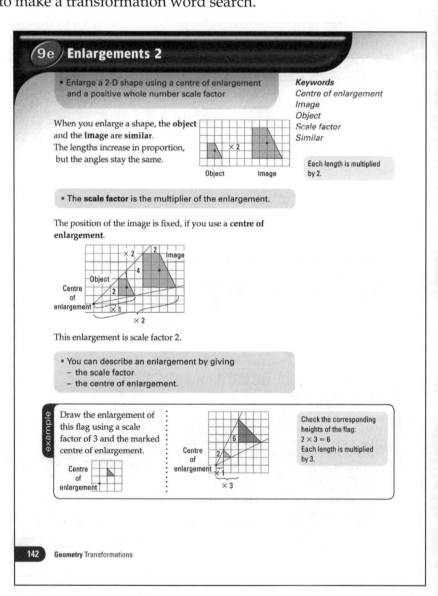

Plenary

Show a diagram with an object and an enlarged image. Students work in pairs to write a bullet point explanation of how to find the centre of enlargement. They develop this so that their explanation is twenty words or less. After a few minutes, each pair shares their explanation with another pair. Which is the best and most concise explanation? (EP)

Simplification

Students will find copying the drawings from the exercise difficult: provide the initial diagrams on a printed sheet.

Exercise 9e

1 The pink triangle ABC is enlarged to give the green triangle $A_1B_1C_1$.
   Copy the diagram on square grid paper.
   a Draw and extend the lines from A_1 to A from B_1 to B and from C_1 to C.
   b At the intersection of the lines, mark the centre of enlargement as O.
   c Measure the lines BC, B_1C_1 and AC, A_1C_1.
   d State the scale factor of the enlargement.
   e Measure the lines OA, OA_1 and OB, OB_1 to check the scale factor.

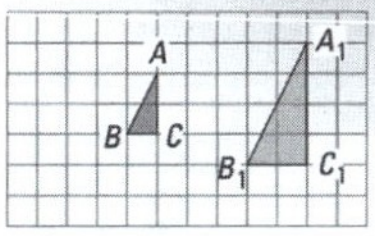

2 Copy these shapes on square grid paper.
   Draw the enlargement of each shape using the dot as the centre of enlargement and the given scale factor.

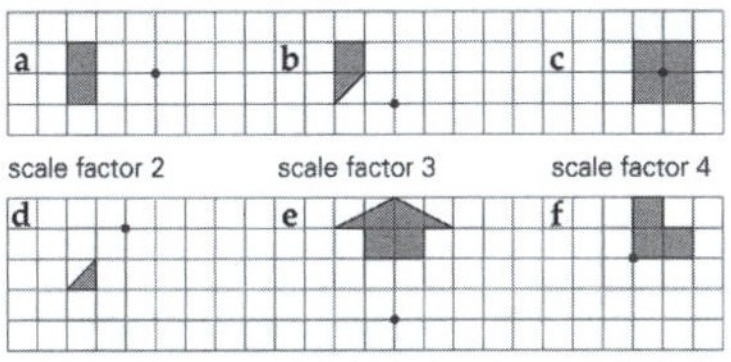

a	b	c
scale factor 2	scale factor 3	scale factor 4

d	e	f
scale factor 2	scale factor 3	scale factor 2

3 The pink triangle is enlarged by scale factor 3 to give the green triangle, which is not all shown here.
   The point (1, 4) moves to (3, 0).
   Find the other two coordinates of the green triangle.

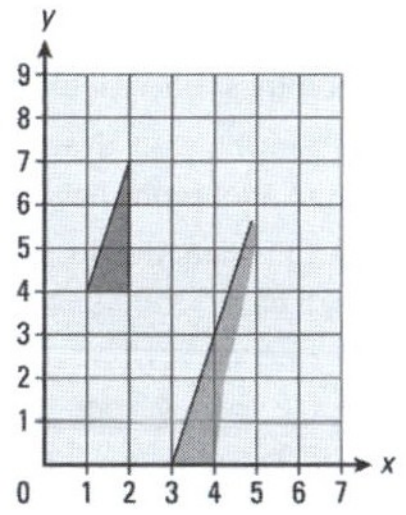

activity

   a i Draw a large rectangle.
     ii Mark a point O inside the rectangle.
    iii Draw lines from O to the vertices.
    iv Find the midpoints of each line and join these four points to form a rectangle.
   This rectangle is similar to the first rectangle.
   b Use this method to draw other similar shapes.

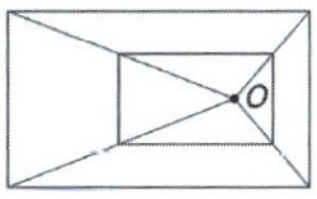

Enlargements 2   **143**

Extension

For those students who have completed the **activity**, look at question 2 parts a, c, d and f again and ask them to draw an enlargement of scale factor $\frac{1}{2}$.

Exercise 9e commentary

Question 1 – A general method for finding a centre of enlargement. The diagram must be drawn accurately otherwise the extended lines will not cross at a unique point. Pairing students may help them to follow the instructions (1.3).

Question 2 – Similar to the example. Check that students place their diagrams in such a way that the image will fit on the page (1.3).

Question 3 – Challenge students to find the centre of enlargement, using logic rather than guess work. As in question **1**, the centre will lie on the line joining the two bottom-left vertices and the distances from any centre should be in the ratio 1 : 3 (IE) (1.2).

Activity – An introduction to an enlargement with a fractional scale factor, giving a reduced-size image. Discuss the confusing language and the relationship between inverse transformations (CT) (1.3).

Assessment Criteria – SSM, level 6: enlarge 2-D shapes, given a center of enlargement and a positive whole-number scale factor.

Links

Pictures can be copied and enlarged or reduced using a device called a pantograph. A pantograph consists of several hinged rods joined together in a parallelogram shape with extended sides. One end is traced over the image and a pencil attached to the other end reproduces the image to the desired scale. Pantographs are often sold as toys. There is more information about pantographs at http://en.wikipedia.org/wiki/Pantograph and an interactive pantograph at http://www.ies.co.jp/math/java/geo/panta/panta.html

• Make scale drawings (L6) ***Useful resources***
Ruler and protractor

● ●

Starter – Shape pairs

Write the following list of shapes, orders of rotation and line symmetries on the board:
regular hexagon, isosceles triangle, rectangle, regular pentagon, equilateral triangle, square,
order 1, 3 lines, 6 lines, order 2, 4 lines, order 5.
Challenge students to match up the pairs in the shortest possible time.
Ask students which shapes out of the six will not tessellate. (pentagon)

Teaching notes

As a whole class introductory
activity, ask students where they
have seen scale factors used. Maps
may be the common answer but
scale factors arc all around us:
drawings in books, photographs,
posters, toy/model cars, *etc*. It is
important to highlight that all
dimensions in scale factor reductions
and enlargements are in proportion
with the original.

As a practical activity for all, ask
students to draw a scale factor
reduction of their desk (table) –
where students are sharing a table
they should work together. To show
this diagram in their book or on
paper what would be a sensible scale
to use? 1 cm for 10 cm? or 1 cm for
20 cm? or ? The scale drawing can
then be used to 'predict' the length
of the desk's diagonal; how accurate
is their prediction compared to an
actual measurement?

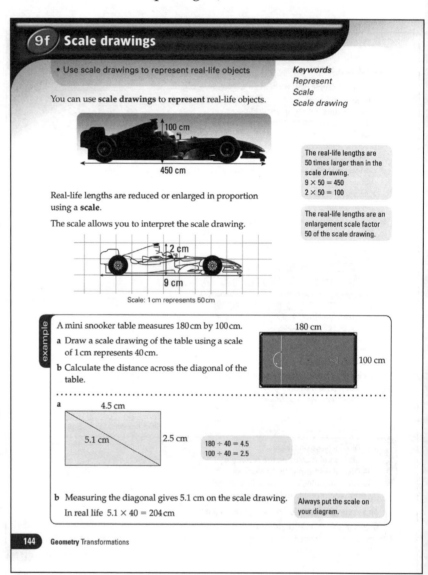

Plenary

Show a poster (or projection) of a real-life scene; it could be of
buildings in New York or London. What do students think is
the scale factor reduction? How can you work it out? How
close was your estimate?

Simplification

Make this a real activity – students find it difficult to relate book work to real life. How would they draw a scale plan of the classroom on a piece of squared paper? If the dimensions of the classroom are 7 m × 6 m what would be a good scale to use (1 cm : 1 m or 1 cm : 50 cm?) (EP).

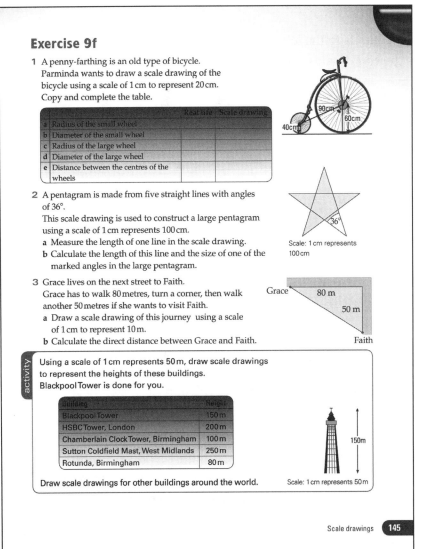

Exercise 9f commentary

Question 1 – Similar to the example part **a**. Clarify the meanings of radius and diameter.

Question 2 – Similar to the example part **b**.

Question 3 – A point for discussion is how accurately to quote the answer. To the nearest metre is sensible but why? Could students give a more accurate answer based on their diagram? How accurately do they think 80 m and 50 m are measured? (1.3).

Activity – Have a ready supply of heights for other tall buildings. What would be a sensible scale for the tallest skyscraper in the world? The Burj Dubai is planned to reach 818 m (IE).

Links

Model railways are available in different gauges. OO gauge means that the model is built to a scale of 1:76, or 1 cm on the model represents a distance of 76 cm in real life. 1 cm on an N gauge model represents a distance of 146 cm. If an N gauge model locomotive is 5 cm long, how long is the real-life locomotive? There are pictures of different gauge model locomotives at http://ngaugesociety.com/modelling/scales/scales.htm

Extension

Give students a plan of the school (or part of it) – can they work out the scale factor reduction? Having found the scale for the plan can they work out the dimensions for some of the corridors/halls/rooms and check. It is likely that these will not be accurate, raising the question of how the plan could be improved (IE).

9a

1 Copy the diagram on square grid paper.
  a Reflect the blue hexagon using the x-axis as the mirror line.
    Label the shape A and give the coordinates of the vertices.
  b Rotate the blue hexagon through 180° about the point (0, 0).
    Label the shape B and give the coordinates of the vertices.
  c Translate the blue hexagon by 4 units to the left.
    Label the shape C and give the coordinates of the vertices.

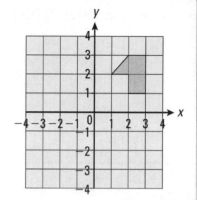

9b

2 a Tessellate a parallelogram using repeated translations.
  b Colour the equal angles in your tessellation.

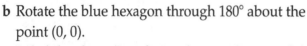

3 Copy the diagram on square grid paper.
  a Reflect the green triangle in the mirror line M_1. Call the image I_1.
  b Reflect I_1 in the mirror line M_2.
    Call the image I_2.
  c Describe the single transformation that moves the green triangle to I_2.

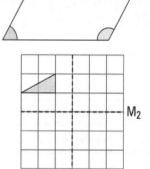

9c

4 Draw these symbols from Steph's calculator.
  Draw any lines of reflection symmetry and state the order of rotation symmetry for each symbol.
    a **b** **c** 4 **d** 8 **e** 0

5 a Draw a polygon with three lines of symmetry and rotational symmetry of order 3.
  b Give the mathematical name of this shape.

9d

6

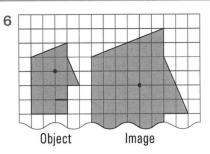

Object   Image

Part of an enlargement of the face is shown.

a Calculate the scale factor of the enlargement.

b Copy and complete the table of measurements.

	Object	Image
Length of the forehead	1 cm	
Slanting length of the nose		
Slanting length of the top of the head		
Thickness of the neck		
Width of the mouth		

c Draw the completed image on square grid paper.

9e

7 a Draw the rectangle on a coordinate grid.

    b Enlarge the rectangle by scale factor 2 using (0, 0) as the centre of enlargement.

    c Write down the coordinates of the vertices of the object and the image.

    d Explain the relationship between these coordinates.

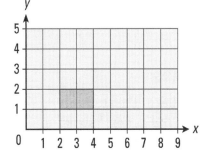

9f

8 A sales brochure shows a scale drawing of a door.
The scale is 1 cm represents 50 cm.

    Calculate   **a** the height

                    b the width

                    c the area of the real door.

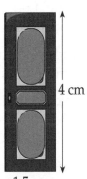

4 cm

←1.5 cm→

Scale: 1 cm represents 50 cm

- Visualise and manipulate dynamic images   (L6)
- Recognise efficiency in an approach   (L6)

Useful resources
Tracing paper
Assorted dotted grid paper
String

Background

Celtic knots are interwoven designs where the lines alternately pass over then under the other lines that they cross. Some of the designs are traversable and could be made from one continuous closed loop whilst others are made from the combination of two or more separate loops that still interweave.

Celtic designs date from early times. However, the type of knot work that is now commonly associated with Celtic designs only came into existence after the Christian influence on the Celts (around A.D. 450), before which time their artwork mainly consisted of geometrical patterns such as spirals and steps but not interwoven knots.

This case study looks at the symmetry found in some Celtic knots and ways of constructing knots (TW).

Teaching notes

Use the examples in the student book to look at the general structure of Celtic knots; noting how the lines alternately pass over then under the

lines they are crossing. Look also at the way that the knots can be made from one continuous line, the quatrefoil knot, or from a number of separate but still interwoven lines, the trefoil knot.

Look at the case study and consider each of the six knots in turn to see if they are made from one continuous line or more than one.

Ask students to look at the six knots and think about what symmetries they have (1.1). After allowing time to study the knots consider each knot as a class. Ask, *does this knot have reflective symmetry?*

If the students think that the knot has reflective symmetry, ask them to show where any lines of symmetry might be positioned. Discuss whether they actually are lines of symmetry and how, whilst the silhouette of the design might have reflective symmetry, the alternating under/over nature of the crossings means that the designs do not actually have reflective symmetry as an 'under' on one side is an 'over' on the other side (EP, RL) (1.2).

Once students have understood the issues in deciding whether a knot has reflective symmetry, challenge then to construct their own single line knot that has reflective symmetry. A knot can be

Teaching notes continued

constructed by using a pencil to draw a self-overlapping closed loop and then rubbing out parts to indicate alternate over- and under-laps

Return to the six knots in the student book and in turn ask *does this knot have rotation symmetry?*

Discuss how this time the alternating under/over does not cause the same problem. Students may find the rotated knots hard to visualise. One option is to use tracing paper and rotate the image. A second option is for small groups of students to use two books and hold one fixed while they rotate the other one.

Then move on to the scroll on the right hand page showing one way of making a Celtic knot design. It may be helpful to have pre-prepared grids for students to use in order to help them construct an accurate drawing. Asking students to work in pairs may also help them support one another when following the instructions (TW, SM) (1.3).

Once they have successfully constructed their knot and answered the questions about the knot, students should be encouraged to experiment with further designs (CT).

Extension

Challenge students to construct designs using the following procedure (IF) (1.4, 1.5).

Draw a symmetric pattern of dots and add a single closed line that loops around the dots. Add a second line, an equal distance from the first, to make a 'rope' and then remove line segments to make an under and over pattern.

The site http://www.entrelacs.net/ shows other examples of this way of forming knot designs. It also includes other methods, such as those based on a plait, and many designs like those that appear as borders in old manuscripts and in jewelery.

Students could also try making their own knots using string or modeling wire.

9 Summary

Assessment criteria

* Reflect simple shapes in a mirror line, translate shapes horizontally or vertically and begin to rotate a simple shape or object about its centre or a vertex   Level 4

* Identify all the symmetries of 2-D shapes   Level 5

* Reason about position and movement and transform shapes   Level 5

* Enlarge 2-D shapes, given a centre of enlargement and a positive whole-number scale factor   Level 6

Question commentary

Example

The example illustrates a typical problem on rotation and rotational symmetry. Encourage the use of tracing paper or turning the page to make visualisation easier. Some students may give the rotational symmetry as order 4 as the rhombus is rotated four times. Emphasise that the order of rotational symmetry is not the number of times the shape is turned, but the number of times that the shape turns onto itself in a complete turn.

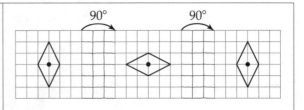

Past question

The question requires students to enlarge a shape on isometric paper using a given centre of enlargement and scale factor. Some students may enlarge the shape but use an incorrect centre of enlargement. Emphasise that the centre of enlargement determines the position of the image while the scale factor only determines the size of the image. Ask probing questions such as "What information do you need to complete a given enlargement" and "If someone has completed an enlargement, how would you find the centre and the scale factor?"

Answer

Level 6

2

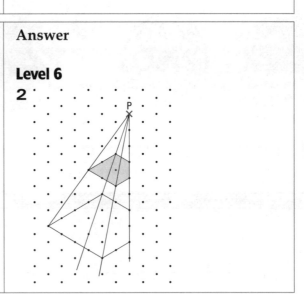

Development and links

The use of a scale factor leads into work on ratio in Chapter 12 and scale drawing is used in work on Loci and Bearings in Chapter 14. Enlargement and Transformations are developed further in Year 9.

Reflection and line symmetry is important in physics when studying the properties of light, and in biology where symmetry occurs in nature. Most rotating mechanical parts used in design technology exploit the property of rotational symmetry and students rotate pictures and drawings using computer programs in ICT. Symmetry and tessellations are important in art and design, particularly in Roman and Islamic art. Students will draw and use scale drawings in technology when creating design plans and in geography when using maps.

Algebra

Objectives

- Use squares, positive, [and negative] square roots, cubes and cube roots... **5**
- Carryout more difficult calculations effectively and efficiently using the function keys for powers and roots **5**
- Generate terms of a linear sequence using term-to-term and position-to-term rules... **6**
- Use linear expressions to describe the nth term of a simple arithmetic sequence, justifying its form by referring to the activity or practical context from which it was generated............ **6**

Introduction

The focus of this chapter is on reviewing and developing Year 7 work on sequences. Students find and use term-to-term and position-to-term rules for numerical sequences and sequences in context. The work on squares and cubes from Chapter 1 is extended to finding square roots and cube roots using inverses, a calculator or trial and improvement.

The student book discusses the first electric traffic light invented in Salt Lake City in 1912. Students will be familiar with light sequences in many contexts. They are used in navigation, for error diagnostics on computer printers, in Christmas lights and advertising signs, for signalling and in programmable LED messages. Each light is turned on or off in a sequence and may change colour. There are examples of signs using light sequences at http://tavmjong.free.fr/INKSCAPE/MANUAL/html/NeonSign.html and at http://www.signsdeluxe.com/

⚡ast-track

All spreads

Level

MPA

1.1	10a, b, c, d, d^2
1.2	10a, b, c, d^2
1.3	10d, d^2
1.4	10a, d, d^2
1.5	10a, c

PLTS

IE	10a, d^2
CT	10a, c
RL	10a, d, d^2
TW	10b
SM	10c, d, d^2
EP	10c, d

Extra Resources

10   Start of chapter presentation
10b Animation: Sequences
10b Animation: Sequences 2
10b Animation: Sequences 3
10b Worked solution: Q3a
10b Consolidation sheet
10c Starter: Arithmetic sequence multichoice
10d Starter: Square and cube matching
10d Worked solution: Q4b

Assessment: chapter 10

- Describe integer sequences (L5)
- Generate terms of a linear sequence using term-to-term [and position-to-term] rules (L6)

Useful resources
Mini whiteboards

Starter – Connections

If $x = -2$ and $y = 5$, ask students to write down 10 rules connecting x and y, for example,
$y = 5x + 15$, $y - x = 7$, $x + 2y = 8$...
Can be extended by changing the values of x and y.

Teaching notes

To make connections with students' prior learning, ensure that the vocabulary of this unit is understood. Ask students to write definitions of 'term', 'rule' and 'sequence' and take feedback.

When providing worked examples, place an emphasis on looking at how to get from one term to the next – this can often be linked to a diagrammatic representation. A useful technique is to look at first differences, which will be most useful for linear sequences such as occur in question **2** parts **a – f**.

Students often forget to state the first term of a sequence when giving a term-to-term rule. To help reinforce the need for this, give the partial rule 'add 5' and challenge students to give first terms that result in a sequence of odd numbers, even numbers, multiples of 4 and non-integer numbers (RL).

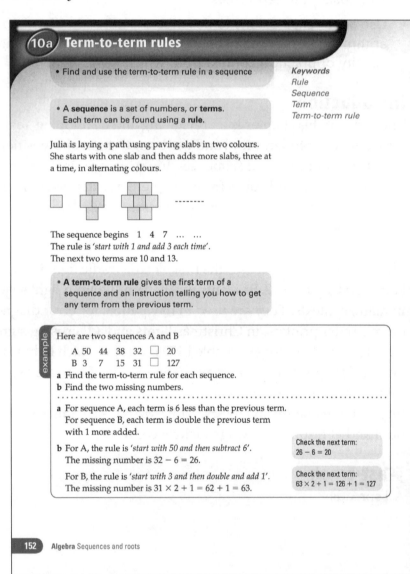

Plenary

Ask students to generate sequences for a friend to find the rule. They could use both numbers and diagrams.

Simplification

If students struggle at this point, offer more visual consolidation with sequences in diagrammatic form.

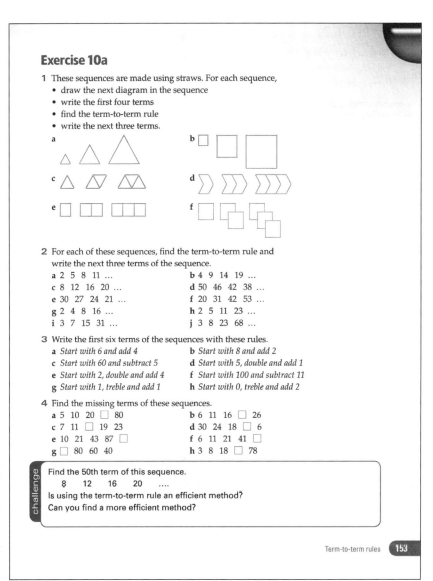

Exercise 10a

1 These sequences are made using straws. For each sequence,
   - draw the next diagram in the sequence
   - write the first four terms
   - find the term-to-term rule
   - write the next three terms.

a

b

c

d

e

f

2 For each of these sequences, find the term-to-term rule and write the next three terms of the sequence.
   a 2 5 8 11 ...
   b 4 9 14 19 ...
   c 8 12 16 20 ...
   d 50 46 42 38 ...
   e 30 27 24 21 ...
   f 20 31 42 53 ...
   g 2 4 8 16 ...
   h 2 5 11 23 ...
   i 3 7 15 31 ...
   j 3 8 23 68 ...

3 Write the first six terms of the sequences with these rules.
   a *Start with 6 and add 4*
   b *Start with 8 and add 2*
   c *Start with 60 and subtract 5*
   d *Start with 5, double and add 1*
   e *Start with 2, double and add 4*
   f *Start with 100 and subtract 11*
   g *Start with 1, treble and add 1*
   h *Start with 0, treble and add 2*

4 Find the missing terms of these sequences.
   a 5 10 20 □ 80
   b 6 11 16 □ 26
   c 7 11 □ 19 23
   d 30 24 18 □ 6
   e 10 21 43 87 □
   f 6 11 21 41 □
   g □ 80 60 40
   h 3 8 18 □ 78

challenge

Find the 50th term of this sequence.
   8   12   16   20   ....
Is using the term-to-term rule an efficient method?
Can you find a more efficient method?

Extension

More able students can progress towards finding the *n*th term of a sequence.

Exercise 10a commentary

The first two questions can be done with the whole class as a mini whiteboard activity to assess prior learning and understanding.

Question 1 – A visual approach to understanding sequences (IE) (1.1, 1.2).

Question 2 – Similar to the example. Parts **a** – **f** are linear and looking at first differences should be encouraged. Part **j** may benefit from whole class discussion (1.2).

Question 3 – Applying a term-to-term rule.

Question 4 – Similar to the example. Students may need to be reminded to find the rule first and to use the next number on as a check if appropriate (1.2).

Challenge – Here students are encouraged towards finding a position-to-term rule (CT) (1.2, 1.4, 1.5).

Assessment Criteria – Algebra, level 6: generate terms of a sequence using term-to-term [and position-to-term] definitions of the sequence, on paper [and using ICT].

Links

The look-and-say sequence is a famous non-linear sequence that sometimes appears in puzzle books. The first seven terms are 1, 11, 21, 1211, 111221, 312211, 13112221, 1113213211. What is the rule for moving from one term to the next? (Describe the previous term in words and then write it in numbers, so one one; two ones; one two, two ones etc). Try starting the sequence with 2 or 3 instead of 1.

- Generate sequences from patterns or practical contexts and describe the general term in simple cases   (L6)
- Generate terms of a linear sequence using [term-to-term and] position-to-term rules   (L6)

Useful resources

Tape measures

Starter – Match point!

Write the following coordinates of points on the board:
(1, -1) (-1, 5) (-1, -3) (-1, 2) (-1, - 4)
Write the following equations of lines on the board:
$y = 4x + 1$, $y = 3x - 4$, $y = -4$, $y = 4 - x$, $y = 2(x + 2)$
Ask students which points lie on which lines and to give the coordinates of another point on each line.

Teaching notes

The key processes in algebra, particularly the development of analysing through using mathematical reasoning, are supported here where students are required to identify and describe numerical patterns.

The worked examples show the process that students need to use and offer clear visual representations which students can apply for themselves in other situations. The use of tables to organise thinking is especially important. When finding the value of any term from its position in the sequence, make links back to work on substitution

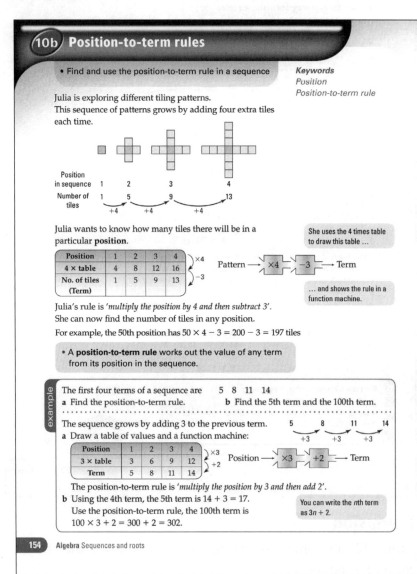

10b Position-to-term rules

- Find and use the position-to-term rule in a sequence

Keywords
Position
Position-to-term rule

Julia is exploring different tiling patterns.
This sequence of patterns grows by adding four extra tiles each time.

| Position in sequence | 1 | 2 | 3 | 4 |
| Number of tiles | 1 | 5 | 9 | 13 |

+4 +4 +4

Julia wants to know how many tiles there will be in a particular **position**.

She uses the 4 times table to draw this table …

Position	1	2	3	4
4 × table	4	8	12	16
No. of tiles (Term)	1	5	9	13

×4
−3

Pattern → ×4 → −3 → Term

… and shows the rule in a function machine.

Julia's rule is 'multiply the position by 4 and then subtract 3'.
She can now find the number of tiles in any position.
For example, the 50th position has 50 × 4 − 3 = 200 − 3 = 197 tiles

- A **position-to-term rule** works out the value of any term from its position in the sequence.

example

The first four terms of a sequence are   5 8 11 14
a Find the position-to-term rule.   b Find the 5th term and the 100th term.
....
The sequence grows by adding 3 to the previous term.   5   8   11   14
a Draw a table of values and a function machine: +3 +3 +3

Position	1	2	3	4
3 × table	3	6	9	12
Term	5	8	11	14

×3
+2

Position → ×3 → +2 → Term

The position-to-term rule is 'multiply the position by 3 and then add 2'.
b Using the 4th term, the 5th term is 14 + 3 = 17.
Use the position-to-term rule, the 100th term is
100 × 3 + 2 = 300 + 2 = 302.

You can write the *n*th term as 3*n* + 2.

Plenary

Use the plenary to link to the research students have done on the Fibonacci sequence. Discuss the golden ratio in nature and give tape measures to students to investigate their own height to belly button to floor ratio. Who is the 'most perfect' student? Or is their teacher perfect?

Simplification

Where students struggle with the later questions in the exercise, refer them back to the worked example and encourage the use of diagrams to represent the sequence in a more concrete way.

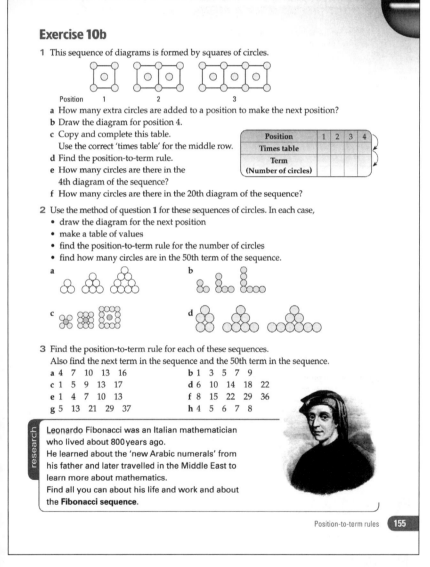

Exercise 10b

1 This sequence of diagrams is formed by squares of circles.

Position   1   2   3

  a How many extra circles are added to a position to make the next position?
  b Draw the diagram for position 4.
  c Copy and complete this table.
    Use the correct 'times table' for the middle row.
  d Find the position-to-term rule.
  e How many circles are there in the 4th diagram of the sequence?
  f How many circles are there in the 20th diagram of the sequence?

Position	1	2	3	4
Times table				
Term (Number of circles)				

2 Use the method of question 1 for these sequences of circles. In each case,
  • draw the diagram for the next position
  • make a table of values
  • find the position-to-term rule for the number of circles
  • find how many circles are in the 50th term of the sequence.

  a   b
  c   d

3 Find the position-to-term rule for each of these sequences.
  Also find the next term in the sequence and the 50th term in the sequence.
  a 4 7 10 13 16   b 1 3 5 7 9
  c 1 5 9 13 17   d 6 10 14 18 22
  e 1 4 7 10 13   f 8 15 22 29 36
  g 5 13 21 29 37   h 4 5 6 7 8

research

Leonardo Fibonacci was an Italian mathematician who lived about 800 years ago.
He learned about the 'new Arabic numerals' from his father and later travelled in the Middle East to learn more about mathematics.
Find all you can about his life and work and about the **Fibonacci sequence**.

Position-to-term rules   155

Extension

Encourage more able students to discuss how to write a formula for finding the nth term of a sequence

Exercise 10b commentary

Question 1 – A structured investigation supported by a diagrammatic representation (1.1, 1.2).

Question 2 – Similar to question **1** but with less scaffolding (IE) (1.1, 1.2).

Question 3 – Similar to the example. Encourage unsure students to draw their own diagrams and follow the procedure set out in the previous questions (1.2).

Research – An opportunity to discuss the historical development of mathematics and how much harder arithmetic was before Arabic numerals. Some students may have seen the film 'The DaVinci Code' – link to this if appropriate (TW).

Assessment Criteria – Algebra, level 6: generate terms of a sequence using [term-to-term and] position-to-term definitions of the sequence, on paper [and using ICT].

Links

Chronophotography is the art of taking a series of photographs of a moving object at regular time intervals. The resulting photographs form a sequence, similar to a flip book. Chronophotography was made popular during Victorian times by photographers such as Étienne-Jules Marey and Georges Demeny. There are examples of chronophotography at http://www. sequences.org.uk/chrono1/0000. html and http://www.elearning-art.net/art-net_courses/Moving_ Images_Workshop_(Eng)/1SHORT HISTORY/1Zoopraxinoscope.htm

- Use linear expressions to describe the *n*th term of a simple arithmetic sequence, justifying its form by referring to the activity or practical context from which it was generated   (L6)

Useful resources

Starter – Strange sequences

Write the following sequences on the board:

50, 48, 44, 38, 30,…   J, F, M, A, M,…   3, -6, 12, -24, 48,…

2, 5, 10, 17, 26,…   S, M, T, W, T,…

Ask students for the next term in each sequence.

Answers: 20, J (June), -96 (multiply previous term by -2), 37, F (Friday).

Can be extended by students making their own 'strange sequences'.

Teaching notes

A focus in this unit is to demonstrate the use of sequences in a real-life context. Students can be encouraged to describe the rules with which they are working in words and using symbols. Collaborative approaches can be encouraged here in order to develop student confidence and to promote purposeful mathematical discussion.

Having initiated whole class discussion through the worked example, invite students to discuss how sequences might appear in real-life contexts.

To become functional in mathematics, students need the confidence to use mathematics as a tool to investigate and solve problems, which leads to better understanding. For this reason, the link to real-life settings is most important (EP).

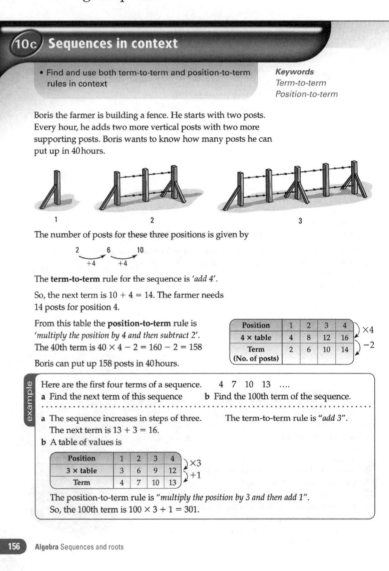

10c Sequences in context

- Find and use both term-to-term and position-to-term rules in context

Keywords
Term-to-term
Position-to-term

Boris the farmer is building a fence. He starts with two posts. Every hour, he adds two more vertical posts with two more supporting posts. Boris wants to know how many posts he can put up in 40 hours.

1   2   3

The number of posts for these three positions is given by

2   6   10
  +4   +4

The **term-to-term** rule for the sequence is '*add 4*'.

So, the next term is 10 + 4 = 14. The farmer needs 14 posts for position 4.

From this table the **position-to-term** rule is '*multiply the position by 4 and then subtract 2*'.
The 40th term is 40 × 4 − 2 = 160 − 2 = 158.

Position	1	2	3	4
4 × table	4	8	12	16
Term (No. of posts)	2	6	10	14

×4
−2

Boris can put up 158 posts in 40 hours.

> **example**
>
> Here are the first four terms of a sequence.   4 7 10 13 ….
> a Find the next term of this sequence   b Find the 100th term of the sequence.
>
> a The sequence increases in steps of three.   The term-to-term rule is "*add 3*".
>   The next term is 13 + 3 = 16.
> b A table of values is
>
> | Position | 1 | 2 | 3 | 4 |
> |---|---|---|---|---|
> | 3 × table | 3 | 6 | 9 | 12 |
> | Term | 4 | 7 | 10 | 13 |
>
> ×3
> +1
>
> The position-to-term rule is "*multiply the position by 3 and then add 1*".
> So, the 100th term is 100 × 3 + 1 = 301.

156   Algebra Sequences and roots

Plenary

Challenge students to write a word-based sequence problem of their own, using the examples in the exercise as a model.

Simplification

Encourage students having difficulty to make full use of diagrams and tables for structure.

Exercise 10c

1 Melissa buys a plant with three leaves on it.
  Each week, it grows another two leaves.
  a Use a term-to-term rule to find the number of leaves on Melissa's plant in week 4 and week 5.
  b Copy and complete this table of values to find the position-to-term rule.
  c How many leaves will be on Melissa's plant after twelve weeks?

Position (week)	1	2	3	4	5
Times table					
Term (leaves)					

2 Wasim has £20 already saved. He takes a Saturday job earning £8 a week.
  This sequence gives the total amount that he has saved after each week.

  20   28   36   44   …   …

  a Write the term-to-term rule.
  b Write the next two terms in the sequence.
  c Copy and complete this table of values and find the position-to-term rule.
  d Find the 40th term of the sequence.
  e How much will Wasim have saved after 40 weeks?

Position (week)	1	2	3	4
Times table				
Term (£)				

3 Migrating geese visit a local pond every winter. On Day 1, fifty geese are living there and then an average of twenty more geese arrive every day for a month.
  a Use the term-to-term rule to write the next two terms of the sequence

  50   70   90   110   …   …

  b Find the position-to-term rule.
  c Find the 20th term of the sequence.
  d How many geese are on the pond on Day 30?

investigation

Everyone in a group shakes hands with everyone else.
If there are just 2 people in the group, there is just 1 handshake.
  a If there are 3 people in the group, how many handshakes will there be?
  b How many handshakes will there be for a group of 4, 5, 6, … people?
  c Find a sequence for the numbers of handshakes.
  d Can you use a term-to-term rule to find the next term?
  e Can you find a position-to-term rule? If not, why not?

Sequences in context **157**

Extension

More able students can be stretched with the investigation of a rule for triangular numbers (as in the handshakes investigation).

Exercise 10c commentary

Question 1 – Invite students to suggest similar situations from nature where a sequence might arise (CT) (1.1, 1.2).

Question 2 – Ask students how their answers would change if Wasim earned £10 a week or £12: can they see how this relates to their position-to-term rule (1.2).

Question 3 – Ensure that students know what are 'migrating geese'. Also the 'average' may need some explaining – invite students to discuss how this will impact on the answer for day 30 (1.2).

Investigation – Handshakes investigation leading to identifying the sequence of triangular numbers. Part **e** will promote mathematical discussion (SM) (1.1, 1.2, 1.5).

Assessment Criteria – Algebra, level 6: write an expression to describe the nth term of an arithmetic sequence.

UAM, level 7: justify generalisations, arguments or solutions.

Links

All human cells contain DNA. DNA stands for Deoxyribonucleic Acid and DNA molecules hold the instructions for building living things. DNA molecules are made up of four building blocks called bases or nucleotides. The order or sequence of the bases along the molecule gives the genetic information for characteristics such as hair and eye colour. There is more information about DNA at http://www.koshland-science-museum.org/exhibitdna/index.jsp

- Recognise the squares of numbers to at least 12 × 12 and the corresponding roots **(L5)**
- Use squares and positive [and negative] square roots **(L5)**
- Use ICT to estimate square roots [and cube roots] **(L6)**

Useful resources
Calculators
Mini whiteboards
A1, 2, 3, 4, 5 size paper

Starter – Power bingo

Ask students to draw a 3 × 3 grid and enter nine numbers that are either square numbers (up to 144) or cube numbers (up to 125). Give possible numbers, for example, 6^2, 2^3

Winner is the first student to cross out all their numbers.

Teaching notes

Make links to students' prior knowledge of squares and square roots by using the first two questions as whole class interactive activities using mini whiteboards. Point out that, as most numbers are not perfect squares so that the square roots of these numbers can't be whole numbers.

Expand this to a discussion of where √30 lies on the number line. How can they improve on between 5 = √25 and 6 = √36? (Knowing square numbers allows you to use efficient starting values for the search.) Use this to introduce the systematic trial-and-improvement approach. A common mistake is not to bound the solution and the need to do this must be emphasised. Making reference to a number line, clarify which numbers between 5 and 6 round up and which round down. This should highlight 5.5 as a suitable point to use as a trial solution. Students should now be able to give √30 to the nearest whole number. Can they tell you how to find √30 to one decimal place? By making constant reference to the number line students should gain an understanding of where to choose subsequent trial values (RL).

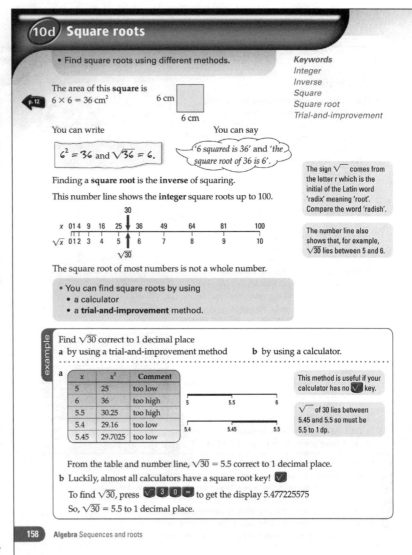

Plenary

Ask students to choose one of the examples from question **4** in the exercise and to write a model trial-and-improvement solution. This will link to future work on equations.

Simplification

Some students will need to use the number line to get a clearer picture when looking at numbers with square roots that are not whole numbers and will need support with trial and improvement.

As a preliminary check that all students are confident in the correct use of their calculator.

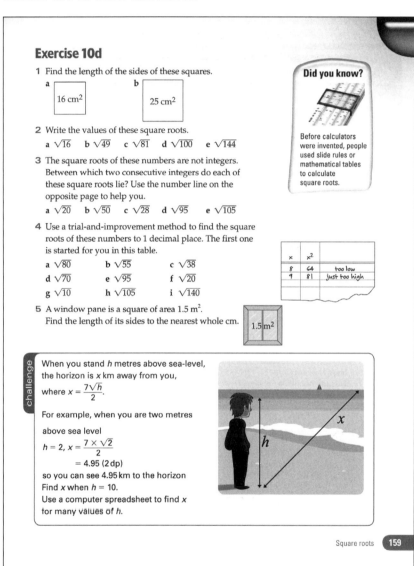

Exercise 10d

1 Find the length of the sides of these squares.

a 16 cm²   b 25 cm²

2 Write the values of these square roots.

a $\sqrt{16}$ b $\sqrt{49}$ c $\sqrt{81}$ d $\sqrt{100}$ e $\sqrt{144}$

3 The square roots of these numbers are not integers. Between which two consecutive integers do each of these square roots lie? Use the number line on the opposite page to help you.

a $\sqrt{20}$ b $\sqrt{50}$ c $\sqrt{28}$ d $\sqrt{95}$ e $\sqrt{105}$

4 Use a trial-and-improvement method to find the square roots of these numbers to 1 decimal place. The first one is started for you in this table.

a $\sqrt{80}$ b $\sqrt{55}$ c $\sqrt{38}$
d $\sqrt{70}$ e $\sqrt{95}$ f $\sqrt{20}$
g $\sqrt{10}$ h $\sqrt{105}$ i $\sqrt{140}$

x	x^2	
8	64	too low
9	81	just too high

5 A window pane is a square of area 1.5 m². Find the length of its sides to the nearest whole cm.

1.5 m²

Did you know?

Before calculators were invented, people used slide rules or mathematical tables to calculate square roots.

challenge

When you stand h metres above sea-level, the horizon is x km away from you, where $x = \frac{7\sqrt{h}}{2}$.

For example, when you are two metres above sea level

$h = 2$, $x = \frac{7 \times \sqrt{2}}{2}$
$= 4.95 \ (2\,dp)$

so you can see 4.95 km to the horizon
Find x when $h = 10$.
Use a computer spreadsheet to find x for many values of h.

Square roots 159

Extension

The most able students could begin to look at Pythagoras' Theorem and how this uses square roots.

Question 1 – Simple instances with whole number answers.

Question 2 – Students need to be encouraged to memorise these.

Question 3 – Use this question to emphasise the need to learn the square numbers.

Question 4 – Similar to the example. Ensure that students know to find upper and lower bounds for the square root, and do not just home in from one side. Also ensure that they know how to establish an answer to 1 dp (SM) (1.3, 1.4).

Question 5 – relate back to question 1 (1.1, 1.3, 1.4).

Challenge – This may be difficult for many students and require further explanation. The activity may be best completed as a whole class.

Assessment Criteria – UAM, level 6: solve problems and carry through substantial tasks by breaking them down into smaller, more manageable tasks, using a range of efficient techniques, methods and resources, including ICT.

Links

Bring in some sheets of A3, A4 and A5 paper and other ISO sizes if available. Ask the class to measure the paper and calculate the ratio of the length to the width for each size. The ratio is √2 : 1 or 1.4142 : 1 in all cases. Now fold a sheet of A4 in half and compare with a sheet of A5 (same size). ISO paper sizes are designed so that A0 has an area of 1 m² but with a length to width ratio of √2 : 1. When a sheet of A0 is cut in half, it makes two smaller sheets size A1, each with a length to width ratio of √2 : 1, *etc.* There is a chart illustrating ISO paper sizes at http://en.wikipedia.org/wiki/Image:A_size_illustration.svg (EP)

- Use cubes and cube roots   (L5)
- Carry out more difficult calculations effectively and efficiently using function keys for powers and roots (L5)
- Use ICT to estimate [square roots] and cube roots   (L6)

Useful resources
Mini whiteboards

Starter – All the nines!

Write the following five calculations on the board:

$99 \times 999, 99^3, \sqrt{999999}, 9(99 + 999), \sqrt{99} \times 99^2$.

Ask students to estimate the answers and arrange them in order of size, smallest first.
Challenge students to explain their methods.

Correct order: $\sqrt{999999}, 9(99 + 999), \sqrt{99} \times 99^2, 99 \times 999, 99^3$.

Teaching notes

Make links here with work on squares and square roots. Assess understanding using the first two questions in the exercise as whole class interactive activities using mini whiteboards (1.2).

This spread offers the opportunity to consolidate finding a solution by trial-and-improvement, a skill that students will use elsewhere.

Make effective links with work in geometry and measures involving volume and use this opportunity to consolidate understanding.

It is important to support the efficient use of a calculator. Many students are caught out by the fact that not all calculator functions are found in the same place on all calculators. This is a good time to look at more than one model/make of calculator to locate the cube root button. Students need to be familiar with their own model of calculator (RL).

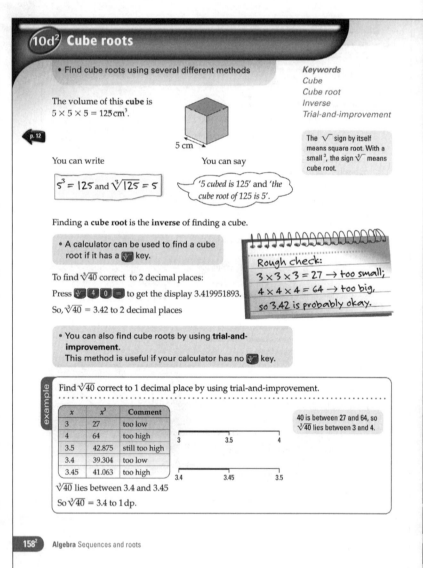

Plenary

Provide a word-based problem involving cubes and/or cube roots along similar lines to the examples in the exercise. To support analysis, discuss the problem as a whole class highlighting key information.

Simplification

Offer additional support with trial-and-improvement where this is needed.

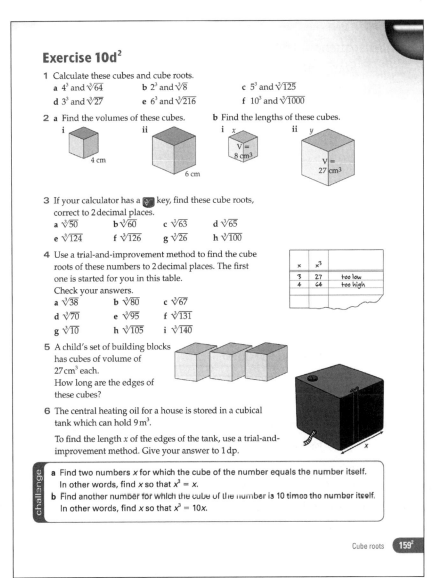

Exercise 10d²

1 Calculate these cubes and cube roots.
   a 4^3 and $\sqrt[3]{64}$ b 2^3 and $\sqrt[3]{8}$ c 5^3 and $\sqrt[3]{125}$
   d 3^3 and $\sqrt[3]{27}$ e 6^3 and $\sqrt[3]{216}$ f 10^3 and $\sqrt[3]{1000}$

2 a Find the volumes of these cubes.   b Find the lengths of these cubes.
   i   ii   i   x   ii   y
      4 cm   6 cm   V =   V =
                                            8 cm3   27 cm3

3 If your calculator has a ⬛ key, find these cube roots, correct to 2 decimal places.
   a $\sqrt[3]{50}$ b $\sqrt[3]{60}$ c $\sqrt[3]{63}$ d $\sqrt[3]{65}$
   e $\sqrt[3]{124}$ f $\sqrt[3]{126}$ g $\sqrt[3]{26}$ h $\sqrt[3]{100}$

4 Use a trial-and-improvement method to find the cube roots of these numbers to 2 decimal places. The first one is started for you in this table. Check your answers.

x	x^3	
3	27	too low
4	64	too high

   a $\sqrt[3]{38}$ b $\sqrt[3]{80}$ c $\sqrt[3]{67}$
   d $\sqrt[3]{70}$ e $\sqrt[3]{95}$ f $\sqrt[3]{131}$
   g $\sqrt[3]{10}$ h $\sqrt[3]{105}$ i $\sqrt[3]{140}$

5 A child's set of building blocks has cubes of volume of 27 cm³ each. How long are the edges of these cubes?

6 The central heating oil for a house is stored in a cubical tank which can hold 9 m³.
   To find the length x of the edges of the tank, use a trial-and-improvement method. Give your answer to 1 dp.

challenge
   a Find two numbers x for which the cube of the number equals the number itself. In other words, find x so that $x^3 = x$.
   b Find another number for which the cube of the number is 10 times the number itself. In other words, find x so that $x^3 = 10x$.

Cube roots 159²

Extension

More able students can be challenged with a volume question in a real-life context, such as designing packaging, to consolidate the work in this spread and understanding of measures.

Exercise 10d² commentary

Not all calculators are the same so students may need help finding/ using the ⬛ʳ key which may require the use of **shift** or **2nd fn** .

Question 1 – All have whole number answers; students should be encouraged to learn to recognise the first few cubes.

Question 2 – Students may need reminding how to calculate the volume of a cuboid.

Question 3 – A chance to confirm that students can use their calculators properly.

Question 4 – Make links here to prior experience of trial-and-improvement (1.3, 1.4, 1.5)

Question 5 – A simple example but in a text based problem; check that units are given (IE) (1.1, 1.3).

Question 6 – A word problem involving solution by trial-and-improvement (SM) (1.1, 1.3, 1.4).

Challenge – Part **a** may require a hint, part **b** has similar solutions but is more difficult.

Assessment Criteria – Algebra, level 6: use systematic trial and improvement methods and ICT tools to find approximate solutions to equations such as $x^3 + x = 20$

Links

In 2005, artist Rachel Whiteread displayed a work named Embankment at the Tate Modern gallery in London. The work comprised 14 000 white plastic cuboids and took five weeks to install. There is more information about Embankment and a video at. http://news.bbc.co.uk/1/hi/entertainment/arts/4326462.stm#

1 This sequence is made using straws.
- **a** Draw the next diagram in the sequence.
- **b** Write the first four terms.
- **c** Find the term-to-term rule.
- **d** Write the next three terms.

2 For each of these sequences, find the term-to-term rule and write the next three terms of the sequence.
- **a** 2   6   10   14   ...
- **b** 3   8   13   18   ...
- **c** 2   5   11   23   ...
- **d** 30   27   24   21   ...

3 Write the first six terms of each of the sequences with these rules.
- **a** *Start with 5. Add 4.*
- **b** *Start with 12. Add 6.*
- **c** *Start with 40. Subtract 6.*
- **d** *Start with 2. Double and add 1.*
- **e** *Start with 3. Double and subtract 2.*
- **f** *Start with 100 and subtract 21.*

4 This sequence of diagrams is made from triangles.
- **a** How many extra triangles are added to each position to make the next position?
- **b** Draw the diagram for position 5.
- **c** Copy and complete this table. Use the correct 'times table' for the middle row.
- **d** Find the position-to-term rule.
- **e** How many triangles are there in the 10th diagram of the sequence?

Position	1	2	3	4
Times table				
Term (No. of triangles)				

Position   1   2   3   4

5 a Find the position-to-term rule for this sequence.
Find the next two terms of the sequence.

Sequence	4	7	10	13	16
Position →	1	2	3	4	5

b Find the position-to-term rule for each of these sequences. Also find the next term and the 20th term in the sequence.
- **i** 1   3   5   7   9
- **ii** 3   6   9   12   15
- **iii** 7   12   17   22   27
- **iv** 19   18   17   16   15

6 Jamie bought three tins of cat food on the day he first owned some cats.
He then bought two tins every day after that.
The sequence for the total number of tins bought is 3   5   7   9   …

Day 1   Day 2   Day 3

a Find the term-to-term rule and the next two terms of this sequence.

b Copy and complete this table and find the position-to-term rule.

Position (day)	1	2	3	4	5
Times-table					
Term (tins)					

c Find the 20th term of the sequence.

d How many tins had he bought altogether by the 100th day?

7 Write the values of these square roots.
  a $\sqrt{36}$ **b** $\sqrt{64}$ **c** $\sqrt{144}$

8 The square roots of these numbers are not integers. Between which two consecutive integers does each of these square roots lie?
  a $\sqrt{30}$ **b** $\sqrt{18}$ **c** $\sqrt{80}$

9 Use a trial-and-improvement method to find the square roots of these numbers to 1 decimal place. The first one is started for you in this table.
  a $\sqrt{70}$ **b** $\sqrt{55}$ **c** $\sqrt{38}$

$\times$	x^2	
8	64	64<70 Too low
9	81	81>70 Too high

10 Calculate these cubes and cube roots.
  a 10^3 and $\sqrt[3]{1000}$ **b** 2^3 and $\sqrt[3]{8}$ **c** 4^3 and $\sqrt[3]{64}$

11 Use a trial-and-improvement method to find the cube roots of these numbers to 1 decimal place. The first one is started for you in this table. Check your answers using a calculator.
  a $\sqrt[3]{40}$ **b** $\sqrt[3]{98}$ **c** $\sqrt[3]{50}$

$\times$	x^3	
3	27	27<40 Too low
4	64	64>40 Too high

12 A cube of ice has a volume of 1 litre.
Find the length of each edge of the cube in centimetres.

10 Summary

Assessment criteria

- Generate terms of a sequence using term-to-term and position-to-term definitions of the sequence on paper and using ICT   Level 6
- Write an expression to describe the nth term of an arithmetic sequence   Level 6
- Use systematic trial and improvement methods and ICT tools to find approximate solutions to equations such as $x^3 + x = 2$ Level 6
- Solve problems and carry through substantial tasks by breaking them into smaller more manageable tasks, using a range of efficient techniques, methods and resources, including ICT   Level 6
- Justify generalisations, arguments or solutions   Level 7

Question commentary

Example

The example illustrates a question on finding a square root by trial and improvement. Use of a number line may help students to visualise the two values between which the root must lie. Some students may confuse square and square root, or multiply the estimate by two instead of squaring. Ask probing questions such as "How do you go about choosing a value of x to start", "How do you use the previous outcomes to decide what to try next?", "How do you know when to stop?" and "Is your solution exact?" Encourage students to check their answer by squaring.

$$8^2 \quad < 70 < 9^2$$
$$8^2 \quad < 70 < 8.5^2$$
$$8^2 \quad < 70 < 8.4^2$$
$$8.3^2 \ < 70 < 8.4^2$$
$$8.35^2 < 70 < 8.4^2$$

$$\sqrt{70} = 8.4 \text{ (1 dp)}$$

Past question

This question asks students to find a position-to-term rule for an arithmetic sequence given a similar sequence as a starting point. In part **a**, relate the need to +1 to changing the constant in the given rule. In parts **b** and **c**, remind students that *both* the step size, coefficient of n, and starting value, constant term, are halved or doubled. Encourage students to test their new rules for some values of n. The question can also be done directly from the new sequence without reference to the old rule.

Answer

Level 6

2   a $(4n + 1) + 14 = n + 2$
     b $\frac{1}{2}(6n + 6) = 3n + 3$
     c $2(5n - 3) = 10n - 6$

Development and links

This topic links with work on graphs and functions in Chapter 13. Powers and roots link to work on indices in algebra. Sequences are developed further in Year 9.

Sequences are found in all subjects of the curriculum: in the study of the natural world in science, in musical scales, in patterns in art and design, in tool sizes in technology, in sequences of dates in history and in sequences of instructions in ICT. Square and cube roots are used to find distances from known areas or volumes, for example the diagonal of a square piece of wood in technology.

11 Collecting and representing data

Objectives

- Decide which data to collect to answer a question, and the degree of accuracy needed 6
- Identify possible sources 6
- Consider appropriate sample size ....
- Plan how to collect the data 6
- Construct frequency tables with equal class intervals for gathering continuous data two-way tables for recording discrete data 6
- Construct pie charts for categorical data 6
- Construct bar charts and frequency diagrams for discrete and continuous data.. 6
- Construct simple line graphs for time series 6
- Construct simple scatter graphs 6

Level

MPA

1.1	11a, b, c
1.2	11a, e
1.3	11c, d, e, f
1.4	11a, b, c, d, f
1.5	11a, c, d, e, f

PLTS

IE	11a, b, c, f
CT	11a, e
RL	11a, c, d, f
TW	11a, d
SM	11b, d, e
EP	11a, c, e, f

Introduction

The focus of this chapter is on methods of collecting and representing data. Students consider the differences between primary and secondary data and plan how to collect data, constructing two-way tables for discrete data and frequency tables for grouped continuous data. They represent data by constructing pie charts, bar charts, frequency diagrams, simple line graphs for time series and scatter graphs.

The student book discusses how advertisers and manufacturers collect information from the public. Loyalty cards can tell a retailer exactly what consumers are buying and when. When this information is combined with the information given on the application form for the card, the retailers can build up a detailed picture of an individual's lifestyle and use this to target goods and services. In return, the consumer is given rewards, usually in the form of money off their shopping. There is more information about loyalty cards at http://en.wikipedia.org/wiki/Loyalty_card

Extra Resources

11   Start of chapter presentation

11d   Worked solution: Q2

        Assessment: chapter 11

⚡Fast-track

All spreads

- Decide which data to collect to answer a question, and the degree of accuracy needed (L5)
- Identify possible sources (L5)
- Consider appropriate sample size (L6)

Useful resources
Access to the internet

Starter – Today's number is ...72

Ask questions based on 'Today's number', for example

What is double 72 and add 3? (147)
What are the factors of 72? ($2^3 \times 3^2$: 1, 2, 3, 4, 6, 8, 9, 12, 18, 24, 36, 72)
How many of the factors are square numbers? (4)
What is 10% of 72? (7.2)
The square root of 72 lies between which two whole numbers? (8, 9)

Teaching notes

As an introduction to the lesson, discuss the differences between primary and secondary data. In pairs, ask students to consider a statement such as, 'Year 8 girls are taller than Year 8 boys'. How could they show whether or not this was true? What data should they use? If they only use data collected from school, it will give information about Year 8 heights in that school but not more generally. Therefore there is a need to use both primary and secondary data.

A useful way to approach issues surrounding the collection of data is from a consideration of the possible sources of bias. If students can first identify ways in which data collection might be biased, then they can try to think of ways to avoid this (RL).

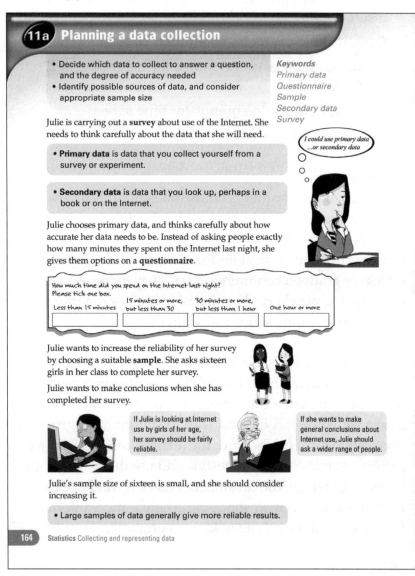

Plenary

Give students two sets of data, the weight of ten Y7 school bags and ten Y10 school bags (the data collected last week). In pairs, ask students to agree what they could do with the data to support investigating 'Y7 school bags are heavier than Y10 school bags'. Allow them a further two minutes to make comments about the data, for example, insufficient, gender, did Y10 have PE or maths (big thick books or…) Discuss the problem in general (EP).

Simplification

To help make the difference between primary and secondary data more concrete and allow students to get a 'feel' for data, use real data for question **1** and allow students to discuss what it shows. In question **2** supply more specific statements, for example, 'men live longer in Scotland than in Japan'.

Students would benefit from whole class discussion of the answers to this exercise (TW) (1.5).

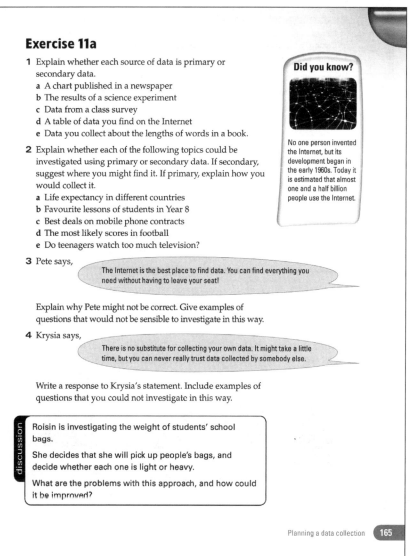

Exercise 11a

1 Explain whether each source of data is primary or secondary data.
   a A chart published in a newspaper
   b The results of a science experiment
   c Data from a class survey
   d A table of data you find on the Internet
   e Data you collect about the lengths of words in a book.

2 Explain whether each of the following topics could be investigated using primary or secondary data. If secondary, suggest where you might find it. If primary, explain how you would collect it.
   a Life expectancy in different countries
   b Favourite lessons of students in Year 8
   c Best deals on mobile phone contracts
   d The most likely scores in football
   e Do teenagers watch too much television?

3 Pete says,

> The Internet is the best place to find data. You can find everything you need without having to leave your seat!

Explain why Pete might not be correct. Give examples of questions that would not be sensible to investigate in this way.

4 Krysia says,

> There is no substitute for collecting your own data. It might take a little time, but you can never really trust data collected by somebody else.

Write a response to Krysia's statement. Include examples of questions that you could not investigate in this way.

discussion

Roisin is investigating the weight of students' school bags.

She decides that she will pick up people's bags, and decide whether each one is light or heavy.

What are the problems with this approach, and how could it be improved?

Planning a data collection 165

Did you know?

No one person invented the Internet, but its development began in the early 1960s. Today it is estimated that almost one and a half billion people use the Internet.

Extension

Use question **2** as the basis for an investigation. First, formulate a statement, for example in part **b**, 'most football games have three goals scored in them' or 'more goals are scored in the second half than the first'. Then plan what and how much data they would need and how they would get it (IE).

Exercise 11a commentary

Question 1 – Pair students and ask them to agree on answers. This could be developed by asking them to think of projects they have completed one using primary data and another using secondary data – they should explain each project and why primary or secondary data was used (CT).

Question 2 – Ask students to think of ways in which the data they collect might be biased (1.1).

Question 3 and **4** – Pair students and ask them to list two or three reasons why the statements may not be correct and to come up with two or three example questions which would require a different approach (1.2, 1.4).

Discussion – Set the challenge for students to plan for this. First they need to agree the statement they are going to investigate, for example, Y7 school bags are heavier than Y10 school bags and then they need to plan how they would investigate this statement (1.4).

Assessment Criteria – HD, level 6: design a survey or experiment to capture the necessary data from one or more sources.

Links

The Internet is a worldwide system of computer networks that allows the user to access information from any other computer (with permission). It was developed by the U.S. government in 1969 to allow researchers to access information on computers at other universities and was first known as the Advanced Research Projects Agency Network or ARPANET. There is a timeline showing the development of the Internet at http://www.webopedia.com/quick_ref/timeline.asp

- Plan how to collect the data   (L5)   *Useful resources*
- Construct frequency tables with equal class intervals for gathering discrete and continuous data and two way tables for recording discrete data   (L6)

Starter – Numbered cubes

Ask students to imagine a bag containing ten cubes numbered 1, 2 or 5.

Cubes are drawn out of the bag and replaced each time.

Write numbers on the board, for example, 5, 1, 5, 5, 2, 2, 2, 5, 2, 5, 2, 2, 2, 2, 5, 2, 2, 2, 5, 2

Ask students to estimate how many of each number there are in the bag? (for example, 1×1, 6×3, 3×5)

What is the mean of the numbers pulled out? (3)

Teaching notes

As an introduction to the lesson students should be made aware of different types of data – many find it difficult to determine the difference between continuous and discrete data. At this level it is very useful to highlight that continuous data tends to occur where you cannot get the exact data, for example, in height you may be 1.57 m or 5 feet 2 inches but you are always 'and a bit' or 'nearly'. Discrete data tends to occur where you are counting (often whole numbers).

Transferring data from a list into a table, as in questions **2** and **3**, is prone to silly mistakes (losing your place, double counting, missing out a category, *etc.*) and it will be instructive to work through an example where this is done systematically. Possible strategies to minimise the risk of errors include:

- Choose small groups of data to tally each time, for example, work through each column, where there are only 2 or 3 pieces of data.
- Tick off data values as they are entered (though not in books!).
- Check the number of entries in the table and in the list are the same.

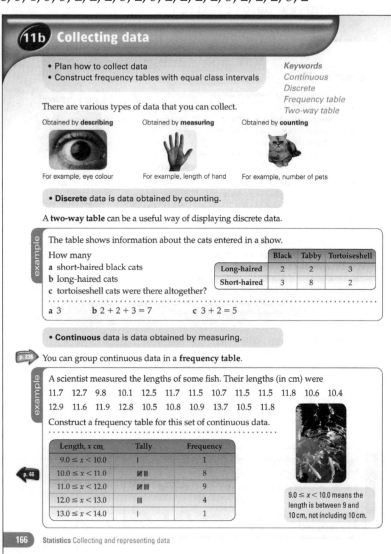

Plenary

Statements such as $9.0 \leq \leq x < 10.0$ are confusing. Give (groups of) students a number of similar mathematical and worded statements. Ask them to match statements (where possible) and supply any missing statements. For example, $9.0 \leq \leq x < 10.0, \ldots x < 10.0, \ 9.0 \leq \leq x, \ldots 9.0 < x \leq \leq 10.0, \ldots 10.0 > x > 9.0$

'x is smaller than 10', 'x is larger than 9 but less than or equal to 10', 'x is larger or equal to 9 and smaller or equal to 10'; 'x is less than 10 but larger than 9'.

Simplification

To support students in completing the frequency tables, reduce the amount of data to just one line. When this has been put into the table, provide the second and then third line of data.

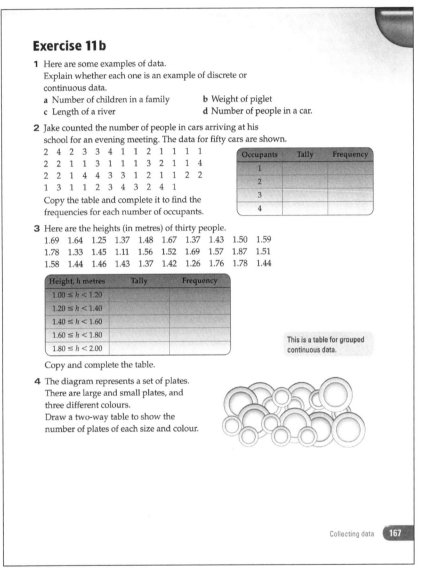

Exercise 11b

1 Here are some examples of data.
  Explain whether each one is an example of discrete or continuous data.
  a Number of children in a family   **b** Weight of piglet
  c Length of a river   **d** Number of people in a car.

2 Jake counted the number of people in cars arriving at his school for an evening meeting. The data for fifty cars are shown.

  2 4 2 3 3 4 1 1 2 1 1 1 1
  2 2 1 1 3 1 1 1 3 2 1 1 4
  2 2 1 4 4 3 3 1 2 1 1 2 2
  1 3 1 1 2 3 4 3 2 4 1

  Copy the table and complete it to find the frequencies for each number of occupants.

Occupants	Tally	Frequency
1		
2		
3		
4		

3 Here are the heights (in metres) of thirty people.

  1.69 1.64 1.25 1.37 1.48 1.67 1.37 1.43 1.50 1.59
  1.78 1.33 1.45 1.11 1.56 1.52 1.69 1.57 1.87 1.51
  1.58 1.44 1.46 1.43 1.37 1.42 1.26 1.76 1.78 1.44

Height, h metres	Tally	Frequency
$1.00 \leq h < 1.20$		
$1.20 \leq h < 1.40$		
$1.40 \leq h < 1.60$		
$1.60 \leq h < 1.80$		
$1.80 \leq h < 2.00$		

  Copy and complete the table.

  This is a table for grouped continuous data.

4 The diagram represents a set of plates. There are large and small plates, and three different colours.
  Draw a two-way table to show the number of plates of each size and colour.

Extension

We use tables to show information more clearly. Ask students to look at the tables they have made in question **2** and **3**. In each case ask them to make four comments about what the table is showing and share the ideas with other students. Then, using twenty words, can they explain why information in a table is more helpful than a list (IE, SM).

Exercise 11b commentary

Question 1 – It may help students to list some possible data values for each part before deciding on the type of data. Whole class discussion could be used to clarify the distinctions (1.4).

Question 2 – A discrete data version of the second example.

Question 3 – Similar to the second example. Check that students are correctly interpreting the intervals.

Question 4 – Constructing a two-way table like that in the first example (1.1).

Assessment Criteria – HD, level 6: construct tables for large discrete and continuous sets of raw data, choosing suitable class intervals. HD, level 6: design and use two-way tables.

Links

http://fishermansview.com/fishing_world_records.htm lists the World records by weight for many different types of fish caught on rod and line. What are the largest and smallest fish on the list? (Grass Pickerel 1 lb and Great White Shark 2664 lb). What is the range of the weights? (2663 lb) What is the range in kg? (about 1200 kg)

- Construct pie charts for categorical data   (L6)   ***Useful resources***
Calculator
Compasses
Protractor

Starter – Student survey

Draw a Venn diagram on the board with three different attributes, for example:
brown eyes, cereal for breakfast, plays football.

Enter possible numbers or enter students' own data and ask questions, for example,
how many students have brown eyes, did not have cereal for breakfast and play football?

Teaching notes

Show students a simple pie chart with three shaded areas ($\frac{1}{4}$, $\frac{1}{3}$, and the remainder) and supply a simple scenario: it represents 24 pupils, the smallest proportion have had one mobile phone, the slighter larger proportion $\left(\frac{1}{3}\right)$ have had two mobile phones, the largest proportion represents students who have had more than two different mobile phones. Take feedback on what the pie chart tells them. How can they be more accurate? (measure angles)

Students should be taken through an example of how to calculate the angles in a pie chart, either as in the second example $\left(\frac{6}{17} \times 360\right)$ or by first calculating the angle associated with one entry $\left(6 \times \frac{360}{17}\right)$. As a check of the calculation the sum of the angles should be 360°. Students may need to be reminded how to use a protractor and could be asked to explain the procedure to be followed: centre the cross, align the base, measure from zero, *etc.* Again, as a check of the drawing, the angle in the final sector should equal that calculated. This calculation can be related to earlier work on ratio and proportion.

Pie charts also arise in other subjects, for example geography, and it may be helpful to ask students how they are used and constructed in those subjects. Other variants of pie charts may also be available for comparison from the media (EP) (1.2).

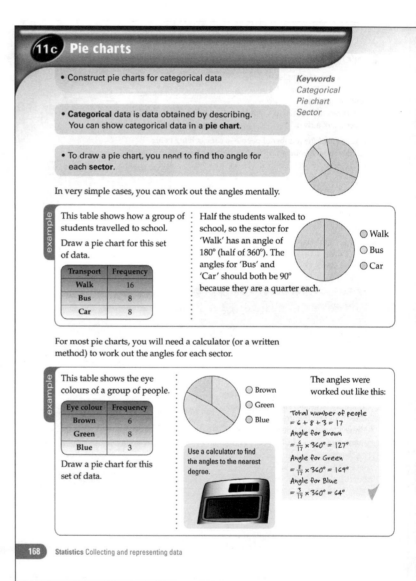

Plenary

Give pairs of students either a frequency table (for 24 or 36 pieces of data) with three of four categories, or a pie chart for a similar amount of data in three or four categories. Students have to draw a pie chart from the data or construct the frequency table from the pie chart. Discuss which was easier to do (1.1).

Simplification

Where students are unable to draw pie charts effectively at first give them two or three pie charts and ask them to explain what they show. Then give them a simple set of data with 36 items in three categories and the angles already worked out. Ask them to draw this as a pie chart and develop from here.

Exercise 11c

1 Sketch a pie chart for the data in each of these tables.
You do not need to calculate angles or draw these accurately.

a

Animal	Frequency
Dogs	12
Cats	11
Birds	13

b

Country	Medals won
Spain	19
Cuba	10
USA	11
Japan	41

c
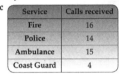

Service	Calls received
Fire	16
Police	14
Ambulance	15
Coast Guard	4

2 Draw the pie charts from question **1** accurately.
You will need to

- calculate the angles (mentally, with a written method or using a calculator), and
- draw the diagram using a protractor, ruler and compasses.

3 Marcus used a spreadsheet to produce this pie chart for the data in the table.

	A	B
1	**Genre**	**Number**
2	Fiction	8
3	History	12
4	Reference	9
5	Biography	7

- Fiction
- History
- Reference
- Biography

a Explain why this pie chart cannot be correct.
b Draw a correct version of the pie chart.

discussion

These two pie charts show the types of housing in two towns.

Make a list of some conclusions that you can draw from these pie charts.

What misconceptions might cause someone to draw incorrect conclusions from these pie charts?

- Terraces - Semi ○ Detached ○ Flat

Dalton   Farley

Extension

The pie chart in the second example shows colours of eyes in a group – ask students to survey their class, draw a similar pie chart and write a short commentary on similarities and differences (IE).

Exercise 11c commentary

Question 1 and **2** – These form a pair. Ask students to compare their sketches and resolve any (significant) differences, identifying where they went wrong, before proceeding (RL) (1.5). Question **2** is similar to the second example; only part **a** involves 'nice' numbers. Weaker students could work together on calculating the angles before drawing and labelling their own pie charts (1.3).

Question 3 – Students should be able to give more than one reason why the pie chart is incorrect. One entry is 10° (1.3).

Discussion – You may need to remind students that the two towns may have very different populations (1.4).

Assessment Criteria – HD, level 5: interpret graphs and diagrams, including pie charts, and draw conclusions. HD, level 6: construct pie charts for categorical data.

Links

Eye colour is an inherited characteristic. Only one of the pair of genes that control eye colour is passed from each parent to a child. The combination of inherited genes determines the child's eye colour. There is an eye colour calculator at http://museum.thetech.org/ ugenetics/eyeCalc/eyecalculator. html

• Construct bar charts and frequency diagrams for discrete and continuous data (L5)

Useful resources

. . . .

Starter – Favourite crisps

A pie chart represents the favourite crisps of 60 students.

The flavours and angles are: onion, chicken, plain, prawn, salt & vinegar; 30°, 36°, 60°, 90° and 144°.

25% of students preferred prawn.
Twice as many preferred salt & vinegar to chicken.

$\frac{1}{10}$ preferred plain.

Ask students to match angles with flavours.

How many students prefer each flavour?

(o=24, ch=5, pl=6, pr–15, s&v=10)

Teaching notes

Students are well aware of 'bar charts' but they rarely appreciate the difference between categorical data, numerical data which is discrete and continuous data and how this affects the type of 'bar chart' that is drawn. As an introduction to the lesson highlight the examples given in the text book. Show that the bars for categorical and numerical discrete data have gaps between them. For continuous data we draw a frequency diagram where there are no gaps between the bars; this shows that the measures are continuous.

To complete questions such as **2** and **3** students will need to be shown how to design a table and systematically fill in the 30 entries, *c.f.*, spread **11b**. In question **2**, the colour order for the bars does not matter, although this is often best done alphabetically. However in question **3**, draw bars from smallest to largest showe size and make sure that there is a gap between each bar.

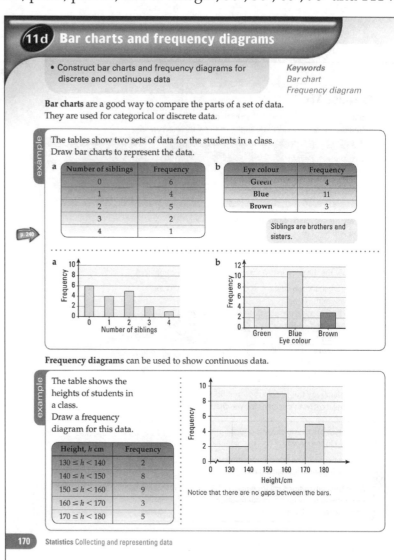

Plenary

Project a bar chart with errors, for example:

• it is not named
• the vertical axis is not marked – bars have numbers at the top that don't relate to their height
• there are no spaces between the bars (which are colours) or there are spaces and the data is continuous
• some of the bars are not in the correct order
• the bars are of different widths

In pairs students have to find the mistakes and then redraw the bar chart accurately (RL) (1.3).

Simplification

Provide blank tables for questions 1 and 2. Alternatively, if time (or a teaching assistant) is available work with the weaker students to fill in the table, and discuss how best to do this systematically. See also spread **11b**.

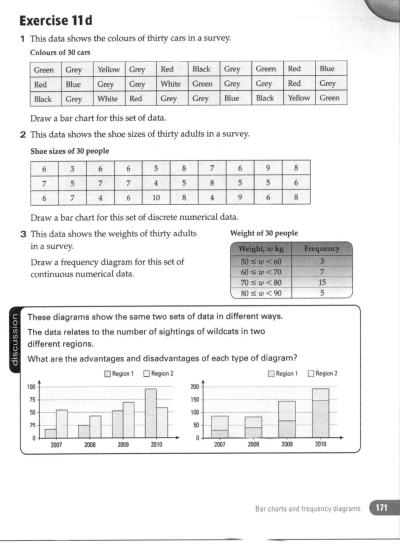

Exercise 11d

1 This data shows the colours of thirty cars in a survey.

Colours of 30 cars

Green	Grey	Yellow	Grey	Red	Black	Grey	Green	Red	Blue
Red	Blue	Grey	Grey	White	Green	Grey	Grey	Red	Grey
Black	Grey	White	Red	Grey	Grey	Blue	Black	Yellow	Green

Draw a bar chart for this set of data.

2 This data shows the shoe sizes of thirty adults in a survey.

Shoe sizes of 30 people

6	3	6	6	5	8	7	6	9	8
7	5	7	7	4	5	8	5	5	6
6	7	4	6	10	8	4	9	6	8

Draw a bar chart for this set of discrete numerical data.

3 This data shows the weights of thirty adults in a survey.

Draw a frequency diagram for this set of continuous numerical data.

Weight of 30 people

Weight, w kg	Frequency
$50 \leq w < 60$	3
$60 \leq w < 70$	7
$70 \leq w < 80$	15
$80 \leq w < 90$	5

> **discussion**
>
> These diagrams show the same two sets of data in different ways.
>
> The data relates to the number of sightings of wildcats in two different regions.
>
> What are the advantages and disadvantages of each type of diagram?

Bar charts and frequency diagrams **171**

Extension

Bar charts are one of the most useful diagrammatic representations because it is easy to understand the results of the diagram and they are also one of the easiest to draw. To challenge students, supply two sets of data (for example, weights of males and weights of females) and ask them to draw a single bar chart to highlight this information – they should draw dual bar charts similar to those in the discussion activity (SM).

Exercise 11d commentary

In each question ask students to write one or two sentences describing what the bar charts show.

Question 1 – Categorical data as in part **b** of the first example. Suggest that the categories are listed alphabetically (1.3).

Question 2 – Discrete numerical data as in part **a** of the first example. Check that the shoe sizes are in ascending order (1.3).

Question 3 – Continuous numerical data as in the second example. Check that the bars now touch (1.3).

Discussion – Challenge pairs of students to find two advantages and two disadvantages for each graph. Pairs of pairs could then agree which are the two best and two worst features of each graph before presenting them to the whole class (TW) (1.4, 1.5).

Assessment Criteria – HD, level 6: construct bar charts and frequency diagrams for discrete and continuous data.

Links

The Scottish Wildcat is the only native member of the cat family in the British Isles and has been extinct in England and Wales since 1862. It is now found only in the Highlands of Scotland, where it is thought that about 500 remain. The Scottish wildcat is larger than a domestic cat and preys mainly on rodents and small mammals. There is more information about the Scottish Wildcat at http://www.scottishwildcats.co.uk/

• Construct simple line graphs for time series. (L6) ***Useful resources***
Squared paper

Starter – Temperatures

Write a list of times and temperatures on the board:

6 a.m., 9°; 9 a.m., 16.5°; 12 noon, 21°; 3 p.m., 23.5°; 6 p.m., 17°; 9 p.m., 14°.

Give students quick-fire questions, for example:

What is the biggest temperature difference?

By how much did the temperature change between 9 a.m. and 12 noon?

Can be differentiated by the choice of temperatures.

Teaching notes

Most students will have seen time-series graphs around them and may have drawn them (or used them) in science and in geography. As an introduction, give students data for the amount of sunshine per day for two weeks in June, for example:

Sun 6.5 hrs; Mon 8.2 hrs; Tues 2.7 hrs; Wed 3.4 hrs; Thurs 3 hrs; Fri 5.8 hrs; Sat 7.5 hrs; Sun 4.8 hrs; Mon 5.5 hrs; Tues 6.4 hrs; Wed 7.2 hrs; Thurs 7.8 hrs; Fri 8.4 hrs; Sat 9.5 hrs

Ask them to show this information on a graph. Allow them five minutes to do this in pairs. Remind students that they need to label their graphs correctly (EP).

Now compare the graphs that students have drawn. Most will have drawn bar charts but if there are some who have drawn a time-series graph, use these as an example. If not, then suggest that for this type of information a quicker way of showing the data is by drawing a time-series graph.

Show them a time-series graph discussing the solid line between the two data points. Explain that it shows a trend rather than an 'in between' value.

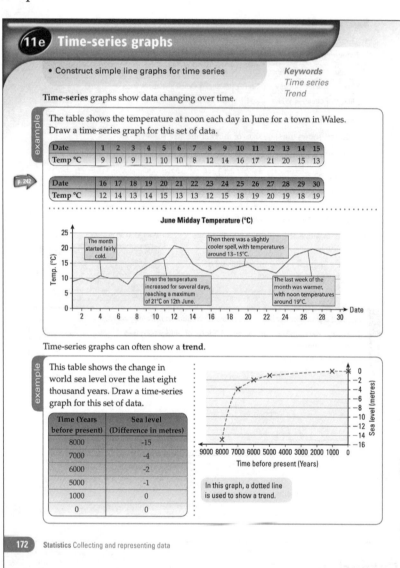

Plenary

Ask students to think about the homework they do during a school week. Can they draw a time-series graph to show the amount of homework that is set each day (students could work together here to agree on what is the usual amount). On the same graph (with Saturday and Sunday!) can they show how much homework they actually do each day! Comment on the differences.

Simplification

For question **1**, provide partly-labelled, pre-prepared axes so that students can concentrate on plotting and interpreting the data. In question **2** students may need support in choosing the scales for the axes for the time-series graph.

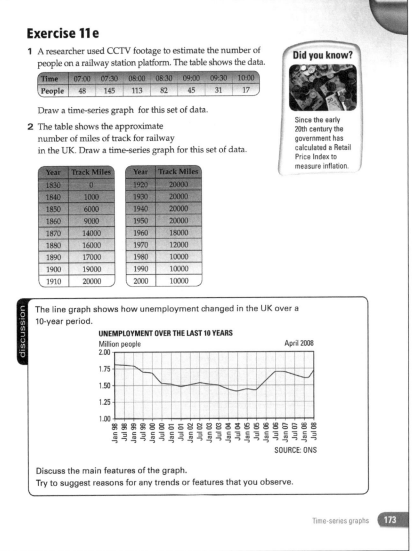

Exercise 11e

1 A researcher used CCTV footage to estimate the number of people on a railway station platform. The table shows the data.

Time	07:00	07:30	08:00	08:30	09:00	09:30	10:00
People	48	145	113	82	45	31	17

Draw a time-series graph for this set of data.

2 The table shows the approximate number of miles of track for railway in the UK. Draw a time-series graph for this set of data.

Year	Track Miles	Year	Track Miles
1830	0	1920	20000
1840	1000	1930	20000
1850	6000	1940	20000
1860	9000	1950	20000
1870	14000	1960	18000
1880	16000	1970	12000
1890	17000	1980	10000
1900	19000	1990	10000
1910	20000	2000	10000

Did you know?

Since the early 20th century the government has calculated a Retail Price Index to measure inflation.

discussion

The line graph shows how unemployment changed in the UK over a 10-year period.

UNEMPLOYMENT OVER THE LAST 10 YEARS

Million people — April 2008

SOURCE: ONS

Discuss the main features of the graph.
Try to suggest reasons for any trends or features that you observe.

Time-series graphs **173**

Extension

Construct an 'effortometer'. On squared paper set up axes for a time-series graph. On the horizontal (time) axis start at the time students got up today and then mark at one hour intervals, for example, 7.15, 8.15, 9:15, *etc*. On the vertical (effort) axis mark from 0-100%. Students then plot points on the 0-100% effort scale for each hour of their day and predict future hours. Although this is largely guesswork, done properly it is an opportunity for students to reflect on their day, their efforts and themselves. They can exchange their 'effortometer' graphs and can describe what they show (SM).

Exercise 11e commentary

Question 1 – Similar to the examples. Ask students to explain the pattern that they see (1.3).

Question 2 – Students may need help to decide on suitable scales to use (1.3).

Discussion – Students may not know about political and economic events in the last decade so ask them to suggest how a war or the cost of oil might affect the numbers.

Ask students to suggest how the government might go about finding the total number of unemployed people. Are they likely to under- or over-estimate the total? Why? (EP, CT) (1.2, 1.5)

Assessment Criteria – HD, level 6: construct simple time graphs for time series.

Links

The Consumer Price Index (CPI) is an official measure of the average price of goods and services including travel costs, food, heating and household goods. The index number is calculated each month by finding the price of a sample of goods that a typical household might buy, and comparing the price to a reference value. The percentage change in the CPI from the same month in the previous year is a measure of inflation. The latest figures for the CPI can be found at http://www.statistics.gov.uk/cci/nugget.asp?ID=19

- Construct simple scatter graphs

(L6) **_Useful resources_**
Squared and graph paper

Starter – Data handling jumble

Write a list of anagrams on the board and ask students to unscramble them.

Possible anagrams are
VURSEY, DOME, TAAD, ATORSIINENQUE, UNNSUITCOO, NAME, SLAMEP, CERTSIDE
(survey, mode, data, questionnaire, continuous, mean, sample, discrete)

Teaching notes

A common mistake in drawing scatter graphs is for students to plot points without any thought about uniform scales on the axes. Start with a graph similar to that in the second example and ask students to interpret it. Then, if necessary, draw attention to the scales and discuss how the points should be plotted. Produce a correctly drawn graph and ask students to describe what they now see. With a little forethought it is simple to remove or even reverse a correlation by misplotting points (RL).

In pairs, ask students to list the information they would require to draw the axes for a scatter diagram showing the number of texts made against the time spent on the mobile phone per day for pupils in the class. What information do they need to be able to label this pair of axes? Take feedback from groups and list the points made. This will support a strategy for drawing appropriate axes. You may like to ask students what they think the scatter graph would look like and also what they think the maximum number of text and minutes used would be (EP).

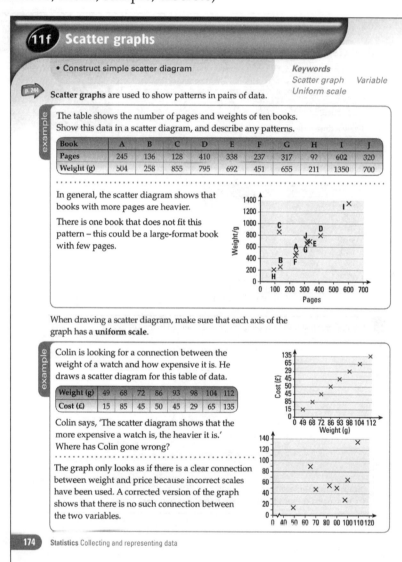

Plenary

Give small groups of students 4–6 paired data topics that can be drawn as scatter graphs. Ask each group to agree the likely trend and then discuss. For example:

- age and hours of sleep
- weight and height
- temperature and the amount of coffees sold at a cafe
- result in english and result in maths
- results in maths and time for 100 m sprint
- amount of money spent on sweets and on clothes.

Simplification

To allow students to get a better understanding of scatter graphs, allow the group to plot real information. For a group of about ten students, measure their height and the length of their shoe, as accurately as possible (this is better than shoe size as this is often incorrect). For best results, place shoe heel against the wall and measure on a sheet of paper. Draw a large pair of exes and get students to plot their own coordinates (they could identify their cross with their initials). Discuss the results.

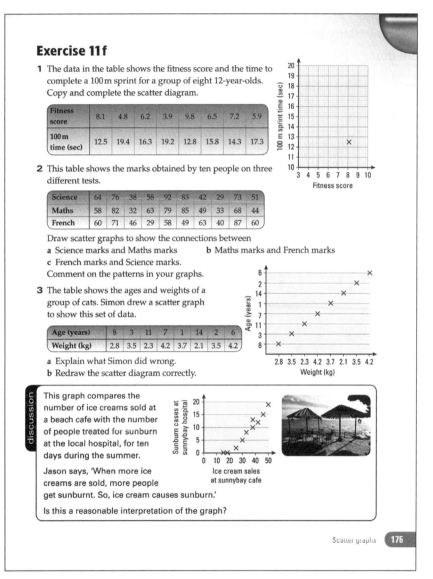

Exercise 11f

1 The data in the table shows the fitness score and the time to complete a 100 m sprint for a group of eight 12-year-olds. Copy and complete the scatter diagram.

Fitness score	8.1	4.8	6.2	3.9	9.8	6.5	7.2	5.9
100 m time (sec)	12.5	19.4	16.3	19.2	12.8	15.8	14.3	17.3

2 This table shows the marks obtained by ten people on three different tests.

Science	64	76	38	58	92	83	42	29	73	51
Maths	58	82	32	63	79	85	49	33	68	44
French	60	71	46	29	58	49	63	40	87	60

Draw scatter graphs to show the connections between
a Science marks and Maths marks   **b** Maths marks and French marks
c French marks and Science marks.
Comment on the patterns in your graphs.

3 The table shows the ages and weights of a group of cats. Simon drew a scatter graph to show this set of data.

Age (years)	8	3	11	7	1	14	2	6
Weight (kg)	2.8	3.5	2.3	4.2	3.7	2.1	3.5	4.2

a Explain what Simon did wrong.
b Redraw the scatter diagram correctly.

discussion

This graph compares the number of ice creams sold at a beach cafe with the number of people treated for sunburn at the local hospital, for ten days during the summer.

Jason says, 'When more ice creams are sold, more people get sunburnt. So, ice cream causes sunburn.'

Is this a reasonable interpretation of the graph?

Scatter graphs **176**

Extension

Ask students if they agree with this statement: Pupils who get up early on a school day also get up early on Saturday and Sunday. How would they check if this was true? Can they design a survey and scatter graph to see if this is true.

Exercise 11f commentary

Question 1 – Similar to the first example. Students could be asked to comment on what they see (1.3).

Question 2 – Students should decide what scales to use for themselves: check that they are sensible. It may help to produce three separate tables for each pair of variables (1.3).

Question 3 – Similar to the second example (1.3).

Discussion – A chance to discuss correlation and causation (necessary but not sufficient) and the significance of 'third' variables which might offer a better/more reasonable explanation (IE) (1.4, 1.5).

Assessment Criteria – HD, level 6: construct scatter graphs.

Links

A Hertzsprung-Russell diagram is a scatter graph showing the connection between the luminosity of a star and its temperature. The position of each star on the chart indicates its age. The diagram is named after the Danish astronomer Ejnar Hertzsprung (1873–1967) and the American astronomer Henry Norris Russell (1877–1957). There is more information about the Hertzsprung-Russell diagram at http://en.wikipedia.org/wiki/Hertzsprung-Russell_diagram and an interactive version at http://aspire.cosmic-ray.org/labs/star_life/hr_diagram.html

11a

1 Give some examples of primary and secondary data that you could use to investigate these topics.
  a Recycling
  b Mobile phone use
  c Exercise and sport

11b

2 Tom asked 30 people in his class how many books they currently had out on loan from the school library. Here are his results.

3 2 2 2 1 0 3 0 1 4
4 3 1 0 0 1 2 3 3 4
1 3 2 2 2 3 3 4 0 1

Organise this data into a frequency table.

3 Sara recorded the size and colour of crayons in a box.
There are three colours of crayon − red, blue and green.
There are two sizes of crayon − large and small.
Sara used upper and lower case letters to represent the crayons.
She used G for a large green, and b for a small blue, and so on.

R r r G r b B r G g
B B G g r R r r b b
g g g R B R G r G B

Draw a two-way table for this set of data.

11c

4 Here are one season's results for two football teams.
United: Won 14, Drew 8, Lost 8
Wanderers: Won 10, Drew 6, Lost 14
Draw a pie chart for each team's results.

5 This set of data shows the number of phone calls received by a shop each day for 20 days.

4   5   5   3   0   2   7   4   6   5
4   4   1   0   3   2   1   5   5   4

Draw a bar chart to show the data.

6 This set of data shows the length (in minutes and seconds) of 20 phone calls received by a shop.

0:35   1:22   2:47   1:26   3:55   2:50   0:15   1:03   3:35   4:09
3:10   0:59   3:09   2:26   3:11   3:28   2:05   3:54   2:12   1:08

Draw a frequency diagram to show the data.

7 In an experiment, Jackie heated up a beaker of water, and then left it to cool. She recorded the temperature every minute (in degrees Celsius). Here are her results.

95, 79, 67, 57, 49, 43, 38, 34, 31, 28, 26, 25, 23, 22

Draw a time-series graph for this set of data.

8 John recorded the length and weight of 10 earthworms.

| Length (cm): | 7 | 8 | 7 | 9 | 11 | 6 | 10 | 5 | 6 | 5 |
| Weight (g): | 22 | 24 | 21 | 28 | 30 | 16 | 28 | 17 | 18 | 16 |

Draw a scatter diagram for this set of data.

11 Summary

Assessment criteria

- Interpret graphs and diagrams, including pie charts, and draw conclusions   Level 5
- Select, construct and modify, on paper and using ICT:
  - pie charts for categorical data   Level 6
  - bar charts and frequency diagrams for discrete and continuous data   Level 6
  - simple line graphs for time series   Level 6
  - scatter graphs   Level 6

and identify which are most useful in the context of the problem

- Design a survey or experiment to capture the necessary data from one or more sources   Level 6
- Construct tables for large discrete and continuous sets of raw data, choosing suitable class intervals   Level 6
- Design and use two-way tables   Level 6

Question commentary

Example

The example illustrates a typical problem on collecting data. Students are asked to identify three reasons why some collected data may be unreliable. Some students may only identify one or two reasons. Encourage further thinking by asking questions such as "Who should Julie ask?", "How many people should Julie ask?" and "When should Julie take her survey?"

- Only people at one shop are asked
- The sample is small
- Only lunchtime shoppers are surveyed.

Past question

The question requires students to interpret a pie chart and to calculate the angle of a sector on a pie chart. Encourage students to estimate the size of each angle before measuring to prevent misreading the protractor. Emphasise the need to show clear working for each step. In part **b**, ask students what information they need to calculate the size of a pie chart angle. Some students may prefer to use fractions to find the angle, that is, $\frac{9}{24}$ of 360°.

Answer

Level 6

2 a $60 \div 5$
$= 12°$ per pupil
$96° \div 12°$
$= 8$ pupils

b $360° \div 24$
$= 15°$ per pupil
$15° \times 9 = 135°$

Development and links

This topic is developed further in Chapter 15 where students will analyse and interpret data presented in the form of tables, charts and graphs.

The methods of collecting and representing data encountered in this chapter are important in geography, where statistical information is presented in many forms of chart and graph. Charts and graphs are used in Science to present and analyse results of experiments. In everyday life, students will be bombarded with requests for their opinion, either online, by telephone, in the street or on printed surveys and questionnaires.

12 Ratio and proportion — Number

Objectives

- Apply understanding of the relationship between ratio and proportion.. **5**
- Simplify ratios, including those expressed in different units, recognising links with fraction notation **5**
- Divide a quantity into two or more parts in a given ratio............. **6**
- Use the unitary method to solve simple problems involving ratio and direct proportion **6**
- Calculate percentages and find the outcome of a given percentage increase or decrease **6**
- Use the equivalence of fractions, decimals and percentages to compare proportions.. **6**

Introduction

This chapter builds on the ideas of ratio and proportion introduced in Year 7 and on the work on percentages in Chapter 4. Students simplify ratios, divide a quantity in a given ratio, use direct proportion, consolidate knowledge of the difference between ratio and proportion and solve simple problems. The topic develops to finding percentage increases and decreases and using the equivalence of percentages, fractions and decimals to compare simple proportions.

The topic of ratio and proportion has many applications in everyday life. The student book discusses the use of ratio and proportion in mixing water, sand, aggregate and cement powder to make concrete. Other applications where ratio and proportion are important include mixing paint colours, mixing screen wash and antifreeze with water in a car, calculating medicine doses in proportion to the weight of a patient, creating metal alloys, adapting recipes for larger or smaller numbers and drawing to scale, including maps. There is an animated explanation of gears and gear ratios at http://www.howstuffworks.com/gears.htm

Fast-track

12a, b, c, d, e

Level

MPA

1.1	12a, b, c, d, e, f, CS
1.2	12c, d, e, CS
1.3	12a, b, c, d, e, f, CS
1.4	12b, c, f, CS
1.5	12e, CS

PLTS

IE	12b, c, d, e, f, CS
CT	12a, b, c, d, e, f, CS
RL	12b, d, e, CS
TW	12e, CS
SM	12a, d, e, f, CS
EP	12a, c, e, f, CS

Extra Resources

12	Start of chapter presentation
12b	Starter: Ratio matching
12b	Worked solution: Q1b
12c	Consolidation sheet
12d	Animation: Ratios
12d	Consolidation sheet
12e	Worked solution: Q2c
12e	Consolidation sheet
12f	Consolidation sheet Maths life: Daily bread
	Assessment: chapter 12

- Solve simple problems involving ratio and proportion using informal strategies (L5)
- Simplify a ratio, including those expressed in different units, recognising links with fraction notation (L5)

Useful resources
Mini whiteboards

Starter – Amazing digits

Ask students to think of a three digit number (3 different digits), for example, 451.
Ask them to reverse the digits (154).
Subtract the smaller number from the larger number ($451 - 154 = 297$).
Add the answer digits together until a single digit is obtained ($2 + 9 + 7 = 18$, $1 + 8 = 9$).
Repeat with another three digit number.
Ask students what they notice. (Always 9)

Teaching notes

Students can be encouraged to make explicit links here between fractions and ratios in terms of cancelling and simplifying. Stress the importance of both parts of the ratio being in the same units, as in the introductory example (CT).

To facilitate assessment of students' prior knowledge at an early stage, use question **1** as a class activity using mini whiteboards.

In the second worked example, for those students who appear uncertain, focus on looking for how many lots of 7 there are in 175 to find the multiplier.

The example involving scales on maps can be linked to work elsewhere in the curriculum such as geography or scale drawing in design and technology. Invite students to suggest other examples. This work provides opportunities for students to demonstrate their capacity to transfer mathematical skills to other contexts, supporting their development as functional mathematicians (EP).

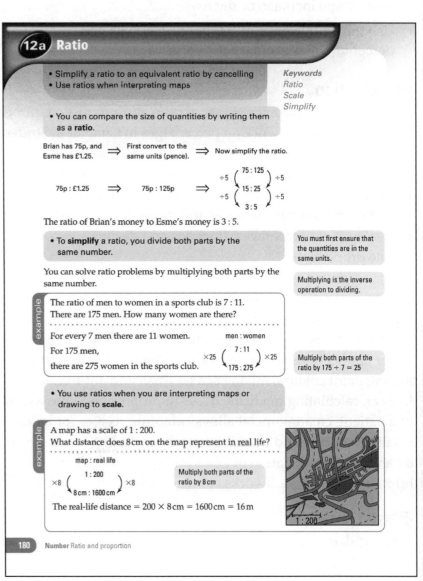

Plenary

Invite students to think of how they use ratio in real life: Food technology? Recipes? Art? Mixing colours? … (EP)

Simplification

Where students have difficulty, focus on supporting the analysis of word problems. This is often a problem and will be a focus for learning through KS3 and into KS4.

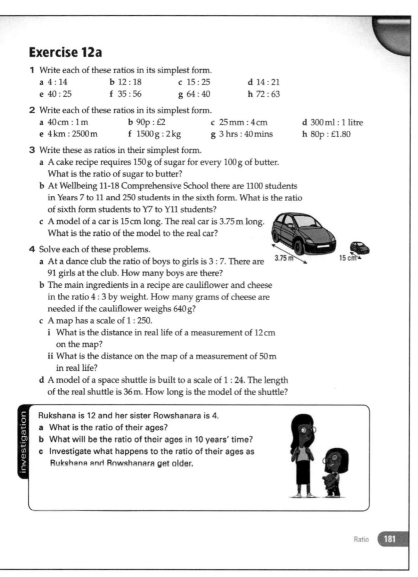

Exercise 12a

1 Write each of these ratios in its simplest form.

a 4 : 14	**b** 12 : 18	**c** 15 : 25	**d** 14 : 21
e 40 : 25	**f** 35 : 56	**g** 64 : 40	**h** 72 : 63

2 Write each of these ratios in its simplest form.

a 40 cm : 1 m	**b** 90p : £2	**c** 25 mm : 4 cm	**d** 300 ml : 1 litre
e 4 km : 2500 m	**f** 1500 g : 2 kg	**g** 3 hrs : 40 mins	**h** 80p : £1.80

3 Write these as ratios in their simplest form.

a A cake recipe requires 150 g of sugar for every 100 g of butter. What is the ratio of sugar to butter?

b At Wellbeing 11-18 Comprehensive School there are 1100 students in Years 7 to 11 and 250 students in the sixth form. What is the ratio of sixth form students to Y7 to Y11 students?

c A model of a car is 15 cm long. The real car is 3.75 m long. What is the ratio of the model to the real car?

4 Solve each of these problems.

a At a dance club the ratio of boys to girls is 3 : 7. There are 91 girls at the club. How many boys are there?

b The main ingredients in a recipe are cauliflower and cheese in the ratio 4 : 3 by weight. How many grams of cheese are needed if the cauliflower weighs 640 g?

c A map has a scale of 1 : 250.
  i What is the distance in real life of a measurement of 12 cm on the map?
  ii What is the distance on the map of a measurement of 50 m in real life?

d A model of a space shuttle is built to a scale of 1 : 24. The length of the real shuttle is 36 m. How long is the model of the shuttle?

investigation

Rukshana is 12 and her sister Rowshanara is 4.
a What is the ratio of their ages?
b What will be the ratio of their ages in 10 years' time?
c Investigate what happens to the ratio of their ages as Rukshana and Rowshanara get older.

Ratio **181**

Extension

Ask more able students to design a worksheet of some simple ratio problems around mixing colours or making bead jewellery that could be used in a year 6 class room as part of a transition package with feeder schools (SM).

Exercise 12a commentary

Question 1 – Make links here with fractions and cancelling, possibly use this as a whole class mini whiteboard activity.

Question 2 – Here students will need to rewrite the examples in the same units first before cancelling. Refer students back to the introduction if necessary.

Question 3 – This example provides the opportunity for students to practice rewriting text-based problems in mathematical form before simplifying. The units may catch some students out in part **c** (IE) (1.1, 1.3).

Question 4 – Similar to the two examples. Word problems to solve in a real-life context to develop functional skills competence (1.1, 1.3).

Investigation – It may help students to write the ratio in the form $n : 1$ (CT).

Assessment Criteria – NNS, Level 5: understand simple ratio.

Links

The aspect ratio of a screen or an image is the ratio of its width to its height. HD televisions and monitors have an aspect ratio of 16 : 9 (also known as 1.78 : 1) but older-style screens have an aspect ratio of 4 : 3 (1.33 : 1) Common cinema film ratios are 1.85 : 1 and 2.35 : 1. When an image filmed in one aspect ratio is displayed on a screen with a different aspect ratio, the image has to be either cropped or distorted.

- Divide a quantity into two [or more] parts in a given ratio   (L6)
- Use the unitary method to solve simple problems involving ratio and direct proportion   (L6)

Useful resources

Starter – Countdown

Ask students for six numbers: five between 1 and 10, and one from 25, 50 and 100.
Write the numbers on the board.
Throw a die three times to generate a three digit target number.
Challenge students to calculate this target number (or get as close as possible to it) using the six numbers and any operations.

Teaching notes

The worked example emphasises the value of a simple checking strategy and students should be encouraged to use this for themselves. This is a means of developing key processes in number through considering the appropriateness and accuracy of numerical results and also, as a paired activity, as a means of assessing themselves and others and reviewing progress (RL).

The real-life, everyday settings used in the worked examples and in the word-based problems in the exercise. based fully support the embedding of functional skills in the curriculum, where the focus is on students learning to access their 'mathematical toolbox' to find the appropriate skill to solve a problem and then to understand how to check a solution so that they know they have a reasonable answer (IE).

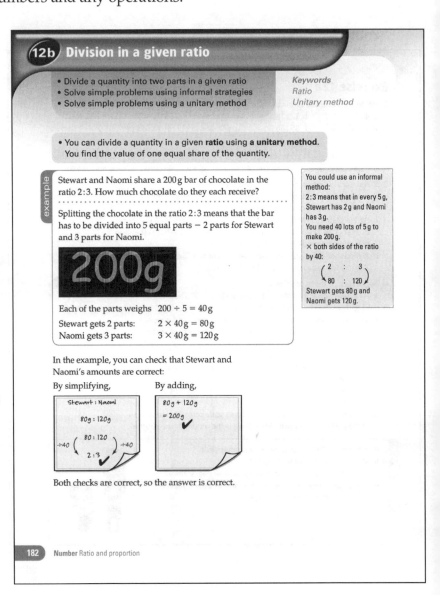

12b Division in a given ratio

- Divide a quantity into two parts in a given ratio
- Solve simple problems using informal strategies
- Solve simple problems using a unitary method

Keywords
Ratio
Unitary method

- You can divide a quantity in a given **ratio** using a **unitary method**. You find the value of one equal share of the quantity.

example

Stewart and Naomi share a 200 g bar of chocolate in the ratio 2:3. How much chocolate do they each receive?

Splitting the chocolate in the ratio 2:3 means that the bar has to be divided into 5 equal parts – 2 parts for Stewart and 3 parts for Naomi.

200g

Each of the parts weighs	$200 \div 5 = 40\,g$
Stewart gets 2 parts:	$2 \times 40\,g = 80\,g$
Naomi gets 3 parts:	$3 \times 40\,g = 120\,g$

You could use an informal method:
2:3 means that in every 5 g, Stewart has 2 g and Naomi has 3 g.
You need 40 lots of 5 g to make 200 g.
× both sides of the ratio by 40:
$$\binom{2\ :\ 3}{80\ :\ 120}$$
Stewart gets 80 g and Naomi gets 120 g.

In the example, you can check that Stewart and Naomi's amounts are correct:

By simplifying,
Stewart : Naomi
80g : 120g
÷40 (80 : 120) ÷40
2 : 3 ✓

By adding,
80g + 120g
= 200g ✓

Both checks are correct, so the answer is correct.

182   **Number** Ratio and proportion

Plenary

Ask students to reflect on the word problems they have been working on and to vote on which was most challenging. Discuss this as a whole class.

Simplification

Support students who are experiencing difficulty with word problems by guiding them in underlining or highlighting the key information needed to solve the problem.

Exercise 12b

1 Divide these quantities in the ratios given.
   a £50 in the ratio 2:3   **b** 60 cm in the ratio 5:7
   c 72 MB in the ratio 4:5   **d** 90p in the ratio 1:5
   e 120 seconds in the ratio 3:5 **f** £240 in the ratio 5:3

2 Sian picks some apples.
   She shares out 15 apples between herself and her mum in the ratio 2:3.
   a Draw a diagram to show how Sian divides the apples.
   b Write the number of apples she gives to her mum.

3 Morgan wins £40.
   He shares out the £40 between himself and his dad in the ratio 3:5.
   a Draw a diagram to show how Morgan divides the money.
   b Write the amount of money he gives to his dad.

4 Solve each of these problems.
   a At a gym club the ratio of boys to girls is 3:4. There are 63 children in total at the club. How many girls are there?
   b Jack and Mona share £84 in the ratio 2:5. How much money does Jack receive?

5 For each of these questions, check that the answer is correct. Explain your reasoning in each question.
   a At a sports club the ratio of boys to girls is 2:3. There are 25 boys and girls in total at the club. How many girls and how many boys are there?
   ANSWER: There are 10 boys and 15 girls.
   b Javed and Oprah share £65 in the ratio 4:9. How much money do they each receive?
   ANSWER: Javed receives £12 and Oprah receives £30.

> **investigation**
>
> Meredith wants to design a flag using two colours — blue and green. The flag must be coloured blue to green in the ratio 3:5.
> **a** Draw a rectangle 5 cm by 16 cm.
>   Colour the flag blue and green in the ratio 3:5.
> **b** Draw a different rectangle 12 cm by 8 cm.
>   Colour the flag blue and green in the ratio 3:5.
> **c** How can you tell which sizes of rectangular flags, drawn on squared paper, can easily be coloured in blue and green in the ratio 3:5?

Division in a given ratio   **183**

Extension

Challenge more able students to develop real-life word-based problems that could be used in an assessment for a parallel group. They must provide a mark scheme.

Exercise 12b commentary

Take opportunities to use the real-life scenarios in this exercise to develop the students as functional mathematicians (CT).

Question 1 – Simple calculations that can be done mentally.

Questions 2 and **3** – Similar to the example (1.1, 1.3).

Question 4 – If students struggle here, support them in identifying key information (1.1, 1.3).

Question 5 – This question highlights the value of the simple checking strategy promoted in the discussion and emphasises the importance of explaining thinking. Students could compare their reasoning with a friend. (RL) (1.1, 1.3, 1.4)

Investigation – Here students investigate links between ratio and fractions of areas.

Assessment Criteria – Calculating, level 6: divide a quantity into two or more parts in a given ratio.

Links

The Golden Ratio occurs in mathematics, art and in nature and is calculated as 1 : 1.618 (to 3 dp). It can be used to divide an object into two parts so that the ratio of the smaller part to the larger part is the same as the ratio of the larger part to the whole object. The ratio is used in architecture to produce buildings of aesthetically pleasing proportions. There are pictures of buildings built on the Golden Ratio at http://goldennumber.net/architecture.htm

- Use direct proportion in simple contexts   (L5)   *Useful resources*
- Use the unitary method to solve simple problems
  involving ratio and direct proportion   (L6)

· ·

Starter – Grandad Bob

Bob wanted to share £5500 between his 4 grandchildren.
He decided to give the money in the ratio of their ages.
Simon was 16, Lucy and Jo were both 12 and Steven was 10.
Ask students how much money each grandchild received. (£1760, £1320, £1320, £1100)

Can be extended by students making up their own ratio problems.

Teaching notes

The material throughout this spread is related to real-life everyday application of the mathematics being learned. Students can be encouraged to look for examples of their own from everyday experience and equally to identify cases where increases are not in direct proportion (EP).

The unitary method of solving problems will need to be supported for some students, although all will be able to use doubling and halving strategies. It may be helpful to use the spider diagram approach to illustrate which values are easy to calculate for any given example. Students will be familiar with this strategy from pervious work on fractions and percentages and this may help to make connections (CT).

To embed functional skills, discuss how this work can be used to make comparisons of such things as nutritional value or value for money.

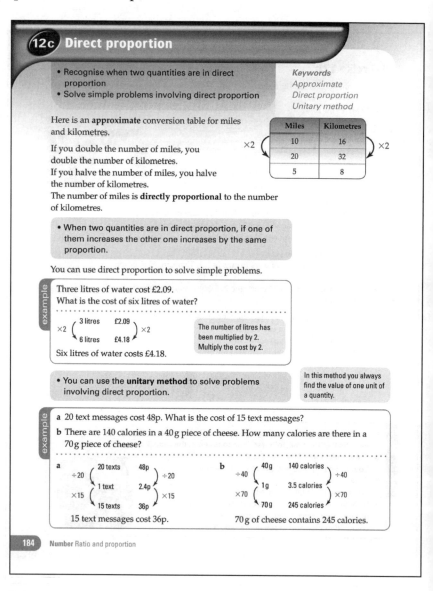

Plenary

Link to real-life situations and invite students to suggest examples where two quantities may not increase in direct proportion. For example, with special offers on food, drinks, text message costs, *etc*. Why would this happen?

Simplification

Use spider diagrams to illustrate how some values can be easily calculated to increase confidence and make links with fractions and percentages (CT).

Exercise 12c

1 48 bags of crisps cost £11.04. Work out the cost of
  a 24 bags of crisps   **b** 12 bags of crisps   **c** 1 bag of crisps
  d 5 bags of crisps   **e** 50 bags of crisps   **f** 6 bags of crisps.

2 Here are three offers for text messages on a mobile phone.
In which of these offers are the numbers in direct proportion?
In each case explain and justify your answers.

D-Mobile

Text messages	Cost (£)
10	£0.40
20	£0.80
40	£1.60

Yellow

Text messages	Cost (£)
5	£0.19
20	£0.72
45	£1.56

Codaphone

Text messages	Cost (£)
20	£0.68
50	£1.70
120	£4.08

3 Use direct proportion to solve each of these problems.
  a 4 apples cost 92p. What is the cost of 12 apples?
  b 28 g of cashew nuts contain 14 g of fat.
    How many grams of fat are there in 42 g of cashew nuts?

> A typical 28 g serving of cashew nuts contains 14 g of fat

  c 200 g of chips contain 500 calories.
    How many calories are there in 40 g of chips?
  d A recipe for 4 people uses 500 g of mushrooms.
    i What weight of mushrooms is needed for 7 people?
    ii How many grams of mushrooms are needed for 13 people?

4 Solve each of these problems.
  a 3 litres of lemonade costs £1.11.
    What is the cost of **i** 9 litres of lemonade   **ii** 8 litres of lemonade?
  b 4 identical pans have a capacity of 14 litres. What is the capacity of 7 of these pans?
  c 16 pencils cost £1.92. What is the cost of 7 pencils?
  d £3 is worth 42 Chinese Yuan. How much is £25 worth in Chinese Yuan?
  e 35 litres of petrol costs £38.15. What is the cost of 40 litres of petrol?
  f A recipe for 4 people needs 300 g of pasta. How much pasta is needed to make the recipe for 7 people?

challenge

Use direct proportion to copy and complete this approximate conversion table for converting between kilograms and pounds.

How many pounds make 1 kg?
How many kilograms make 1 lb?

Kilograms (kg)	Pounds (lb)
1	
	4
5	11
10	
23	
	110

Extension

Invite more able students to apply their skills to investigate value for money between various mobile phone packages (IE).

Exercise 12c commentary

Question 1 – This question could be used as a whole class activity. Parts **c**–**e** naturally suggest using the unitary method (1.3).

Question 2 – The explanation of how students find their answers is particularly important (CT) (1.2, 1.4).

Questions 3 and **4** – Real-life settings for word problems. Students may need to be reminded to use the unitary method.

Challenge – Ask students to explain the strategy they used to decide which parts of the table to complete first.

Assessment Criteria – Calculating, level 6: solve problems involving ratio and direct proportion.

Links

As part of the design process for a product, manufacturers draw up a list of all the components incorporated in the assembled product. This is called a parts list. The manufacturer decides how many of the product he is going to build, and then orders the number of parts required. The number he needs is in direct proportion to the number on the parts list. There are examples of parts lists at http://www.turbocharged.com/catalog/parts_list.html and at http://www.wellsdental.com/Techbull/U801/u801.htm

- Understand the relationship between ratio and proportion (L5)
- Apply understanding of the relationship between ratio and proportion (L6)

Useful resources

Starter – Emergency!

Ask students to arrange the digits 1 to 9 to make 3 three digit numbers that will add up to 999. Challenge students to see who can find the greatest number of different ways this can be done.

(One possible way is $498 + 375 + 126 = 999$)

Teaching notes

This spread clarifies the terms ratio and proportion providing illustration through worked examples. The emphasis in the spread is on real-life use of this learning and students can be encouraged to look for or generate examples of their own to provide further illustration and to develop increasing responsibility for their learning.

Students can be encouraged to work collaboratively here and share ideas to clarify understanding. This also provides the opportunity to explore issues or problems from different perspectives (IE) and to generate ideas and explore possibilities (CT).

Again, emphasise connections with students' knowledge of fractions throughout.

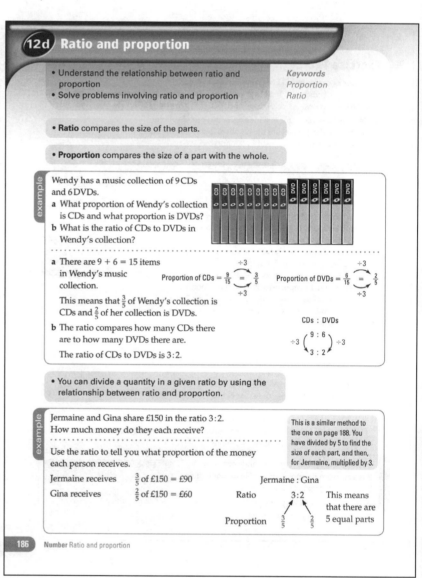

Plenary

Take whole class feedback on the challenge activity to discuss examples that students have generated in pairs.

Simplification

Make further use of diagrams to offer visual clues, as in the first question of the exercise, to support students who struggle with this work.

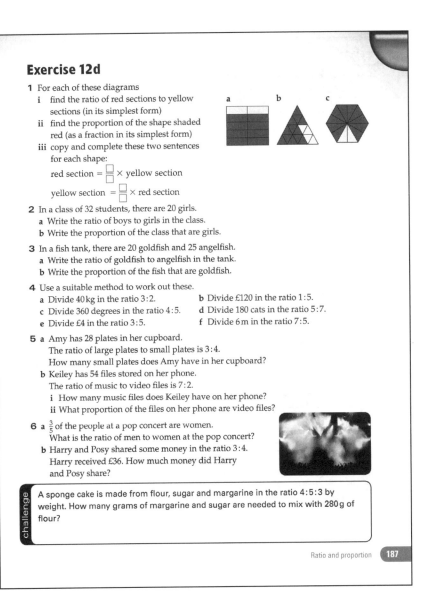

Exercise 12d

1 For each of these diagrams
  i find the ratio of red sections to yellow sections (in its simplest form)
  ii find the proportion of the shape shaded red (as a fraction in its simplest form)
  iii copy and complete these two sentences for each shape:

  red section $= \dfrac{\square}{\square} \times$ yellow section

  yellow section $= \dfrac{\square}{\square} \times$ red section

2 In a class of 32 students, there are 20 girls.
  a Write the ratio of boys to girls in the class.
  b Write the proportion of the class that are girls.

3 In a fish tank, there are 20 goldfish and 25 angelfish.
  a Write the ratio of goldfish to angelfish in the tank.
  b Write the proportion of the fish that are goldfish.

4 Use a suitable method to work out these.
  a Divide 40 kg in the ratio 3 : 2.   **b** Divide £120 in the ratio 1 : 5.
  c Divide 360 degrees in the ratio 4 : 5.   **d** Divide 180 cats in the ratio 5 : 7.
  e Divide £4 in the ratio 3 : 5.   **f** Divide 6 m in the ratio 7 : 5.

5 a Amy has 28 plates in her cupboard.
    The ratio of large plates to small plates is 3 : 4.
    How many small plates does Amy have in her cupboard?
  b Keiley has 54 files stored on her phone.
    The ratio of music to video files is 7 : 2.
    i How many music files does Keiley have on her phone?
    ii What proportion of the files on her phone are video files?

6 a $\frac{3}{5}$ of the people at a pop concert are women.
    What is the ratio of men to women at the pop concert?
  b Harry and Posy shared some money in the ratio 3 : 4.
    Harry received £36. How much money did Harry and Posy share?

> **challenge**
> A sponge cake is made from flour, sugar and margarine in the ratio 4 : 5 : 3 by weight. How many grams of margarine and sugar are needed to mix with 280 g of flour?

Ratio and proportion **187**

Extension

Involve more able students in a mini project using these skills, possibly involving some aspect of school life, such as comparing exam results for boys and girls (SM).

Exercise 12d commentary

Question 1 – Similar to the first example. The diagrams should support understanding and clarify the distinction between ratio and proportion (1.2).

Questions 2 and **3** – Similar to question **1** but without a diagram supplied.

Question 4 – Students may need to be reminded that they will need to change the units in parts **e** and **f** (1.3).

Question 5 – Similar to the second example. Students may need support to establish a method of approach. This may be used effectively as a whole class discussion activity (EP) (1.1, 1.3).

Question 6 – Use this question to develop student dialogue around misunderstandings. Tell students that some said the answer to part **a** was 3 : 5. Is this correct? (RL) (1.1, 1.3).

Challenge – A triple ratio problem from everyday, real life. Ask students to add some challenges of their own and share these with a friend (1.1, 1.3).

Assessment Criteria – Calculating, level 6: divide a quantity into two or more parts in a given ratio and solve problems involving ratio and direct proportion.

Links

Proportional representation is a system of election where the number of seats given to a particular party is proportional to the number of votes that it receives. In an election in the UK, only the winning candidate in each constituency becomes a Member of Parliament. Smaller parties might win a sizeable proportion of the vote without winning any seats.

- Use the equivalence of fractions, decimals and percentages to compare proportions   (L6)
- Calculate percentages and find the outcome of a given percentage increase or decrease   (L6)

Useful resources

Starter – Percentage pairs

Write the following list of percentage calculations on the board:

50% of 28, 30% of 75, 15% of 70,   90% of 25, 20% of 75, 10% of 350, 56% of 25, 40% of 30, 25% of 60, 100% of 35, 10% of 120, 25% of 42.

Challenge students to match up the pairs in the shortest possible time. Hint: 90% of 25 = 25% of 90

Can be differentiated by the choice of percentages.

Teaching notes

The worked examples here illustrate a variety of methods that students should discuss in order to support learning. The mental method will be worth particular attention as students need to have a solid understanding of this in order to fully develop understanding.(CT)

Explicit links are made throughout to the equivalence of fractions, decimals and percentages and these need to be highlighted to students. This particularly supports the key processes in number, particularly encouraging analysing through the use of appropriate mathematical procedures.

The work in which students are engaged and the suggested approaches foster opportunities for the exploration of issues or problems from different perspectives (IE) and to collaborate with others to work towards common goals (TW).

The work in this exercise focuses on skills that strongly support the students in becoming functional mathematicians (EP).

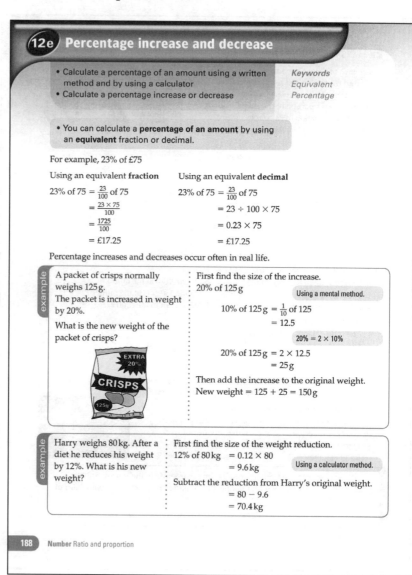

12e Percentage increase and decrease

- Calculate a percentage of an amount using a written method and by using a calculator
- Calculate a percentage increase or decrease

Keywords
Equivalent
Percentage

- You can calculate a **percentage of an amount** by using an **equivalent** fraction or decimal.

For example, 23% of £75

Using an equivalent fraction

23% of $75 = \frac{23}{100}$ of 75

$= \frac{23 \times 75}{100}$

$= \frac{1725}{100}$

$= £17.25$

Using an equivalent decimal

23% of $75 = \frac{23}{100}$ of 75

$= 23 \div 100 \times 75$

$= 0.23 \times 75$

$= £17.25$

Percentage increases and decreases occur often in real life.

example

A packet of crisps normally weighs 125 g. The packet is increased in weight by 20%.

What is the new weight of the packet of crisps?

First find the size of the increase.
20% of 125 g

Using a mental method.

10% of $125\,g = \frac{1}{10}$ of 125
$= 12.5$

20% = 2 × 10%

20% of $125\,g = 2 \times 12.5$
$= 25\,g$

Then add the increase to the original weight.
New weight = 125 + 25 = 150 g

example

Harry weighs 80 kg. After a diet he reduces his weight by 12%. What is his new weight?

First find the size of the weight reduction.
12% of $80\,kg = 0.12 \times 80$
$= 9.6\,kg$

Using a calculator method.

Subtract the reduction from Harry's original weight.
$= 80 - 9.6$
$= 70.4\,kg$

188   **Number** Ratio and proportion

Plenary

Ask students to write three word problems involving percentage increase and decrease and to swap with a friend. They need to know the answers so that they can swap back for marking (RL). By observation collect a few of the most interesting examples for use as a starter activity next lesson.

Simplification

Support students experiencing difficulty here to consolidate mental methods using spider diagrams and involving simple quantities of which to calculate percentages.

Exercise 12e

1 Calculate these percentages using a mental or informal written method.

   a 10% of 80   **b** 20% of £60   **c** 5% of 180 g
   d 15% of £300   **e** 40% of 320 N   **f** 25% of 160 mm
   g 55% of £2800   **h** 65% of £88

2 Calculate these using a suitable method.

   a 13% of £50   **b** 42% of 67   **c** 50% of 86 kg
   d 16% of $24   **e** 37% of £45   **f** 17.5% of 80 km
   g 65% of £230   **h** 95% of 19.2 kg

3 **a** Increase £50 by 15%.   **b** Decrease £50 by 15%.
   c Increase 160 m by 25%.   **d** Decrease 360° by 10%.
   e Increase 60 kg by 5%.   **f** Decrease £1700 by 13%.
   g Increase £240 by 45%.   **h** Decrease 240 J by 36%.

4 **a** Archenal football stadium has 43 400 seats. It is rebuilt with an increased seating of 45%. How many seats are there at the new stadium?
   b A top secret file of size 208 kB is saved on a computer. After some information is stolen and deleted from the file, it is decreased in size by 5%. What is the new size of the file?

   c A bottle of olive oil holds 560 ml. It is increased in capacity by 15%. What is the new capacity of the bottle?

investigation

Here are five items for sale in Cheapos shop with their original prices.

In a 21-day sale, Cheapos reduce all their prices by 21%.
a Calculate the sale price of each of the items.
b How could the sales assistants work out the sale price of an item after a 21% reduction using just a single multiplication?
c Investigate increasing and decreasing the prices by different percentages. Try to find a single multiplication which works out the

Extension

Encourage more able students to use a calculator with more complex calculations to find and use a multiplier for calculating percentage increase and decrease.

Exercise 12e commentary

Question 1 – If necessary use spider diagrams to remind students how to calculate simple percentages. In part **e**, N stands for Newton, the SI unit of force.

Question 2 – Some of these percentage calculations are more challenging than in question **1** but it is best to persist with mental strategies as far as possible because of the embedding of understanding that this promotes.

Question 3 – Again students should use mental methods to find the change. They may need to be reminded to add on or subtract this change. In part **h**, J stands for Joule the SI unit of energy.

Question 4 – Similar to the two examples. Ask students to work in pairs and to explain their approach to each other for each example.

Investigation – A practical scenario that guides students towards an efficient way of calculating (SM).

Assessment Criteria – Calculating, level 6: calculate percentages and find the outcome of a given percentage increase or decrease.

Links

A map showing the percentage change in the number of people over the state pension age (65 for men and 60 for women) by area of the UK can be found at http://www.statistics. gov.uk/cci/nugget.asp?id=875

Which colour indicates areas where the proportion of people of pensionable age has decreased? (dark grey). What two factors could cause this proportion to increase in an area? (More people over the state pension age move into the area, perhaps to retire, or younger people leave the area, perhaps to find work elsewhere).

- Use the equivalence of fractions, decimals and percentages to compare proportions

Useful resources

(L6)

Starter – Paper round

Sam earns £20 each week doing a paper round.
As a bonus Sam was offered a choice of three options.

an extra lump sum of £8 for one week

an extra £2 each week for four weeks

a pay rise of 50% for one week followed by a pay cut of 50% the following week.

Ask students what choice Sam should make and why?

See also the plenary of spread **12e**.

Teaching notes

The worked examples in this unit make explicit the links between fractions, decimals and percentages and their place in comparing proportions, thereby supporting the development of the process skills in number, representing, where knowledge of equivalent forms is essential.

Students should always be encouraged to explain and justify their thinking, both to the teacher and to each other. The development of student dialogue and purposeful mathematical talk fosters confidence in all abilities of students and facilitates learning effectively (EP).

Comparison of proportions is a vital skill in becoming functional in mathematics. Encourage students to take ownership and responsibility for their own learning through inviting them to suggest examples from their own life experiences where this skill is needed (CT, SM).

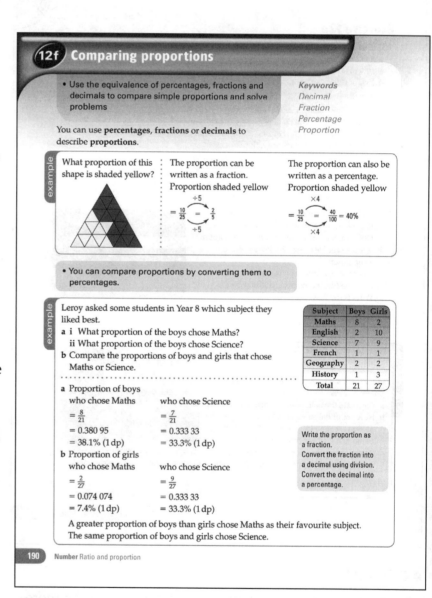

Plenary

Refer students back to the **problem solving** question in the exercise and initiate whole class discussion around explaining how to tackle this activity. Allow students to listen to each others' justifications and to contribute to a whole class 'best answer'.

Simplification

Where students experience difficulty, scaffold understanding with additional examples that can be supported with diagrams as in question **2** of the exercise.

Exercise 12f

1 Convert these fractions into percentages using a calculator where appropriate.

  a $\frac{7}{10}$ **b** $\frac{23}{50}$ **c** $\frac{14}{25}$ **d** $\frac{5}{4}$ **e** $\frac{17}{40}$

  f $\frac{5}{12}$ **g** $\frac{33}{35}$ **h** $\frac{5}{7}$ **i** $\frac{4}{5}$ **j** $\frac{13}{20}$

2 Write the proportion of each of these shapes that is shaded red. Write each of your answers as
  i a fraction in its simplest form
  ii a percentage (to 1 dp where appropriate).

  a   **b**   **c**

3 Express each of your answers to these problems
  i as a fraction in its simplest form
  ii as a percentage (to 1 dp where appropriate).
  a Morton scores 49 out of 70 in his English test.
    What proportion of the test did he answer correctly?
  b Class 8X3 has 33 students. 21 of these students are boys.
    What proportion of the class are girls?

4 a Hilary McGoalmachine has scored 23 goals in 35 games for her club.
    Jodie Goalpoacher has scored 20 goals in 29 games for the same club. Who is the better goal scorer? Explain and justify your answer.
  b Tina put £120 into a savings account. After one year the interest was £8. Harriet put £90 into a savings account. After one year the interest was £7. Who had the better deal?

> **Did you know?**
>
> As of 2008, the top goalscorer in the football Premier League since it started in 1992 is Alan Shearer with 260 goals in 441 appearances. However that doesn't give him the top scoring rate!

problem solving

Here are the Summer exam results of some students in History, Geography and Religious Studies.
For each student, show clearly in which subject they did the best.
Explain and justify your answers.

Name	History (60 marks)	Geography (70 marks)	Religious Studies (80 marks)
Zak	24	30	33
Wilson	20	14	16
Yvonne	45	50	52
Ulf	55	64	73
Veronica	30	38	33

Comparing proportions   **191**

Extension

Support learning in more able students through use of investigations such as in the **problem solving** question. For example, invite students to look at their own test results (IE).

Exercise 12f commentary

Question 1 – Practice converting fractions to percentages: a suitable method is used in the first example. Invite comment as to which ones are most efficiently converted without a calculator (1.3, 1.4).

Question 2 – Similar to the first example (1.4).

Question 3 – Similar to question **2** but with the removal of the visual scaffolding (1.3, 1.4).

Question 4 – Here, real-life examples to investigate develop students' functional mathematics. Again, the key element of the question is 'how do you know?' (1.1)

Problem Solving – Students may need their attention drawn to the fact that the marks are out of different totals for each subject. The explanation and justification of answers is very important (EP).

Assessment Criteria – NNS, level 6: use the equivalence of fractions, decimals and percentages to compare proportions.

Links

For every 100 girls born in the UK, there are around 105 boys. What ratio is this of male to female? What percentage of all babies born in the UK are boys? In 2001 the population of the UK was 28.6 million males and 30.2 million females. What percentage of the population is male? Why is this different to the percentage at birth? (life expectancy for females is longer than for males)

1 Write each of these ratios in its simplest form.

a 6 : 18   b 5 : 15   c 8 : 12   d 6 : 15
e 35 : 28   f 32 : 56   g 63 : 90   h 70 : 60
i 24 : 100   j 24 : 104   k 128 : 256   l 64 : 176

2 Write each of these ratios in its simplest form.

a 20 cm : 3 m   b 50 p : £4   c 65 mm : 8 cm   d 70 cl : 2 litres
e 7 km : 700 m   f 1900 g : 4 kg   g 1 hr 20 mins : 80 mins   h 70 p : £1.70

3 Solve these problems.

a In a fishing club the ratio of men to women is 7 : 2. There are 84 men in the club. How many women are there?

b In a school the ratio of boys to girls is 8 : 9. If there are 624 boys at the school, how many girls are there?

c A map has a scale of 1 : 10 000.
    i What is the distance in real life of a measurement of 3 cm on the map?
    ii What is the distance on the map of a measurement of 5 km in real life?

4 Divide these quantities in the ratios given.

a Divide 65 km in the ratio 6 : 7.   b Divide £225 in the ratio 8 : 7.
c Divide 256 MB in the ratio 3 : 5.   d Divide 4500 N in the ratio 4 : 5.
e Divide 3 minutes in the ratio 4 : 5.   f Divide £2 in the ratio 7 : 13.

5 Solve these problems.

a In a running club the ratio of boys to girls is 5 : 3. There are 96 children in total at the club. How many girls are there?

b Sam and Siobhan share £2400 in the ratio 7 : 5. How much money does Sam receive?

c A pizza is made with dough and vegetables in the ratio 3 : 5. The total weight of the pizza is 320 g. What weight of dough has been used to make the pizza?

6 Here are three offers for different types of bread. In which of these offers are the numbers in direct proportion? In each case explain and justify your answers.

a **Wholemeal loaves**

Weight of bread	Cost
300 g	£0.45
400 g	£0.60
800 g	£1.20

b **Croissants**

Weight of bread	Cost
50 g	£0.32
125 g	£0.75
200 g	£1.25

c **Currant teacakes**

Weight of bread	Cost
40 g	£0.24
100 g	£0.60
240 g	£1.44

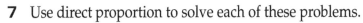

12c

7 Use direct proportion to solve each of these problems.

    a 5 pears cost 82p. What is the cost of 15 pears?

    b 150 g of crisps contain 240 calories. How many calories are there in 50 g of crisps?

    c A recipe for 6 people uses 420 g of flour.

      i What weight of flour is needed for 7 people?

      ii How much flour is needed for 3 people?

    d 5 litres of water costs £1.45.

      i What is the cost of 7 litres of water?

      ii What is the cost of 17 litres of water?

12d

8 **a** Steve and Jenny break a 120 g chocolate bar into two pieces. Steve has $\frac{3}{8}$ of the bar and Jenny has the rest. What is the ratio of Jenny's piece of the bar to Steve's piece of the bar?

    b $\frac{8}{9}$ of the people who attended a football match were men. What was the ratio of men to women at the football match?

    c Shirley and Hanif share some money. Shirley receives $\frac{2}{5}$ of the money and Hanif receives £66. How much money did Shirley and Hanif share?

12e

9 **a** Increase £30 by 10%.   **b** Decrease 700 euros by 5%.

    c Increase 8 miles by 20%.   **d** Decrease 180° by 15%.

    e Increase 280 kg by 25%.   **f** Decrease £100 000 by 3%.

    g Increase 250 rabbits by 30%.   **h** Decrease 2500 kJ by 22%.

    i Increase 70 g by 9%.   **j** Decrease £1.80 by 35%.

10 **a** Chelski football stadium has 71 440 seats. It is rebuilt with an increased seating of 15%. How many seats are there at the new stadium?

    b A jar of jam holds 370 g. It is increased in capacity by 23%. What is the new capacity of the jar? (Give your answer to the nearest gram.)

    c A memory stick normally cost £36. It is reduced in price in a sale by 17.5%. What is the sale price of the memory stick?

12f

11 **a** Dan scores 62% in his Maths exam, $\frac{37}{60}$ in his Science exam and $\frac{29}{50}$ in his English exam. In which subject did Dan do the best?

    b Megan and Jane play tennis. Last week Megan played 7 matches and won 5 of them. Last month Jane played 20 matches and won 14 of them. Who is the better tennis player? Explain and justify your answer.

- Give accurate solutions appropriate to the context or problem (L6)
- Relate the current problem and structure to previous situations (L5)

Useful resources
Bread and cake recipes

•••

Background

Many students probably give little thought to all the things that happen before loaves of bread appear for sale in the supermarket or local bakers. This case study considers a few of the processes involved in running a small bakery, in particular looking at sales figures and related income, baking times and staff working hours (EP).

Teaching notes

Look at the case study and have a general discussion about the running of a small bakery, focusing to begin with on the things the bakery will have to think about to ensure that it makes a profit:

Where does the bakers income come from?

What costs will there be in running a bakery?

Make sure that students think about things such as the cost of ingredients, wages, rent for the premises, electricity bills and so on (TW).

Then move on to the details in the case study, starting with the sales record. Give the students a few minutes to work out the weekly totals for the sales and to consider the ratio of sales of white loaves to wholemeal loaves (1.3).

When complete, look together at the question on the sticky note below the pie chart and, checking that they understand what the pie chart is showing, discuss how the ratio of 2 white to 1 wholemeal becomes 2/3 white and 1/3 wholemeal when expressed in fractions. Describing the sales of wholemeal loaves as 1 in every 3 loaves sold and the sales of white loaves as 2 in every 3 loaves sold might be helpful with this (1.1).

The ready reckoner for the ingredients of the white loaf also gives plenty of opportunities for looking at proportions, for example

If 10 loaves need 20g of salt, how much salt will 5 loaves need? 15 loaves? 30 loaves?

If 2 loaves need 960ml of water, how much will 20 loaves need? 10 loaves? 5?

Maths life — Daily bread

There are lots of things to think about when running a business. Here are just some of the things to consider when running a small bakery.

THE BAKERY
1234 123 456, 123 THIS ROAD

OPENING TIMES
Mon - Fri: 9:00am to 5:30pm, Sat: 9:30am to 4:00pm
Sun: Closed

PRICES
White loaf: 98p, Wholemeal: £1.20

You need to think about your sales and income

Total number of loaves sold each day

Day	White	Wholemeal
Mon	78	42
Tues	84	38
Weds	102	66
Thurs	72	66
Fri	90	48
Sat	186	54
Weekly total		66

Percentage of white and wholemeal loaves sold

white 66%
wholemeal 34%

Why does the pie chart show wholemeal as about 1/3?

Roughly how many white loaves are sold for every wholemeal loaf?

- How much money do the sales of the white loaf raise in the week?
- How much money do the wholemeal loaves raise in the week?
- Roughly what percentage of the income is from wholemeal bread? Why is this higher than 34%?

194   MathsLife

Teaching notes continued

Ask the students to complete the ready reckoner (1.3). If ICT is available this is an activity that is suited to being done using a spread sheet. This would provide an opportunity to investigate other recipes and to also discuss the use of 'algebraic' formulae in computing (CT).

The question about when the bakers will need to start making the first batch of bread on a Saturday is a useful one for considering the working hours of the bakery. Discuss how, with shops such as a clothes shop, members of staff will only need to arrive a short time before opening time as all the stock will already be in the shop. Compare this with a bakers shop that makes its own bread where some members of staff will have to start considerably earlier than the opening time in order to have the bread ready for the first customers. Look at the spread and ask (IE) (1.4):

How can you work out when they will need to start making the bread? What information will you need?

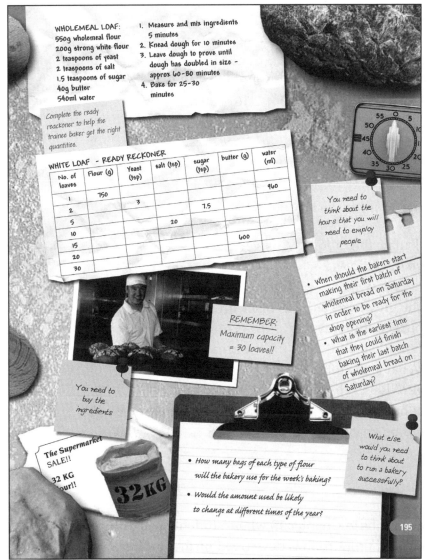

Establish that you need to know when the shop opens and how long it takes to make the bread.

Where is the information about how long it takes to make the bread?

Look briefly at the various processes described in the recipe for wholemeal bread. Give the students a few minutes to work out the time that the bakers will have to start and, when they have finished, ask them how they worked it out. Some may have started at the opening time and worked backwards one process at a time to reach the start time. Others might have found the total time for all the processes and subtracted that from the start time. Ask how they dealt with processes that had a spread of time such as 'bake for 25 - 30 minutes'. A sensible decision would be to take the longer of the two times to make sure that the bread is ready in time (TW, RL) (1.2, 1.5).

Extension

The sales figures given in the case study are real and were taken from a local baker in a small market town. The students could research sales figures in different types of stores - small bakers and supermarkets and look at aspects such as the balance of sales between white bread and wholemeal or the daily figures to see if there is any pattern to the daily sales, for example, do Saturdays always have most sales? (SM)

12 Summary

- Understand simple ratio   Level 5

- Divide a quantity into two or more parts in a given ratio and
  solve problems involving ratio and direct proportion   Level 6

- Calculate percentages and find the outcome of a given percentage
  increase or decrease   Level 6

- Use the equivalence of fractions, decimals and percentages
  to compare proportions   Level 6

Question commentary

Example	
The example illustrates a typical problem on percentage decrease. Encourage students to estimate the new price before calculating. Some students may calculate the reduction rather than the final price. Ask questions such as "Is there a single number that you could multiply the original price by to find the new price?"	18% of £12 = 0.18 × 12 = £2.16 12 − 2.16 = £9.84

Past question	
The question requires students to calculate the ratio of two areas. Some students may give the ratio as 2 : 2 as there are two parts of each. Others may have difficulty in calculating the areas of the grey parts as the dimensions are not explicit. Encourage students to make a rough visual estimate of the ratio before calculating. Ask questions such as "What is the total length of each side of the large square?" Remind students that units are not required when writing ratios and that the final answer should be given in its simplest form.	**Answer** **Level 6** **2** Black area $= 9 \times 9 + 3 \times 3$ $= 90 \text{ cm}^2$ Grey area $= 9 \times 3 + 9 \times 3$ $= 54 \text{ cm}^2$ Black area : Grey area $= 90 : 54$ $= 10 : 6$ $= 5 : 3$

Development and links

This chapter links with work on real life graphs in Chapter 13 where students will draw and use conversion graphs. The topic of ratio and proportion is developed further in Year 9.

Ratio and proportion are important in many areas of the curriculum. Students will use ratio and proportion when scaling formulae in science, converting recipe quantities in food technology, mixing paint colours in art and working with maps in geography. Percentage increase is used in business studies and economics to calculate price rises and VAT.

Objectives

- Simplify or transform linear expressions by collecting like terms **5**
- Multiply a single term over a bracket **5**
- Use formulae from mathematics and other subjects **5**
- Substitute integers into simple formulae, including examples that lead to an equation to solve **5**
- Derive simple formulae **5**
- Construct and solve linear equations with integer coefficients (unknown on either or both sides, without and with brackets) using appropriate methods **6**
- Generate points in all four quadrants and plot the graphs of linear functions, where y is given explicitly in terms of x, on paper and using ICT **6**
- Recognise that equations of the form $y = mx + c$ correspond to straight-line graphs **6**
- Construct linear functions arising from real-life problems and plot their corresponding graphs **6**
- Discuss and interpret graphs arising from real situations e.g. distance-time graphs **6**
- Find the midpoint of the line segment AB, given the coordinates of points A and B **6**

Introduction

The focus of this chapter is on consolidating, developing and extending algebra topics introduced in earlier chapters. Students simplify expressions, derive and use simple formulae and solve further equations including those involving brackets and directed numbers. They plot graphs of linear functions, recognising when the equation result in a straight line. They interpret linear and non-linear real-life graphs and draw and use linear real-life graphs. The topic concludes by finding the gradient and midpoint of a straight line.

The student page discusses Ohm's Law. Scientists and engineers use formulae to explain what is happening around them and to predict what will happen under given conditions. Ohm's Law was discovered by the German physicist George Ohm who published the results of his work in 1827. There is more information about Georg Ohm at http://en.wikipedia.org/wiki/Georg_Ohm

Fast-track
13c, d, e, e², f, g, h

Level

MPA

1.1	13a, b, c, d
1.2	13b, e, f, g, h
1.3	13a, b, c, d, e, e², f, h
1.4	13c, d, e, f, h
1.5	13d, g

PLTS

IE	13a, b, f, h
CT	13b, c, e, e², f, g, h
RL	13a, c
TW	13d, g
SM	13a, b, e, e², f
EP	13a, b ,d, f, g

Extra Resources

13	Start of chapter presentation
13c	Consolidation sheet
13d	Worked solution: Q1c
13e	Starter: BIDMAS multichoice
13e	Worked solution: Q2a
13e	Consolidation sheet
13f	Animation: Interpreting graphs
13f	Consolidation sheet
13g	Animation: Real-life graphs
13g	Consolidation sheet
13h	Starter: Coordinates time challenge
13h	Animation: Gradients
13h	Consolidation sheet

Assessment: chapter 13

- Simplify or transform linear expressions by collecting like terms (L5)
- Multiply a single term over a bracket (L5)

Useful resources
Dictionaries

• •

Starter – Calculate 100

Ask students to calculate 100 using the digits 1 to 9 and any operation(s), for example, $123 - 4 - 5 - 6 - 7 + 8 - 9 = 100$.

Students score a point for each different calculation.
Award bonus points if the digits are kept in numerical order as in the example!

Teaching notes

The work in this spread links with students' previous work on expressions and some of the examples in the exercise offer ideal recapitulation activities that could be used for oral discussion.

A key point to discuss with students is the difference between adding terms and multiplying them – this appears in question **1**.

The text addresses dividing expressions. This merits discussion with and between students as a frequent mistake is to divide one part of the expression only. The use of brackets in the illustration draws attention to this (RL).

The example showing expansion of more than one pair of brackets might be extended to consider one of the examples from question **4**, where there is a negative number after simplification.

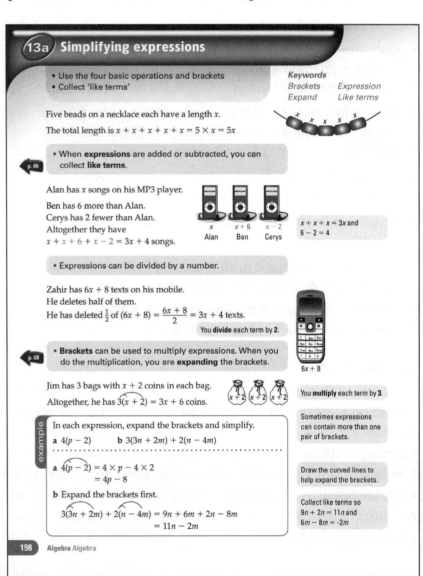

Plenary

Initiate class discussion about the second part of the challenge activity. Here, the aim would be to explain and justify the strategy used. Ask students to discuss and compare their thinking in pairs and then take feedback (EP).

Simplification

Where students struggle with this work, they are often overwhelmed by the number of examples and lose confidence as they repeatedly make errors. Encourage more self-reliance by making question **4**, for example, a paired activity where students answer three questions each and then talk through their work with their friend. This discussion process often helps to clarify points about which students are unsure (RL).

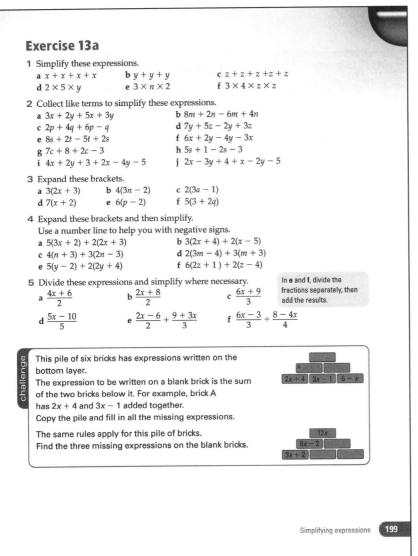

Extension

Consider providing examples that have negative numbers before the brackets for students who are confident here (IE).

Exercise 13a commentary

Question 1 – A recap of basic simplification; possibly best completed orally. (1.3)

Question 2 – An extension of question **1** using addition and subtraction in the same expression together with a mixture of symbols and numbers (1.3).

Question 3 – Similar to the first part of the example. The subtractions will catch out some students (1.3).

Question 4 – Similar to the second part of the example. Again, check that negative numbers are correctly treated (1.3).

Question 5 – Check that students divide both terms and do not write, for example, $(4x + 6) \div 2 = 2x + 6$ instead of $2x + 3$. In parts **e** and **f**, students will need to perform the two divisions before adding (1.3).

Challenge – This activity offers additional practice in collecting like terms. The second part offers the further challenge of finding the value of the missing brick to make the expression true (SM) (1.1, 1.3).

Links

Bring in some dictionaries for the class to use. The word *term* is a noun and can have several different meanings. Ask the class to look up the word *term* and list the meanings. In which curriculum subjects would each meaning be most relevant?

- Substitute integers into simple formulae, including examples that lead to an equation to solve **(L5)**
- Use formulae from mathematics and other subjects **(L5)**
- Derive simple formulae **(L5)**

Useful resources

Starter – Algebraic sums

Draw a 4 × 4 table on the board to form 16 cells.

Label the columns with the terms: x, $3x$, $2y$, $-x$ and the rows with the terms: $2x$, $3y$, $5x$, y.

Ask students to fill in the table by adding the terms corresponding to each cell, for example, the top row in the table would read $3x$, $5x$, $2x + 2y$, x.

Can be differentiated by the choice of terms.

Teaching notes

This spread focuses on the application of students' skills in using and deriving formulae, with explicit links to other areas of mathematics and numerous real-life examples. The problems are word-based and students should, throughout, be encouraged to discuss and develop purposeful mathematical talk, sharing their thinking. To consolidate understanding, it is helpful to talk through how the formula connects with the information provided. This also supports students in connecting their own and others' ideas (CT).

The second example links the derivation of formulae with work on measures. Students are frequently faced with situations similar to this throughout KS3 and into KS4 and often find this difficult. It will be particularly beneficial to generate discussion here.

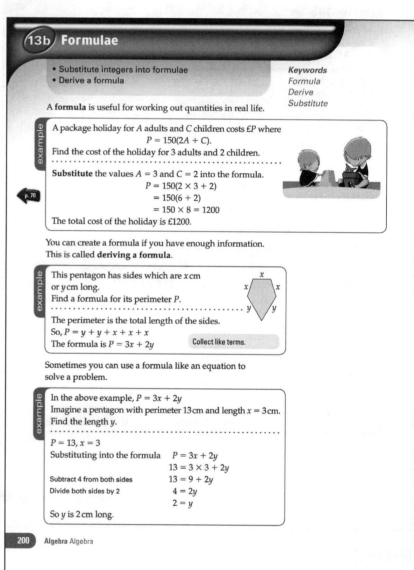

13b Formulae

- Substitute integers into formulae
- Derive a formula

Keywords
Formula
Derive
Substitute

A **formula** is useful for working out quantities in real life.

example

A package holiday for A adults and C children costs £P where
$$P = 150(2A + C).$$
Find the cost of the holiday for 3 adults and 2 children.
....

Substitute the values $A = 3$ and $C = 2$ into the formula.
$$P = 150(2 \times 3 + 2)$$
$$= 150(6 + 2)$$
$$= 150 \times 8 = 1200$$
The total cost of the holiday is £1200.

p. 70

You can create a formula if you have enough information. This is called **deriving a formula**.

example

This pentagon has sides which are x cm or y cm long.
Find a formula for its perimeter P.
....
The perimeter is the total length of the sides.
So, $P = y + y + x + x + x$
The formula is $P = 3x + 2y$ Collect like terms.

Sometimes you can use a formula like an equation to solve a problem.

example

In the above example, $P = 3x + 2y$
Imagine a pentagon with perimeter 13 cm and length $x = 3$ cm.
Find the length y.
....
$P = 13$, $x = 3$
Substituting into the formula $P = 3x + 2y$
$$13 = 3 \times 3 + 2y$$
Subtract 4 from both sides $13 = 9 + 2y$
Divide both sides by 2 $4 = 2y$
$$2 = y$$
So y is 2 cm long.

Plenary

Encourage students to reflect on their learning. Ask them to write down two things of which they were previously unsure but are now confident about and one thing of which they are still unsure. Discuss as a class and use this to inform future teaching.

Simplification

If students experience difficulty, establish whether this is related to their insecurity with formulae or, say, the knowledge of perimeter and/or area. Scaffold if necessary.

The questions in this exercise make explicit links with other areas of mathematics, and with real-life situations (EP).

Question 1 – A simple case similar to the first example. Ask students to describe the formula in words, using the information in the question to help.

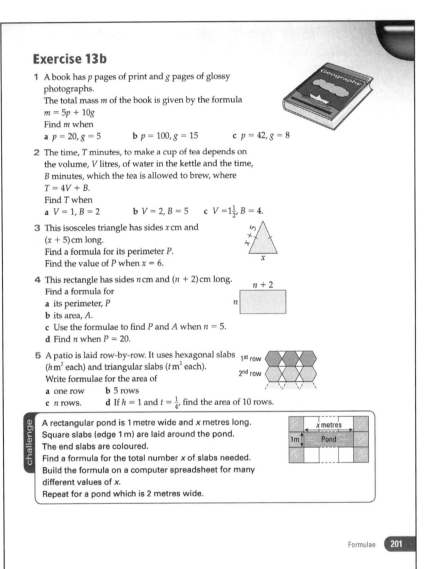

Exercise 13b

1 A book has p pages of print and g pages of glossy photographs.
The total mass m of the book is given by the formula
$m = 5p + 10g$
Find m when
a $p = 20, g = 5$ b $p = 100, g = 15$ c $p = 42, g = 8$

2 The time, T minutes, to make a cup of tea depends on the volume, V litres, of water in the kettle and the time, B minutes, which the tea is allowed to brew, where
$T = 4V + B$.
Find T when
a $V = 1, B = 2$ b $V = 2, B = 5$ c $V = 1\frac{1}{2}, B = 4$.

3 This isosceles triangle has sides x cm and $(x + 5)$ cm long.
Find a formula for its perimeter P.
Find the value of P when $x = 6$.

4 This rectangle has sides n cm and $(n + 2)$ cm long.
Find a formula for
a its perimeter, P
b its area, A.
c Use the formulae to find P and A when $n = 5$.
d Find n when $P = 20$.

5 A patio is laid row-by-row. It uses hexagonal slabs ($h\,\text{m}^2$ each) and triangular slabs ($t\,\text{m}^2$ each).
Write formulae for the area of
a one row b 5 rows
c n rows. d If $h = 1$ and $t = \frac{1}{4}$, find the area of 10 rows.

challenge
A rectangular pond is 1 metre wide and x metres long.
Square slabs (edge 1 m) are laid around the pond.
The end slabs are coloured.
Find a formula for the total number x of slabs needed.
Build the formula on a computer spreadsheet for many different values of x.
Repeat for a pond which is 2 metres wide.

Formulae **201**

Extension

More able students can be challenged with more complex examples involving measures to strengthen their understanding of links within mathematics.

Exercise 13b commentary

Question 2 – Also similar to the first example. Draw attention to the word description and ensure that students link this to the formula.

Question 3 – Similar to the second example. Ensure that students understand the significance of the fact that the triangle is isosceles (1.1).

Question 4 – Combines elements of all three examples. The question assumes prior knowledge of measures and will merit a particular focus, including whole class feedback (1.2).

Question 5 – Students are required to apply their skills in a real-life setting, supporting their development as functional mathematicians (IE) (1.1, 1.3).

Challenge – An additional real-life problem which will also develop students' competence in effective use of ICT (SM) (1.3).

Assessment Criteria – Algebra, level 5: construct, express in symbolic form, and use simple formulae involving one or two operations.

Links

The largest published book in the world is *Bhutan: A Visual Odyssey Across the Last Himalayan Kingdom* by Michael Hawley. Each copy weighs over 133 lbs and measures $60 \times 44 \times 6$ inches. What are the dimensions in metric units? (60 kg, approx. $152 \times 112 \times 15$ cm). Each book uses enough paper to cover a football field and over a gallon of ink. There is more information at http://news.bbc.co.uk/1/hi/world/south_asia/3324797.stm

- Construct and solve linear equations with integer coefficients (unknown on either or both sides, without [and with] brackets) using appropriate methods (e.g. inverse operations, transforming both sides the same way) (L6)

Useful resources

Starter – Five, one, four

Write FIVE − ONE = FOUR on the board.

Ask students to find what digits the letters E, F, I, N, O, R, U and V represent so that the calculation is true.

(One possible solution is E = 7, F = 1, I = 4, N = 5, O = 2, R = 0, U = 3, V = 8)

Challenge students to find as many ways as possible in a given time.

Discuss strategies used.

Teaching notes

The worked examples in the spread combine opportunities to connect with prior learning and to build on what students already know. The key processes in algebra are developed here, with particular emphasis on analysis, using appropriate mathematical procedures, as students gain confidence in manipulating expressions into different equivalent forms.

A major hurdle for many students is to understanding that it doesn't matter which side of the equation the x values are collected on and to appreciate that $4x = 5x − 13$ is the same as $5x − 13 = 4x$, as in the third worked example. Time spent discussing this and similar examples is worth the investment.

Similarly, as suggested in the plenary, allow time to look at deriving equations from word problems, such as those in questions **4** and **5**,.

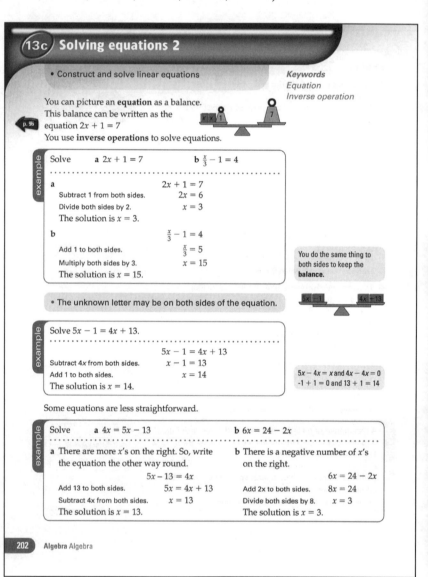

13c Solving equations 2

- Construct and solve linear equations

Keywords
Equation
Inverse operation

p. 96

You can picture an **equation** as a balance. This balance can be written as the equation $2x + 1 = 7$
You use **inverse operations** to solve equations.

example

Solve   **a** $2x + 1 = 7$ **b** $\frac{x}{3} - 1 = 4$

a
                            $2x + 1 = 7$
Subtract 1 from both sides. $2x = 6$
Divide both sides by 2. $x = 3$
The solution is $x = 3$.

b
                            $\frac{x}{3} - 1 = 4$
Add 1 to both sides. $\frac{x}{3} = 5$
Multiply both sides by 3. $x = 15$
The solution is $x = 15$.

You do the same thing to both sides to keep the **balance**.

- The unknown letter may be on both sides of the equation.

example

Solve $5x - 1 = 4x + 13$.
                          $5x - 1 = 4x + 13$
Subtract 4x from both sides. $x - 1 = 13$
Add 1 to both sides. $x = 14$
The solution is $x = 14$.

$5x - 4x = x$ and $4x - 4x = 0$
$-1 + 1 = 0$ and $13 + 1 = 14$

Some equations are less straightforward.

example

Solve   **a** $4x = 5x - 13$ **b** $6x = 24 - 2x$

a There are more x's on the right. So, write the equation the other way round.
                $5x - 13 = 4x$
Add 13 to both sides. $5x = 4x + 13$
Subtract 4x from both sides. $x = 13$
The solution is $x = 13$.

b There is a negative number of x's on the right.
                 $6x = 24 - 2x$
Add 2x to both sides. $8x = 24$
Divide both sides by 8. $x = 3$
The solution is $x = 3$.

Plenary

Refer students back to questions **4** and **5**. Consolidate understanding here as this is often a stumbling block for many students. Provide three or four simple equations presented using symbols and ask students to rewrite these as word sentences. When they are successful, invite them to generate an example of their own. Take feedback and select three examples to use as a starter activity next lesson.

Simplification

Encourage students to use a step-by-step approach to finding a solution and not to make jumps, as this often leads to difficulties. Support them by offering completed solutions written in the wrong order or as cards to be sorted. Students should work in pairs to reorder the solution.

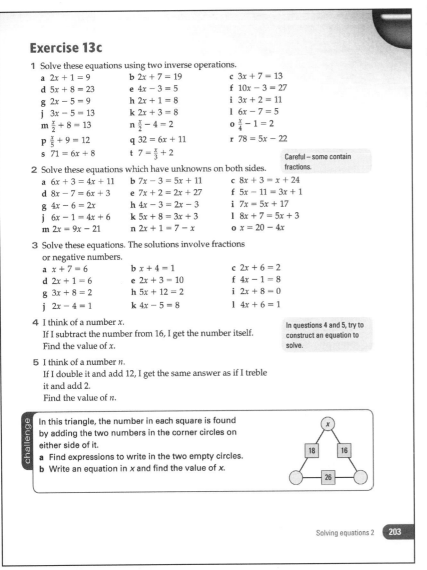

Exercise 13c

1 Solve these equations using two inverse operations.
  a $2x + 1 = 9$ b $2x + 7 = 19$ c $3x + 7 = 13$
  d $5x + 8 = 23$ e $4x - 3 = 5$ f $10x - 3 = 27$
  g $2x - 5 = 9$ h $2x + 1 = 8$ i $3x + 2 = 11$
  j $3x - 5 = 13$ k $2x + 3 = 8$ l $6x - 7 = 5$
  m $\frac{x}{2} + 8 = 13$ n $\frac{x}{2} - 4 = 2$ o $\frac{x}{4} - 1 = 2$
  p $\frac{x}{5} + 9 = 12$ q $32 = 6x + 11$ r $78 = 5x - 22$
  s $71 = 6x + 8$ t $7 = \frac{x}{3} + 2$

Careful – some contain fractions.

2 Solve these equations which have unknowns on both sides.
  a $6x + 3 = 4x + 11$ b $7x - 3 = 5x + 11$ c $8x + 3 = x + 24$
  d $8x - 7 = 6x + 3$ e $7x + 2 = 2x + 27$ f $5x - 11 = 3x + 1$
  g $4x - 6 = 2x$ h $4x - 3 = 2x - 3$ i $7x = 5x + 17$
  j $6x - 1 = 4x + 6$ k $5x + 8 = 3x + 3$ l $8x + 7 = 5x + 3$
  m $2x = 9x - 21$ n $2x + 1 = 7 - x$ o $x = 20 - 4x$

3 Solve these equations. The solutions involve fractions or negative numbers.
  a $x + 7 = 6$ b $x + 4 = 1$ c $2x + 6 = 2$
  d $2x + 1 = 6$ e $2x + 3 = 10$ f $4x - 1 = 8$
  g $3x + 8 = 2$ h $5x + 12 = 2$ i $2x + 8 = 0$
  j $2x - 4 = 1$ k $4x - 5 = 8$ l $4x + 6 = 1$

4 I think of a number x.
  If I subtract the number from 16, I get the number itself.
  Find the value of x.

In questions 4 and 5, try to construct an equation to solve.

5 I think of a number n.
  If I double it and add 12, I get the same answer as if I treble it and add 2.
  Find the value of n.

challenge

In this triangle, the number in each square is found by adding the two numbers in the corner circles on either side of it.
a Find expressions to write in the two empty circles.
b Write an equation in x and find the value of x.

Extension

To consolidate understanding, provide some examples of the particularly challenging equations from the exercise that have been answered to include some errors. Invite more able students to 'mark' these and to provide helpful feedback (RL).

Exercise 13c commentary

Question 1 – Similar to the first example. Ensure that students are confident with the meaning of inverse (1.3).

Question 2 – Similar to the second example. Encourage students to write step-by-step solutions to make errors less likely and easier to correct. Students will need to think on which side to collect the xs. Some solutions involve both negative and fractional numbers (1.3).

Question 3 – Similar to question **1** but now with fractional and negative solutions (1.3).

Question 4 and **5** – Two word-problems that lead to equations similar to those in questions **1** and **3** respectively (1.1, 1.3).

Challenge – This is a stretching activity where students have to generate expressions to make the calculations true, to construct an equation and to find a value for x. Some students will need scaffolding to support their attempts. Encourage talking through their thinking, model if necessary (CT) (1.1, 1.3, 1.4).

Assessment Criteria – Algebra, level 6: construct and solve linear equations with integer coefficients, using an appropriate method.

Links

The human sense of balance is called equilibrioception. The brain collects information from a series of organs in the inner ear called the labyrinth, and combines it with information from the other senses such as sight and touch, to help prevent the body from falling over. A disturbance to the sense of balance can cause the person to feel giddy or unsteady. There is more information about equilibrioception at http://en. wikipedia.org/wiki/ Equilibrioception

- Construct and solve linear equations with integer coefficients (unknown on either side, with and without brackets) using appropriate methods   (L6)

Useful resources

Starter – Think of a number

Ask students to write equations for 'Think of a number' problems and find the starting numbers, for example,

     I multiply by 3 and subtract 7. My answer is 20. (9)
     I double my number, add 14 and divide by 2. My answer is 19. (12)
     I add 5 and multiply by 7. My answer is 21. (-2)

See also the plenary of spread **13c**.

Teaching notes

Give students a few expressions to simplify which include brackets and like terms. This will allow students to review expanding brackets and collecting like terms before beginning the new topic of solving equations with directed numbers. The worked examples emphasise the method of checking through substitution and this is hugely valuable in encouraging students to take increasing responsibility for their learning and to gain confidence (RL).

Wherever possible encourage students to work collaboratively and to discuss their thinking and methods, possibly by sharing out the examples in the exercise with the expectation that students will work as a team. Mathematically this is an effective learning strategy and also supports students in reaching agreements, managing discussions to achieve results (TW) (1.5).

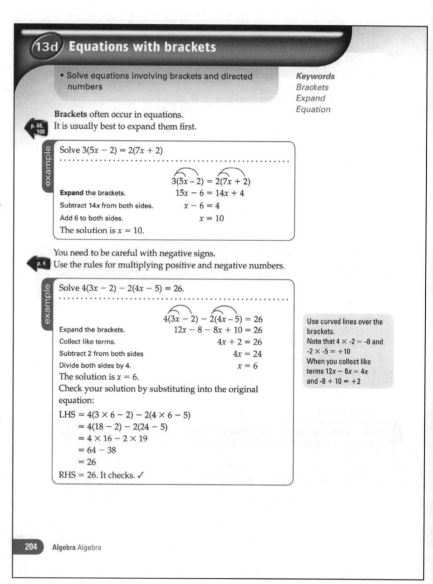

13d) Equations with brackets

- Solve equations involving brackets and directed numbers

Keywords
Brackets
Expand
Equation

Brackets often occur in equations.

p. 68, 100   It is usually best to expand them first.

Solve $3(5x - 2) = 2(7x + 2)$

$$3(5x - 2) = 2(7x + 2)$$

Expand the brackets. $15x - 6 = 14x + 4$
Subtract $14x$ from both sides. $x - 6 = 4$
Add 6 to both sides. $x = 10$
The solution is $x = 10$.

You need to be careful with negative signs.
p. 4   Use the rules for multiplying positive and negative numbers.

Solve $4(3x - 2) - 2(4x - 5) = 26$.

$$4(3x - 2) - 2(4x - 5) = 26$$

Expand the brackets. $12x - 8 - 8x + 10 = 26$
Collect like terms. $4x + 2 = 26$
Subtract 2 from both sides $4x = 24$
Divide both sides by 4. $x = 6$
The solution is $x = 6$.
Check your solution by substituting into the original equation:

LHS $= 4(3 \times 6 - 2) - 2(4 \times 6 - 5)$
       $= 4(18 - 2) - 2(24 - 5)$
       $= 4 \times 16 - 2 \times 19$
       $= 64 - 38$
       $= 26$
RHS $= 26$. It checks. ✓

Use curved lines over the brackets.
Note that $4 \times -2 = -8$ and $-2 \times -5 = +10$
When you collect like terms $12x - 8x = 4x$ and $-8 + 10 = +2$

204   Algebra Algebra

Plenary

To consolidate this work, use three 'solved' equations that involve the multiplication of positive and negative numbers (possibly examples from the exercise that have caused difficulty) and that include some common errors. Ask students to work in pairs to identify the errors and to offer advice to help the 'student' improve (RL).

Simplification

Support weaker students by focusing on generating equations from word problems. Use examples such as those in the exercise in order to consolidate the meaning of equations and thereby scaffold understanding of finding a solution. Provide some model solutions as paired card-sorting activities for additional help.

Students could be asked to work in pairs, each student answering alternate examples and then explaining their reasoning. This is an effective learning strategy (EP).

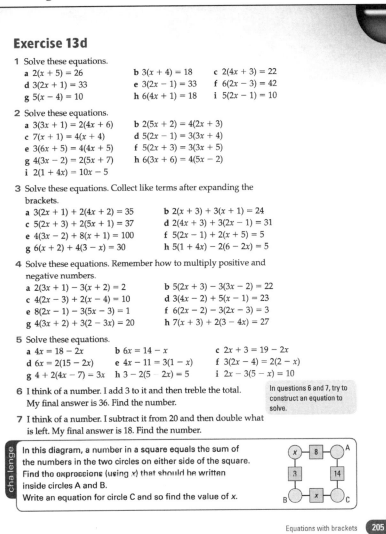

Exercise 13d

1 Solve these equations.
   a $2(x + 5) = 26$ b $3(x + 4) = 18$ c $2(4x + 3) = 22$
   d $3(2x + 1) = 33$ e $3(2x - 1) = 33$ f $6(2x - 3) = 42$
   g $5(x - 4) = 10$ h $6(4x + 1) = 18$ i $5(2x - 1) = 10$

2 Solve these equations.
   a $3(3x + 1) = 2(4x + 6)$ b $2(5x + 2) = 4(2x + 3)$
   c $7(x + 1) = 4(x + 4)$ d $5(2x - 1) = 3(3x + 4)$
   e $3(6x + 5) = 4(4x + 5)$ f $5(2x + 3) = 3(3x + 5)$
   g $4(3x - 2) = 2(5x + 7)$ h $6(3x + 6) = 4(5x - 2)$
   i $2(1 + 4x) = 10x - 5$

3 Solve these equations. Collect like terms after expanding the brackets.
   a $3(2x + 1) + 2(4x + 2) = 35$ b $2(x + 3) + 3(x + 1) = 24$
   c $5(2x + 3) + 2(5x + 1) = 37$ d $2(4x + 3) + 3(2x - 1) = 31$
   e $4(3x - 2) + 8(x + 1) = 100$ f $5(2x - 1) + 2(x + 5) = 5$
   g $6(x + 2) + 4(3 - x) = 30$ h $5(1 + 4x) - 2(6 - 2x) = 5$

4 Solve these equations. Remember how to multiply positive and negative numbers.
   a $2(3x + 1) - 3(x + 2) = 2$ b $5(2x + 3) - 3(3x - 2) = 22$
   c $4(2x - 3) + 2(x - 4) = 10$ d $3(4x - 2) + 5(x - 1) = 23$
   e $8(2x - 1) - 3(5x - 3) = 1$ f $6(2x - 2) - 3(2x - 3) = 3$
   g $4(3x + 2) + 3(2 - 3x) = 20$ h $7(x + 3) + 2(3 - 4x) = 27$

5 Solve these equations.
   a $4x = 18 - 2x$ b $6x = 14 - x$ c $2x + 3 = 19 - 2x$
   d $6x = 2(15 - 2x)$ e $4x - 11 = 3(1 - x)$ f $3(2x - 4) = 2(2 - x)$
   g $4 + 2(4x - 7) = 3x$ h $3 - 2(5 - 2x) = 5$ i $2x - 3(5 - x) = 10$

6 I think of a number. I add 3 to it and then treble the total. My final answer is 36. Find the number.

7 I think of a number. I subtract it from 20 and then double what is left. My final answer is 18. Find the number.

In questions 6 and 7, try to construct an equation to solve.

challenge
In this diagram, a number in a square equals the sum of the numbers in the two circles on either side of the square. Find the expressions (using x) that should be written inside circles A and B.
Write an equation for circle C and so find the value of x.

Equations with brackets **205**

Extension

Provide three challenging examples (possibly taken from the exercise) Ask more able students to put the examples in rank order of difficulty and to explain and justify their decisions.

Exercise 13d commentary

Question 1 – Simple examples allowing practice of expanding a bracket. It is also sensible to solve these by dividing both sides by the number in front of the brackets (1.3).

Question 2 – Similar to the first example. Ensure that all steps are shown: this helps to identify where any errors or misunderstandings occur (1.3).

Question 3 – Similar to the second example (1.3).

Question 4 – Similar to question **3** but now requiring the product of two negative numbers. Students may find this hard, provide scaffolding and refer them to spread **1b** (1.3).

Question 5 – A mix of the different types of equations that students have experienced: aids consolidation (1.3).

Question 6 and **7** – Word problems that produce equations similar to those in question **1** (1.1, 1.3).

Challenge – A stretching activity where students are challenged to generate expressions to make the calculations true, to construct an equation and to find a value for x. Some students will need scaffolding to support their attempts. Encourage students to talk through their reasoning (1.1, 1.3, 1.4, 1.5).

Assessment Criteria – Algebra, level 6: construct and solve linear equations with integer coefficients, using an appropriate method.

Links

Equations are used in Chemistry to describe reactions. The chemicals that react are on the left of the equation and the products are on the right. A balanced equation has equal numbers of each type of atom on each side of the equation. There is a calculator to balance chemical equations at www.webqc.org/balance.php

- Plot the graphs of linear functions, where *y* is given explicitly in terms of *x*, on paper and using ICT.   (L6)

Useful resources

Starter – Twins

Siobhan and Rachel are twins.

Siobhan multiplied her age by 3 and subtracted 6.

Rachel reached the same answer when she multiplied her age by 2 and added 6.

How old are the twins. (12 yrs)

Challenge students to make up their own puzzles.

Teaching notes

Initially, students make links to prior learning through a recapitulation of functions and through constructing a table of values, plotting coordinates and drawing a graph. The worked example offers an opportunity to encourage student dialogue. Students should explain to each other how they would know whether a given point lies on a given line by referring to the equation and without drawing a graph.

There are occasions when the 'student voice' tells us that a friend's explanation is 'better' than the teacher's. This is one of them! This also helps to encourage students to ask questions to extend their thinking (CT).

When students are involved in the exercise, invite them to think about what is going to be different about a straight line graph where the equation contains a –*x* value. To establish stronger cross-curricular links ask students when they have used straight-line graphs in science (EP).

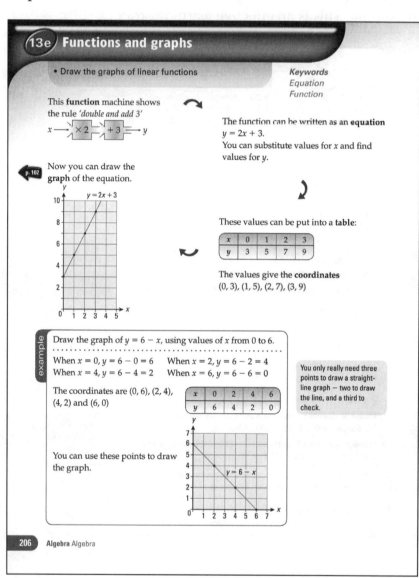

Plenary

Challenge students to work in pairs to produce a concise, clear explanation of how they would know whether a point lies on a given line just by using the equation. This activity will strengthen literacy skills whilst consolidating understanding. Take feedback from the class and agree a whole class 'best' explanation.

Simplification

Remind students who are not confident to look for a regular pattern in the numbers in the table of values. This will help them identify their own errors and thereby take increasing responsibility for their own learning.

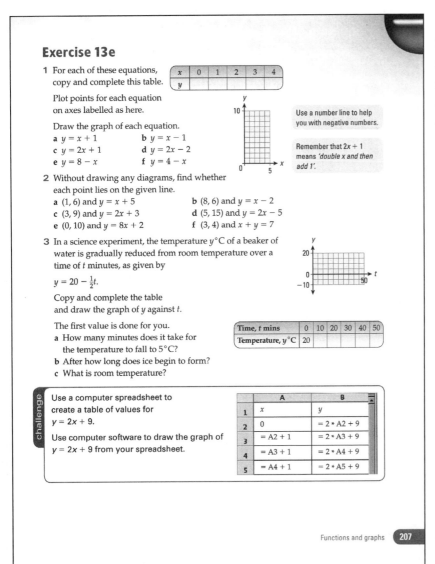

Exercise 13e

1 For each of these equations, copy and complete this table.

x	0	1	2	3	4
y					

Plot points for each equation on axes labelled as here.

Draw the graph of each equation.

a $y = x + 1$ **b** $y = x - 1$
c $y = 2x + 1$ **d** $y = 2x - 2$
e $y = 8 - x$ **f** $y = 4 - x$

Use a number line to help you with negative numbers.

Remember that $2x + 1$ means 'double x and then add 1'.

2 Without drawing any diagrams, find whether each point lies on the given line.

a $(1, 6)$ and $y = x + 5$ **b** $(8, 6)$ and $y = x - 2$
c $(3, 9)$ and $y = 2x + 3$ **d** $(5, 15)$ and $y = 2x - 5$
e $(0, 10)$ and $y = 8x + 2$ **f** $(3, 4)$ and $x + y = 7$

3 In a science experiment, the temperature $y°C$ of a beaker of water is gradually reduced from room temperature over a time of t minutes, as given by

$y = 20 - \frac{1}{2}t$.

Copy and complete the table and draw the graph of y against t.

The first value is done for you.

Time, t mins	0	10	20	30	40	50
Temperature, $y°C$	20					

a How many minutes does it take for the temperature to fall to $5°C$?
b After how long does ice begin to form?
c What is room temperature?

challenge

Use a computer spreadsheet to create a table of values for $y = 2x + 9$.

Use computer software to draw the graph of $y = 2x + 9$ from your spreadsheet.

	A	B
1	x	y
2	0	= 2 * A2 + 9
3	= A2 + 1	= 2 * A3 + 9
4	= A3 + 1	= 2 * A4 + 9
5	= A4 + 1	= 2 * A5 + 9

Functions and graphs **207**

Extension

Challenge more able students to contribute examples of the use of straight-line graphs either from other areas of the curriculum or from real life situations, for example, delivery costs, *etc.*

Exercise 13e commentary

Question 1 – Similar to the structured approach used in the example. Ask why parts **e** and **f** look different (1.3).

Question 2 – This question requires students to identify, from their knowledge of functions only, whether a given point lies on a given line. Ask for explanations of how they would know. Part **f** involves an implicit equation (IE) (1.4).

Question 3 – Here, explicit links are made with the importance of graphs in science experiments. Students are required to look at the results of the experiment through the table of values and to connect generally with the science curriculum (1.2, 1.3).

Challenge – Links the effective use of ICT to this topic (SM).

Assessment Criteria – Algebra, level 6: plot the graphs of linear functions where y is given explicitly in terms of x.

Links

An oscilloscope is a test instrument often used to troubleshoot electrical equipment that is malfunctioning. The instrument has a screen which can display a graph of voltage against time for the part of the circuit that is being tested. There is more information about oscilloscopes at http://en.wikipedia.org/wiki/Oscilloscopes

- Recognise that equations of the form $y = mx + c$ correspond to straight-line graphs (L6)

Useful resources

....

Starter – Stamps

A package costs 60 pence to post.

Ask students how many ways 60 pence can be made using 5p and 7p stamps.

($12 \times 5p$, $5 \times 5p + 5 \times 7p$)

Challenge students to find the largest postage amount that cannot be made using 5p and 7p stamps. (23p)

Teaching notes

Here, students consider how they will know what a straight-line graph will look like from their understanding of equations. Students will need to link to previous experience of the equations for horizontal and vertical lines and it is useful to ensure that this is secure. Many students still express confusion into KS4.

A key point to emphasise is that, if an equation has both x and y values, then it will have a slope of some sort.

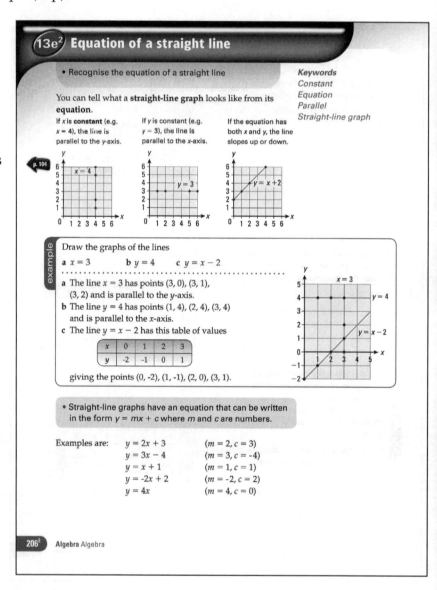

13e² Equation of a straight line

- Recognise the equation of a straight line

Keywords
Constant
Equation
Parallel
Straight-line graph

You can tell what a **straight-line graph** looks like from its **equation**.

If x is **constant** (e.g. $x = 4$), the line is parallel to the y-axis.

If y is constant (e.g. $y = 3$), the line is parallel to the x-axis.

If the equation has both x and y, the line slopes up or down.

Draw the graphs of the lines

a $x = 3$ b $y = 4$ c $y = x - 2$
....

a The line $x = 3$ has points (3, 0), (3, 1), (3, 2) and is parallel to the y-axis.
b The line $y = 4$ has points (1, 4), (2, 4), (3, 4) and is parallel to the x-axis.
c The line $y = x - 2$ has this table of values

x	0	1	2	3
y	-2	-1	0	1

giving the points (0, -2), (1, -1), (2, 0), (3, 1).

- Straight-line graphs have an equation that can be written in the form $y = mx + c$ where m and c are numbers.

Examples are:

$y = 2x + 3$	($m = 2, c = 3$)
$y = 3x - 4$	($m = 3, c = -4$)
$y = x + 1$	($m = 1, c = 1$)
$y = -2x + 2$	($m = -2, c = 2$)
$y = 4x$	($m = 4, c = 0$)

206² Algebra Algebra

Plenary

Show a set of graphs similar to those in question **5** together with their equations and challenge students to match them up. They must also give a reason for matching each pair.

Simplification

Allow students to create tables of values and draw graphs for all the questions. Use the graphs to emphasise the important points to look for in the corresponding equations.

Exercise 13e²

1 a Copy and complete the table for the equation $y = 2x + 3$.

x	0	1	2	3
y				

   b Then draw its graph. Label the x axis from 0 to 4, and the y axis from 0 to 10.

2 a Copy and complete the table for the equations $y = 3x - 2$.

x	0	1	2	3
y				

   b Then draw its graph. Label the x axis from 0 to 4, and the y axis from -3 to 8.

3 Without first completing a table, draw the graphs of these equations.

   a $x = 2$ b $y = 5$ c $x = -2$ d $y = 0$

4 For each equation, state whether its graph is:
   i a horizontal straight line   iii a sloping straight line
   ii a vertical straight line   iv not a straight line
   a $x = 4$ b $y = -3$ c $y = 2x$ d $y = 2x^2 + 1$

   You do not need to draw the graphs.

5 Match the equations to the graphs. Which is the odd equation out?

   $y = x + 1$
   $y = 1$
   $x = -1$
   $y = 2x - 3$
   $y = 2x - 2$

Extension

Challenge students to plot the graph corresponding to $y = 2x^2 + 1$. This could first be done by constructing a table for positive x values and then extending to negative x values. Explain that this is a general method that will work for any equations (SM).

Exercise 13e² commentary

Question 1 and **2** – Similar to part **c** of the example. Allows consolidation of the basic procedure for plotting any graph (1.3).

Question 3 – Horizontal and vertical graphs. Check that students do not reason '$x=2$, x-axis horizontal so line is horizontal' (1.3).

Question 4 – Ask students to write one sentence explaining their reasoning. Part **d** may require some explanation (CT).

Question 5 – Develop question **5**, by requiring students to also think about the significance of the gradient and intercept.

Assessment Criteria – Algebra, level 6: recognize that equations of the form $y = mx + c$ correspond to straight-line graphs.

Links

Beams of light travel in straight lines, but when they meet an obstacle, they may bend around the obstacle or spread out. This effect is called diffraction. There is more information about diffraction at http://www.olympusmicro.com/primer/lightandcolor/diffraction.html and a demonstration of diffraction and the resulting patterns when light is passed through two slits at http://phys.educ.ksu.edu/vqm/html/doubleslit/index.html

- Construct linear functions arising from real-life
  problems and plot their corresponding graphs **(L6)**
- Discuss and interpret graphs arising from real
  situations, e.g. distance-time graphs **(L6)**

Useful resources

Starter – Pigeons and rabbits

Gareth was watching some pigeons and rabbits.

He counted the number of heads and feet. There were 19 heads and 52 feet.
Ask students how many pigeons and how many rabbits there were.

(12 pigeons, 7 rabbits)

Can be extended by asking students to make up their own
bird and animal puzzles.

Teaching notes

The key processes in algebra are developed here with the focus on interpretation and evaluation, as students interpret the shape and other features of graphs. The skills required for students to become functional in mathematics are highlighted and students are required to transfer their skills to new or different situations (CT).

There is an emphasis on real-life graphs and on the use of graphs in other curriculum areas, especially science. Encourage students to identify other information that is shown in the graphs beyond that which is specifically asked for in the question. Discuss other real-life situations that could be represented with straight line-graphs (EP).

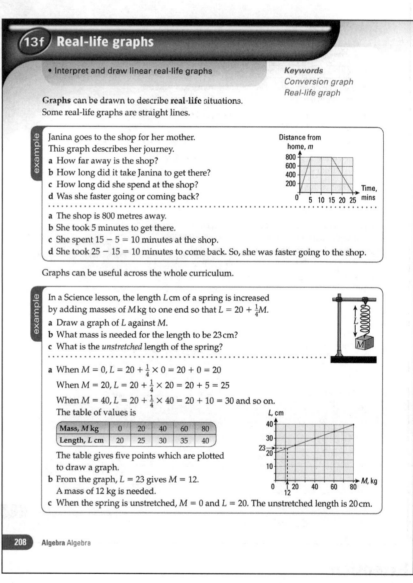

13f Real-life graphs

- Interpret and draw linear real-life graphs

Keywords
Conversion graph
Real-life graph

Graphs can be drawn to describe **real-life** situations.
Some real-life graphs are straight lines.

example

Janina goes to the shop for her mother.
This graph describes her journey.
a How far away is the shop?
b How long did it take Janina to get there?
c How long did she spend at the shop?
d Was she faster going or coming back?

a The shop is 800 metres away.
b She took 5 minutes to get there.
c She spent $15 - 5 = 10$ minutes at the shop.
d She took $25 - 15 = 10$ minutes to come back. So, she was faster going to the shop.

Graphs can be useful across the whole curriculum.

example

In a Science lesson, the length L cm of a spring is increased
by adding masses of M kg to one end so that $L = 20 + \frac{1}{4}M$.
a Draw a graph of L against M.
b What mass is needed for the length to be 23 cm?
c What is the *unstretched* length of the spring?

a When $M = 0$, $L = 20 + \frac{1}{4} \times 0 = 20 + 0 = 20$
When $M = 20$, $L = 20 + \frac{1}{4} \times 20 = 20 + 5 = 25$
When $M = 40$, $L = 20 + \frac{1}{4} \times 40 = 20 + 10 = 30$ and so on.
The table of values is

Mass, M kg	0	20	40	60	80
Length, L cm	20	25	30	35	40

The table gives five points which are plotted
to draw a graph.
b From the graph, $L = 23$ gives $M = 12$.
A mass of 12 kg is needed.
c When the spring is unstretched, $M = 0$ and $L = 20$. The unstretched length is 20 cm.

208 Algebra Algebra

Plenary

Ask students to draw a travel graph and swap with a friend.
The friend should then describe, in words, his or her
interpretation of the depicted journey.

Simplification

To support students who are having difficulty with this work, provide some true and some false statements from the interpretation of a travel graph. Ask students to work in pairs to sort the statements into those that are true and those that are false. Ask how do you know?

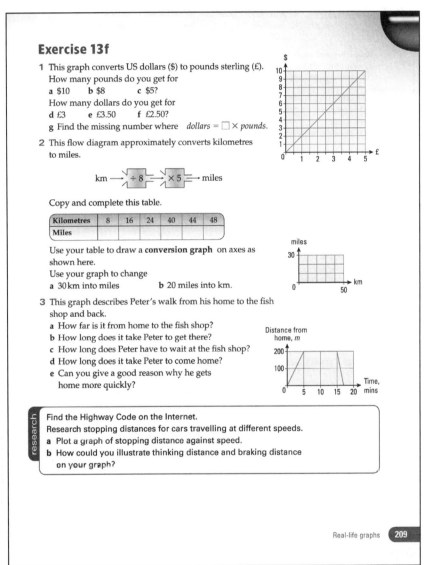

Exercise 13f

1 This graph converts US dollars ($) to pounds sterling (£).
   How many pounds do you get for
   a $10   **b** $8   **c** $5?
   How many dollars do you get for
   d £3   **e** £3.50   **f** £2.50?
   g Find the missing number where *dollars* = □ × *pounds*.

2 This flow diagram approximately converts kilometres to miles.

   km ──→ ÷ 8 ──→ × 5 ──→ miles

   Copy and complete this table.

Kilometres	8	16	24	40	44	48
Miles						

   Use your table to draw a **conversion graph** on axes as shown here.
   Use your graph to change
   a 30 km into miles   **b** 20 miles into km.

3 This graph describes Peter's walk from his home to the fish shop and back.
   a How far is it from home to the fish shop?
   b How long does it take Peter to get there?
   c How long does Peter have to wait at the fish shop?
   d How long does it take Peter to come home?
   e Can you give a good reason why he gets home more quickly?

   research
   Find the Highway Code on the Internet.
   Research stopping distances for cars travelling at different speeds.
   a Plot a graph of stopping distance against speed.
   b How could you illustrate thinking distance and braking distance on your graph?

Extension

Use a travel graph such as the one in question **3**. Ask students to redraw or alter it to depict new, changed criteria. Ask for explanations.

Exercise 13f commentary

The questions in this exercise make explicit links with real-life situations and other areas of the curriculum. Encourage students to think of their own examples (EP) (1.2).

Question 1 – The second example illustrates how to read information from the graph. Check accuracy and encourage students to compare their answers as a check.

Question 2 – Similar to the second example, linking graphs to formulae. Discuss the use of a sensible scale and make sure that students can read the graph both ways in parts **a** and **b** (1.3).

Question 3 – Similar to the first example. To develop functional skills, ask what other questions could be asked about this graph (IE) (1.4).

Research – Here, a real-life example is highlighted. Invite students to identify other real-life examples (SM) (1.2).

Assessment Criteria – Algebra, level 6: interpret graphs arising from real-life situations.

Links

In finance, the exchange rate between two currencies is the price at which one country's currency can be exchanged for another currency, for example, pounds to euros. The rates change daily. Current exchange rates and a conversion calculator can be found at http://uk.finance.yahoo.com/currency-converter?u and at http://www.x-rates.com/calculator.html How many Euros can be bought with £10 at today's exchange rate?

- Discuss and interpret graphs arising from real situations, e.g. distance-time graphs

Useful resources

(L6)

Starter – Containers

Ask students to imagine two unmarked containers.
One holds exactly three litres; the other holds exactly five litres.
Jamie says he can use these containers to get exactly four litres.
Ask students to work out how this can be done.

What if the capacities are 5 litres and 7 litres?

Teaching notes

Again, the emphasis on real-life situations in this spread works towards developing functional mathematics and students ability to interpret and sketch graphs (EP).

Students will learn most effectively by collaborating on tasks and discussing their interpretations. Through this, they will develop the ability to support conclusions using reasoned arguments and evidence (IE).

Whole class discussion of the worked examples can lead to asking students to write a 'storyline' of their own, perhaps about their own school day or to describe a journey they have made. Students can then challenge each other to sketch a graph to depict this.

Alternatively, give students a graph on unlabelled axes and be ask them to write a possible 'storyline' for the graph.

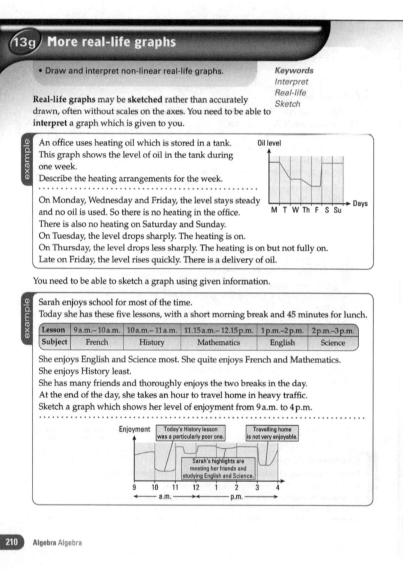

Plenary

Ask students to draw a sketch graph of how they felt during the course of this maths lesson. Ask them to write a written description to accompany this.

Simplification

Students who find this work difficult could be given a sketch graph and a set of cards identifying 'landmarks' in the story of the graph. They should identify where these landmarks should be located on the graph.

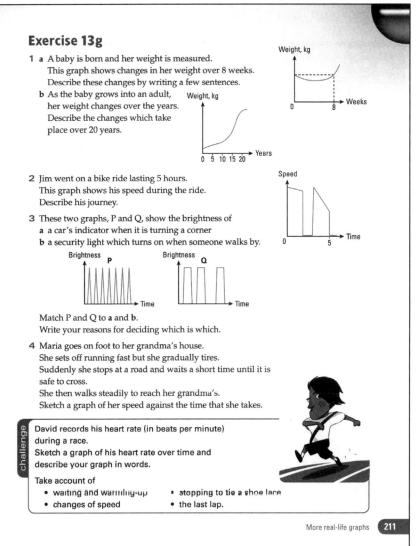

Exercise 13g

1 a A baby is born and her weight is measured. This graph shows changes in her weight over 8 weeks. Describe these changes by writing a few sentences.

   b As the baby grows into an adult, her weight changes over the years. Describe the changes which take place over 20 years.

2 Jim went on a bike ride lasting 5 hours. This graph shows his speed during the ride. Describe his journey.

3 These two graphs, P and Q, show the brightness of
   a a car's indicator when it is turning a corner
   b a security light which turns on when someone walks by.

   Match P and Q to a and b.
   Write your reasons for deciding which is which.

4 Maria goes on foot to her grandma's house. She sets off running fast but she gradually tires. Suddenly she stops at a road and waits a short time until it is safe to cross. She then walks steadily to reach her grandma's. Sketch a graph of her speed against the time that she takes.

challenge

David records his heart rate (in beats per minute) during a race. Sketch a graph of his heart rate over time and describe your graph in words.

Take account of
• waiting and warming-up
• changes of speed
• stopping to tie a shoe lace
• the last lap.

More real-life graphs **211**

Extension

Ask more able students to research real-life graphs in the media, such as those depicting currency fluctuations or the rise and fall of profits in the retail industry (EP).

Exercise 13g commentary

Questions 1–3 – These are similar to the first example. Students should be encouraged to share interpretations and compare ideas (TW, CT) (1.5).

Question 4 – Similar to the second example. Emphasise that a sketch graph is not accurate (1.5).

Challenge – This heart rate graph activity could be carried out as a genuine real-life example – make links with sport. There is not a unique answer to this task (1.2, 1.5).

Assessment Criteria – Algebra, level 6: interpret graphs arising from real-life situations.

Links

Human beings grow fastest before they are born – the fastest growth rate is at about the fourth month of pregnancy. There is a chart showing the average weight and height of an unborn baby during pregnancy at http://www.babycentre.co.uk/pregnancy/fetaldevelopment/chart/

Ask students to plot the graph of weight against age. Is the graph steeper or less steep than the graph in question **1**?

- Generate points in all four quadrants   (L5)   ***Useful resources***
- Find the midpoint of the line segment AB, given the
  coordinates of points A and B   (L6)

Starter – Quadrilaterals

Draw a set of axes from -6 to +6. The following sets of points specify three vertices of a given shape, challenge students to find the coordinates of the fourth point

Square	(0, 3), (2, 0), (-1, -2)	(-3, 1)
Parallelogram	(-4, -5), (-2, -3), (3, -3)	(1, -5)
Rhombus	(-6, 1), (-5, 4), (-4, 1)	(-5, -2)

Can be extended by challenging students to make up their own examples.

Teaching notes

Draw a number of straight-lines with different gradients and ask students to describe their slopes. Using the students' words describe a line (through the origin) and ask them to draw it. Do they all look the same? (Hopefully not!) Agree the need for a precise language. Choose a specific line and work towards a definition of its 'gradient'. Ask, 'if I went one unit along, what should I do next?' Emphasise the need to specify up (+) or down (−). Check understanding by asking students to draw a line with gradient −1 (CT).

Next, reverse the problem by drawing a line and asking students to find its gradient. Allow them to tell you how to proceed and then check their understanding with a further example.

Draw a line segment with specified end coordinates and ask students 'where is the midpoint?' Ask them to imagine looking down from above and agree that the x-value of the midpoint is given by the mean of the two end values. Check that this average value is the same distance from each end. Ask students to explain how to find the y-coordinate of the midpoint. Use a second example to check understanding.

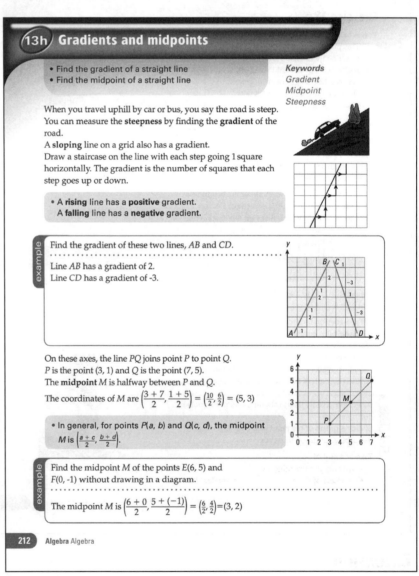

Plenary

Challenge students to find the midpoint and gradient of the line segment joining A (1, 2) and B (5, 5). The midpoint can be found using a formula, can they find a similar formula for finding the gradient?. Allow students to use a drawing to find the gradient but then lead them through an investigation similar to the **extension** activity (IE).(1.2)

Simplification

For both the gradient and midpoint questions, suggest that students make accurate copies on graph paper and then use 'counting squares' methods. This can be used to establish the formula for the midpoint of a line segment.

Exercise 13h

1 Write down the gradient of these five sloping lines.

2 Imagine a staircase of steps on each of the lines *KL, MN, PQ* and *RS*. Find the gradient of each line.

3 Write the coordinates of the midpoint *M* of
  a line *PQ*   **b** line *RS*   **c** line *TU*   **d** line *VW*.

4 Calculate the midpoints of the lines *AB* and *PQ*.

5 Without drawing a diagram, find the midpoints of the lines joining these pairs of points.
  a (2, 1), (8, 7)   **b** (3, 2), (7, 4)   **c** (8, 1), (0, 5)
  d (6, 0), (2, 5)   **e** (12, 41), (18, 19)   **f** (12, 6), (-2, 0)
  g (-1, 8), (5, -2)   **h** (9, 7), (-1, -2)

> **challenge**
> • *M* is the midpoint of *AB*. If *M* is the point (7, -1) and *B* is the point (8, 4), find the point *A*.
> • On axes labelled from 0 to 8, draw two lines from the point (4, 6) to the x-axis with gradients of 2 and -2. Join their midpoints. What shape have you drawn?

Extension

Ask students to investigate finding the gradient of a line using a number of triangles, not necessarily with base one. Encourage them to try using the formula: gradient equals increase in *y* over increase in *x*. Can they justify this result using similar triangles? (IE)

Exercise 13h commentary

Question 1 – Similar to the first example. Encourage students to find the magnitude of the gradient first and then give it the correct sign.

Question 2 – Similar to the first question but without the 'staircase' added (1.3).

Question 3 – Students should be encouraged to use the formula rather than rely on reading off a value from the coordinate grid (ensure that x and y are not mixed up).

Question 4 – Similar to question **3** but now drawn without a grid so that coordinates can not be read off (1.3).

Question 5 – Similar to the second example (1.3).

Challenge –The first part is an inverse problem that can be solved by trial-and-improvement, algebraically or by drawing a graph. The second part relies on the two lines with gradients ±2 being drawn correctly (1.3).

Assessment Criteria – Algebra, level 5: use and interpret coordinates in all four quadrants.

Links

Stairs feature in many buildings throughout the world. The vertical portion between successive steps is known as the 'riser' whilst the horizontal step is known as the 'tread'. Formulae are used to decide on appropriate dimensions for the riser and tread so that stairs are suitable for people to climb. What values are used in students' homes and in the school?

For more information and images, see, http://en.wikipedia.org/wiki/Stairway

13a

1 Simplify

a $y + y + y + y$ 　　　　**b** $y \times y \times y \times y$

c $3 \times 5 \times y$ 　　　　**d** $3 \times 5 \times y \times y$

2 Simplify by collecting like terms

a $2x + 3y + 4x + 5y$ 　　**b** $5p + 2q + 3p - 3q$ 　　**c** $3z + 4 + z - 6$

3 Expand and simplify

a $3(2x + 4) + 4(x + 2)$ 　　**b** $5(y + 3) + 2(3y - 7)$

c $2(4 - 2y) + 3(3y - 1)$ 　　**d** $3(x + 4) + 2(3 - 2x)$

13b

4 Find the values of R, S, T and U when $x = 4$, $y = 2$ and $z = 5$.

$$R = 2(x + 3y) \qquad S = y(2z - x) \qquad T = \frac{4z + 2x}{y} \qquad U = z(x^2 + y^2)$$

5 A rectangle is x cm long and $x - 3$ cm wide.

Find a formula for 　**a** its area, A

　　　　　　　　　　b its perimeter, P.

Use the formulae to find A and P when $x = 5$.

x cm

$x - 3$ cm

13c

6 Find the value of x in this balance.

7 Solve these equations with unknowns on both sides.

a $4x + 1 = 2x + 9$ 　　　**b** $6x + 2 = 2x + 14$

c $8x - 2 = 6x + 4$ 　　　**d** $7x - 2 = 4x + 7$

8 Solve these equations. Take care with the fractions.

a $2x + 5 = 8$ 　　　　**b** $4x - 1 = 8$ 　　　　**c** $\frac{x}{2} + 2 = 7$

13d

9 Solve these equations.

a $2(3x + 2) = 22$ 　　　**b** $3(4x + 1) = 15$ 　　　**c** $4(2x - 1) = 20$

d $2(5x + 1) = 3(3x + 2)$ 　　**e** $3(4x + 3) = 2(5x + 8)$ 　　**f** $4(3x - 2) = 2(4x + 2)$

10 Expand the brackets and solve the equations.

a $5(x + 2) + 3(x + 4) = 30$ 　　　**b** $4(2x + 1) + 3(2x + 3) = 41$

c $3(2x - 1) + 2(4x + 3) = 31$ 　　　**d** $2(3x - 2) + 4(x + 1) = 50$

11 I think of a number. I subtract it from 2 and then treble what is left.

My final answer is 18. Find the number.

12 For each of these equations, copy and complete this table.

x	0	1	2	3	4
y					

Plot points for each equation on axes, both labelled from -2 to 8.
Draw the graph of each equation and label it.
a $y = x + 1$ **b** $y = x - 1$ **c** $y = 2x - 1$ **d** $y = 8 - x$

13 Label both axes from 0 to 10. Draw the graphs of these lines
     a $x = 6$ **b** $y = 5$ **c** $y = x + 1$ **d** $y = 2x$

14 a You can convert euros € into pounds £ with this
formula: $£ = \dfrac{3 \times €}{4}$.

€	0	8	12	20
£				

Copy and complete this table using the formula.

   b Draw a graph to convert euros to pounds,
with both axes labelled from 0 to 20.
     i How many pounds will you get for €16?
     ii How many euros will you get for £3?

15 Khaleda walks the 500 m to the shop in 5 minutes. She
spends 3 minutes in the shop and then returns home the
same way in 2 minutes.
   a Draw a graph to show her visit to the shop on axes as
shown here.
   b What was her speed (in metres per minute) going
from home to shop?

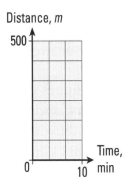

16 Sam heats some soup in the microwave for 2 minutes.
He takes it out and finds it is not warm enough.
He straightaway reheats it for another minute and
then leaves it for a further minute before starting to eat it.

Sketch a graph of the temperature of the soup during
this time.

17 Draw axes and label both from 0 to 8.
Find the gradient of the line through each of these pairs of points.
   a (2, 1), (3, 3)   **b** (1, 3), (2, 6)   **c** (1, 2), (4, 5)   **d** (4, 5), (6, 1)
   e (4, 6), (6, 4)   **f** (4, 0), (6, 1)   **g** (6, 1), (8, 7)   **h** (1, 7), (5, 7)

13 Summary

Summary

Assessment criteria

- Use and interpret coordinates in all four quadrants   Level 5
- Construct, express in symbolic form, and use simple formulae involving
  one or two operations   Level 5
- Construct and solve linear equations with integer coefficients, using
  an appropriate method   Level 6
- Plot the graphs of linear functions, where y is given explicitly in terms of x Level 6
- Recognise that equations of the form $y = mx + c$ correspond to
  straight-line graphs   Level 6
- Interpret graphs arising from real situations   Level 6

Question commentary

Example

The example illustrates a level 5 question on substituting values into a formula. Part **b** results in an equation to solve. Encourage a step-by-step approach and emphasise the need to show working. Ask questions such as "How can you check that your answer is correct?" Emphasise that the question is asking for a weight and a price so units should be given with each answer.

a $p = \frac{1}{2} \times 200 + 50$
  $= 150$ pence
  $= £1.50$
b $300 = \frac{1}{2} \times w + 50$
  $250 = \frac{1}{2} \times w$
  $w = 2 \times 250$
  $= 500$ g

Past question

The question requires students to interpret a distance-time graph. Some students may confuse distance with speed and think that the speed is increasing. Ask questions such as "What information does the graph give you about my walk?" and "How do you know how fast I am walking?"

Answer

Level 6
2 I was walking at a
  steady speed

Development and links

Students will interpret time series graphs in Chapter 15. The topics covered in this chapter are developed further in Year 9.

The ability to solve equations and derive formulae is important in science and engineering, particularly in physics. Students will derive and use formulae when creating spreadsheets in ICT. They will use the properties of straight line graphs to analyse the results of experiments in science. Graphs are also important in other curriculum areas, especially geography.

14 Construction and 3-D shapes

Objectives

- Use straight edge and compasses to construct
  - the midpoint and perpendicular bisector of a line segment 6
  - the bisector of an angle 6
  - the perpendicular from a point to a line 6
  - the perpendicular from a point on a line 6
  - a triangle, given three sides (SSS) 6
- Find simple loci, both by reasoning and by using ICT, to produce shapes and paths.. 6
- Use bearings to specify direction 6
- Know and use the formula for the volume of a cuboid................. 6
- Calculate volumes and surface areas of cuboids and shapes made from cuboids.................................... 6

Level

MPA

1.1	14a, b, d
1.2	14e, f, g
1.3	14a, b, c, c², d, e, f, g
1.4	14b, e
1.5	

PLTS

IE	14b, g
CT	14a, b, d, f, g
RL	14a, c, e
TW	14c
SM	14a, b, c², f
EP	14I, b, d

Introduction

The main focus of this chapter is on accurate construction techniques. Students develop their skills in constructing triangles from Year 7 and extend these to drawing triangles given all three sides (SSS), quadrilaterals, angle bisectors and the perpendicular to and from a point on a line. Other topics developed in the chapter are using bearings to specify direction and finding simple loci. Finally, students consider the net of a cuboid and find the surface area and volume of a cuboid.

The student page discusses the use of triangles in building strong structures. A rectangular shape tends to flatten when a force is applied, but a triangular shape cannot be deformed without changing the length of one of the sides. This means that triangles are very strong and so engineers and architects use triangular shapes in roof trusses, bridges and many other structures. There are pictures of triangles in building structures at http://www.pennridge.org/works/otherstruct.html (EP).

⚡ast-track

14a, b, c, c², d, f, g

Extra Resources

14	Start of chapter presentation
14a, c	Starter: Missing angle bingo
14a	Animation: Drawing triangles
14b	Animation: Drawing triangles 2
14b	Animation: Constructing triangles
14c	Animation: Bisectors
14c	Consolidation sheet
14c²	Animation: Finding perpendicular points
14d	Animation: Loci
14d	Consolidation sheet
14e	Animation: Bearings
14f	Animation: Surface area
14f	Worked solution: Q3a
14f	Consolidation sheet
14g	Animation: Volume of a cuboid
14g	Worked solution: Q2a
14g	Consolidation sheet
	Assessment: chapter 14

* Construct a triangle, given two sides and the included angle (SAS) or two angles and the included side (ASA)   (L6)

Useful resources
Ruler and protractor

• •

Starter – Estimating angles

Ask students to draw angles of different sizes without using a protractor, for example, 35°, 140°, 245°.

Students should then measure the angles and score points accordingly, for example, six points for an exact answer, four points for within 10° and two points for within 15°.

Teaching notes

Before starting, ensure that all students have the correct equipment and a sharp pencil.

Discuss with students the minimum amount of information required to draw a triangle without any ambiguity. Highlight the ASA and SAS cases.

Work through the ASA and SAS constructions, as a whole class activity, taking the opportunity to check that students are following the instructions. Emphasis should be placed on the correct use of a protractor. A common error is to choose the scale that does not start at zero and thereby confuse, for example, 60° and 120°. Encourage students to consider whether the angle is acute or obtuse and whether their measurement makes sense.

Check the constructions by measuring the missing angles and sides: A = 50°, AC = 1.6 cm, AB = 4.3 cm; DF = 4.9 cm, D = 60° and F = 50° (RL).

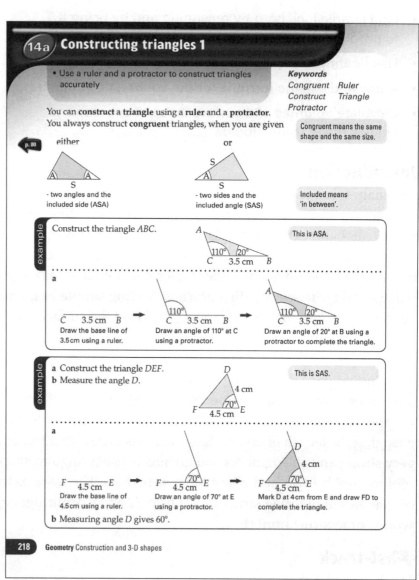

Plenary

Ask students to draw two different triangles, each with one right angle, one side 4 cm long and another 5 cm. long What length is the other side in each of the triangles? (3 cm, 6.4 cm) Can they explain why the triangles are different?

Simplification

Where students are finding this work difficult ask them to following the procedure given in the text. If students are paired together they can check each stage of their constructions as they work through them

As an aid to checking the accuracy of constructions, ask students to measure the missing angles and lengths. Ideally angles should be accurate to $\pm1^\circ$ and lengths to ±1 mm.

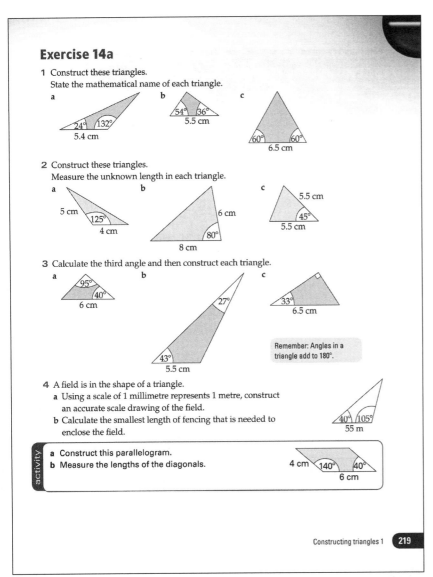

Extension

Challenge students to explain why being given the size of all three angles is insufficient to construct a triangle (CT).

Exercise 14a commentary

Question 1 – Similar to the first example. Remind students not to rub out any construction lines (1.3).

Question 2 – Similar to the second example (1.3).

Question 3 – AAS triangles: help students to realise that they will need to calculate the third angle in order to obtain an ASA triangle as in question **1**. In part **c**, check that students recognise the right-angle symbol (1.3).

Question 4 – A practical example based on a scale drawing; check that students correctly convert lengths between real life and the drawing (1.1, 1.3).

Activity – There are several ways to do this construction. Ask students to write a list of instructions explaining the process for drawing the parallelogram; the two examples provide a model for how to do this (SM).

Assessment Criteria – SSM, level 5: measure and draw angles to the nearest degree, when constructing models and drawing or using shapes. SSM, level 6: use straight edge and compasses to do standard constructions.

Links

A Geodesic Dome is a structure comprised of a network of triangles that form a surface shaped like a piece of a sphere. Geodesic domes are very strong but are lightweight and can be built very quickly. There are famous geodesic domes at the Eden Project in Cornwall and at the Epcot Centre in Florida (a complete sphere) There are pictures of geodesic domes at http://www.geo-dome.co.uk/ and at http://en.wikipedia.org/wiki/Geodesic_dome

- Use straight edge and compasses to construct a triangle, given three sides (SSS) (L6)

Useful resources
Ruler
Protractor
Compasses
Scissors and sticky tape

Starter – Take three

Ask students to make as many triangles as they can from six rods that are exactly 1 cm, 2 cm, 3 cm, 4 cm, 5 cm and 6 cm long (only one of each rod).

How many different triangles can they find? (7)

Which rod does not get used at all? (1 cm)

Can they explain why?

Teaching notes

Insist on the use of sharp pencils and do not allow students to draw with the pencil whilst in the compass. For best results, emphasise that the pencil point should be extended just longer than the point of the compass leg.

When measuring angles made with short lines it is difficult to read the size of the angle using a protractor. As an initial activity in the lesson ask students to draw two lines 4 cm long to make an angle. Ask them to measure that angle; ask, why it is difficult to measure. There are two good strategies here: either extend the length of the line so that it reaches the (circumference) edge of the protractor or use a ruler on top of the protractor to extend the line.

A useful way to demonstrate the SSS construction and to illustrate how to complete question **2** would be to construct a quadrilateral as a whole class activity.

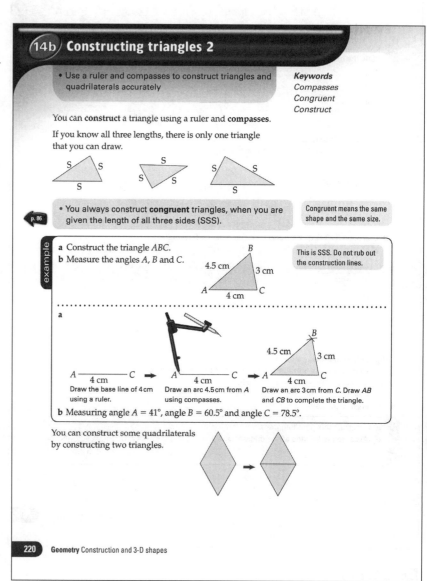

Plenary

The first two nets in the **activity** were constructed from four equilateral triangles and both nets folded into a tetrahedron. How many different nets can students draw for these solids? A cube, a cuboid and an octahedron.

Simplification

A number of students find using compasses difficult; allow them to draw some patterns using compasses to become more confident.

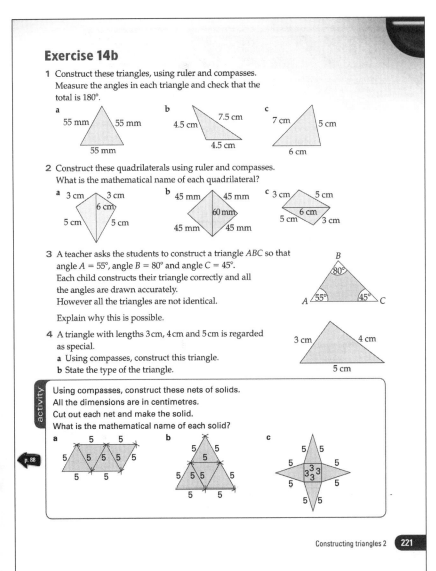

Exercise 14b

1 Construct these triangles, using ruler and compasses. Measure the angles in each triangle and check that the total is 180°.

a
55 mm 55 mm
55 mm

b
4.5 cm 7.5 cm
4.5 cm

c
7 cm 5 cm
6 cm

2 Construct these quadrilaterals using ruler and compasses. What is the mathematical name of each quadrilateral?

a
3 cm 3 cm
6 cm
5 cm 5 cm

b
45 mm 45 mm
60 mm
45 mm 45 mm

c
3 cm 5 cm
6 cm
5 cm 3 cm

3 A teacher asks the students to construct a triangle ABC so that angle A = 55°, angle B = 80° and angle C = 45°.
Each child constructs their triangle correctly and all the angles are drawn accurately.
However all the triangles are not identical.

Explain why this is possible.

B 80°
A 55°   45° C

4 A triangle with lengths 3 cm, 4 cm and 5 cm is regarded as special.
   a Using compasses, construct this triangle.
   b State the type of the triangle.

3 cm 4 cm
5 cm

activity

Using compasses, construct these nets of solids.
All the dimensions are in centimetres.
Cut out each net and make the solid.
What is the mathematical name of each solid?

a
5 5
5 5 5 5 5
5 5

b
5 5
5
5 5 5
5 5

c
5 5
5 5
3 3 3
5 5
5 5

p. 88

Constructing triangles 2 **221**

Extension

Invite students to make more elaborate nets in the activity or to combine tetrahedra to make more complex shapes (IE).

Exercise 14b commentary

Question 1 – Similar to the example. Remind students to draw faint construction lines and not to rub them out. The measured angles provide a check of accuracy (1.3).

Question 2 – Drawing a quadrilateral using two SSS constructions on a common base. (CT) (1.3)

Question 3 – Take the opportunity to remind students of the difference between similar and congruent (1.4).

Question 4 – Do any students know what is the relationship between the lengths of the sides? (EP) (1.3)

Activity – This activity could be extended to include octahedrons and icosahedrons, *etc.*, to make display items for the class room (SM) (1.1, 1.3).

Assessment Criteria – SSM, level 6: use straight edge and compasses to do standard constructions.

Links

A hexaflexagon is a flat hexagon-shaped paper toy that can be folded or flexed along its folds to reveal and conceal its faces alternately. It was invented in 1939 in the USA by Arthur Stone and its construction is based on equilateral triangles. There are instructions to make a hexaflexagon at http://hexaflexagon. sourceforge.net/ and at http:// www. flexagon.net/flexagons/ hexahexaflexagon-c.pdf

14c Bisectors

- Use straight edge and compasses to construct
  - the midpoint and perpendicular bisector of a line segment (L6)
- the bisector of an angle (L6)

Useful resources
Compasses
Ruler

Starter – Quad bingo

Ask students to draw a 3 × 3 grid and enter nine angles from the following list:

30°, 35°, 40°, 45°, 50°, 55°, 60°, 65°, 70°, 75°, 80°, 85°, 90°, 95°, 100°, 105°, 110°, 115°, 120°, 125°.

Ask questions, for example,

One of the angles in a rhombus is 135°, what is the smallest angle? (45°)

Winner is the first student to cross out all nine angles.

Teaching notes

Construction often proves a difficult topic to share with students as some are very capable and others struggle. Rather than showing a construction stage by stage (with students copying) allow them to follow instructions such as in the examples in the student book. They are then able to work at their own speed. Students who are confident can either help those who struggle or complete a further example of their own. As a teacher you will have more time to help those who genuinely need it (TW, RL).

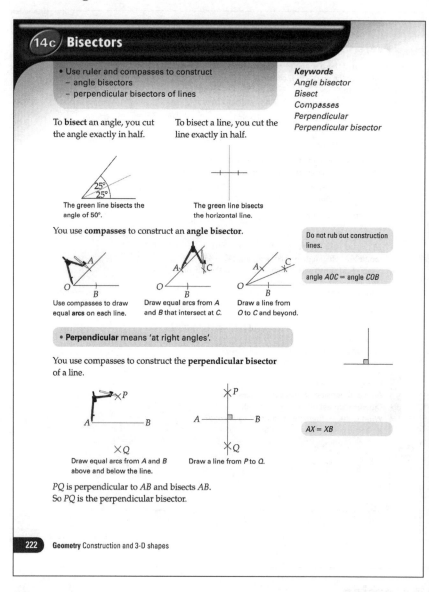

Plenary

In pairs, ask students to agree three or four tips on how to produce the most accurate work when using a compass. Take tips from each pair to share with the class.

Simplification

Group together the students who are finding construction difficult. This will allow you the opportunity to support their needs whilst other students work through other questions. In pairs ask one student to explain to the other how he or she is completing the bisection of the angle. The second student can make comments and highlight errors. Reverse the roles for constructing a perpendicular line (RL).

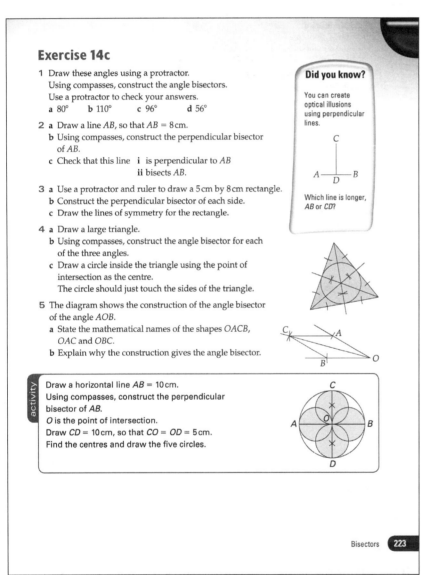

Exercise 14c

1 Draw these angles using a protractor.
Using compasses, construct the angle bisectors.
Use a protractor to check your answers.
   a 80°   b 110°   c 96°   d 56°

2 a Draw a line AB, so that AB = 8 cm.
   b Using compasses, construct the perpendicular bisector of AB.
   c Check that this line   i is perpendicular to AB
                           ii bisects AB.

3 a Use a protractor and ruler to draw a 5 cm by 8 cm rectangle.
   b Construct the perpendicular bisector of each side.
   c Draw the lines of symmetry for the rectangle.

4 a Draw a large triangle.
   b Using compasses, construct the angle bisector for each of the three angles.
   c Draw a circle inside the triangle using the point of intersection as the centre.
     The circle should just touch the sides of the triangle.

5 The diagram shows the construction of the angle bisector of the angle AOB.
   a State the mathematical names of the shapes OACB, OAC and OBC.
   b Explain why the construction gives the angle bisector.

Did you know?

You can create optical illusions using perpendicular lines.

Which line is longer, AB or CD?

activity

Draw a horizontal line AB = 10 cm.
Using compasses, construct the perpendicular bisector of AB.
O is the point of intersection.
Draw CD = 10 cm, so that CO = OD = 5 cm.
Find the centres and draw the five circles.

Bisectors 223

Extension

Give students a reflex angle, can they construct an angle bisector? Using their answer to question 5 as a guide, can they explain why their construction works?

Exercise 14c commentary

Remind students that all construction lines should be left to be seen. All questions (1.3).

Questions 1 and 2 – The previous page models how to do these constructions.

Question 3 – Ensure that students do not guess at right-angles but use a protractor at each corner to draw the initial rectangle. In an accurate diagram, the perpendicular bisectors of opposite sides should coincide to form the lines of symmetry.

Question 4 – If the construction is accurate the three bisectors will pass through a single point – the centre of the inscribed circle. The radius will have to be established by 'inspection'.

Question 5 – If the compasses are reset a kite and scalene triangles will result, otherwise a rhombus and isosceles triangles. Spread **6d** discusses properties of quadrilaterals. Suggest that the mathematical explanation is given as a series of bullet points.

Activity – To draw the small circles, it is only necessary to bisect OA, as OB, OC and OD are all the same length.

Assessment Criteria – SSM, level 6: use straight edge and compasses to do standard constructions.

Links

There is an interactive optical illusion involving parallel lines at http://demonstrations.wolfram.com/ParallelLinesOpticalIllusion/ and at http://www.sapdesignguild.org/resources/optical_illusions/geometrical.html

- Use straight edge and compass to construct
  - the perpendicular from a point to a line (L6)
  - the perpendicular from a point on a line (L6)

Useful resources
Ruler
Compasses
Protractor
Squared paper

Starter – Clock angles

Ask students to give the angle between the hour and minute hands at the following times:

7:00, 4:00, 9:30, 1:30, 4:30 and 3:15

Hint: The hour hand moves as well as the minute hand!
(Answers: 150° or 210°, 120°, 105°, 135°, 45° and 7.5°)

Teaching notes

If students have completed the previous spread they should be confident with the use of a pair of compasses. Organise students to work in pairs and set this challenge before they look in the text book. Draw a diagonal line and mark a point about 5 cm away – can they find the shortest distance from this point to the line? At first, use a ruler only. Then explain that you need to develop a mathematical method to be exactly correct and they can use a ruler and compasses. Anticipate that at this point some will draw an arc that is approximately a tangent with the line; develop this work so that students are able to complete the construction as shown in the example. This method will allow for discussion and an understanding of constructing a perpendicular to a line from a point.

Note that, when constructing a perpendicular to a line, it is not strictly necessary to draw intersecting arcs both above and below the line as one pair and the point will suffice.

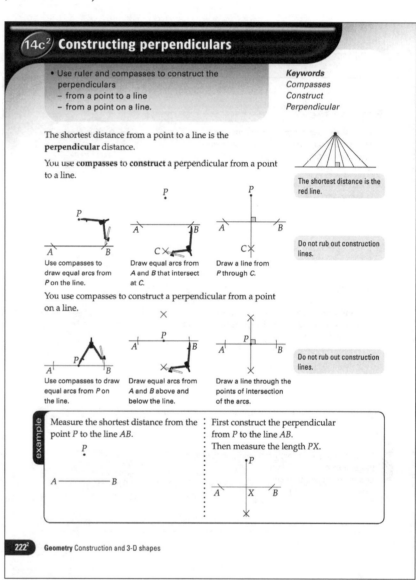

Plenary

Draw a triangle with perpendiculars drawn from two vertices to the opposite sides. Supply lengths for the sides of the triangle and one of the perpendiculars. Can the students work out the length (height) of the other perpendicular line? (Hint at the area of the triangle). Can they say what is the length of the third perpendicular line?

Simplification

Students who are unable to complete this construction should be given more practice at constructing triangles and quadrilaterals from given lengths and angles.

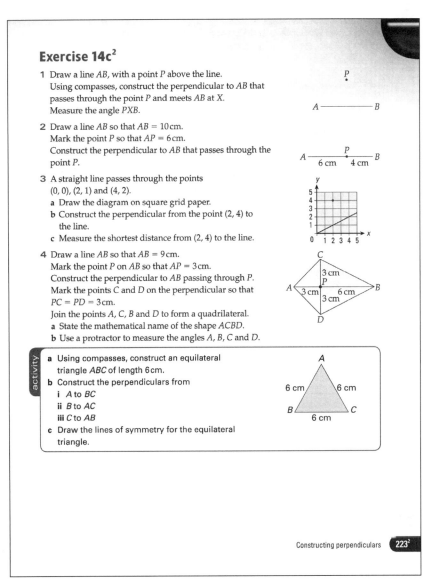

Exercise 14c²

1 Draw a line *AB*, with a point *P* above the line.
Using compasses, construct the perpendicular to *AB* that passes through the point *P* and meets *AB* at *X*.
Measure the angle *PXB*.

2 Draw a line *AB* so that *AB* = 10 cm.
Mark the point *P* so that *AP* = 6 cm.
Construct the perpendicular to *AB* that passes through the point *P*.

3 A straight line passes through the points (0, 0), (2, 1) and (4, 2).
a Draw the diagram on square grid paper.
b Construct the perpendicular from the point (2, 4) to the line.
c Measure the shortest distance from (2, 4) to the line.

4 Draw a line *AB* so that *AB* = 9 cm.
Mark the point *P* on *AB* so that *AP* = 3 cm.
Construct the perpendicular to *AB* passing through *P*.
Mark the points *C* and *D* on the perpendicular so that *PC* = *PD* = 3 cm.
Join the points *A*, *C*, *B* and *D* to form a quadrilateral.
a State the mathematical name of the shape *ACBD*.
b Use a protractor to measure the angles *A*, *B*, *C* and *D*.

activity
a Using compasses, construct an equilateral triangle *ABC* of length 6 cm.
b Construct the perpendiculars from
   i *A* to *BC*
   ii *B* to *AC*
   iii *C* to *AB*
c Draw the lines of symmetry for the equilateral triangle.

Constructing perpendiculars **223²**

Extension

Ask students to draw a (scalene) triangle about 10 cm across, then construct the perpendiculars from each vertex to the opposite side. In an accurate construction all three lines will pass through a single point: the orthocentre. (If the triangle is obtuse, this will be outside the original triangle). Students should be able to measure the 'altitudes' and find the area of the triangle in three different ways (SM).

Exercise 14c² commentary

All questions (1.3).

Questions 1 and **2** – The previous page models how to do these constructions.

Question 3 – Use a 2 cm square grid to avoid a tiny diagram. As a further challenge, ask students to find the gradients of the two lines; do they notice anything?

Question 4 – In an accurate diagram *AC* = *AD* = 4.2 cm and *BC* = *BD* = 6.7 cm. Spread **6d** gives the defining properties of a kite.

Activity – In an accurate diagram all three perpendiculars (which coincide with the angle bisectors) will pass through a single point, the centre of the inscribing and circumscribing circles.

Assessment Criteria – SSM, level 6: use straight edge and compasses to do standard constructions.

Links

Perpendicular recording is a new technology that increases the storage capacity of hard drives. It is predicted that perpendicular recording will allow information densities of up to around 1000 Gb/in² compared with 100–200 Gb/in² using conventional technology. There is an amusing video describing perpendicular recording at http://www.hitachigst.com/hdd/research/recording_head/pr/PerpendicularAnimation.html

- Find simple loci, both by reasoning and using ICT, to produce shapes and paths (L6/7)

Useful resources
Metre rule
Ruler
Compasses
Squared paper

Starter – Triangles from squares!

Ask students to find a triangle where all the angles are square numbers. (100°, 64°, 16°)

Can they find any quadrilaterals where all the angles are square numbers?
(144°, 100°, 100°, 16° and others.)

Teaching notes

It is important for students to realise that a locus of points is actually a mathematical way of describing the path of a moving object and the examples given should be discussed. It is also vital that they are allowed to consider the answer before they draw it accurately and, where appropriate, students should be asked to sketch a solution in their books.

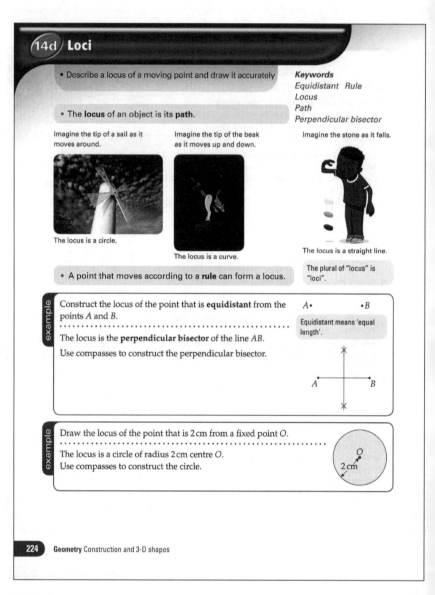

Plenary

Clear a space in your classroom – move a few desks – a people maths activity (EP).

- Have one person stand in the middle. Ask ten others to stand one metre away (they will make a circle of types)
- Have a 1 metre ruler – students have to stand 1 metre away from it – they can stand either side but they will also need to create semicircles around the end point
- Have two students stand 3 m apart – other students have to stand equal distances away from both – this creates a line of students across the middle.

Simplification

Where students are unable to construct the exact answer to the questions, allow them to sketch the answer. This will allow them to understand the concept of loci and, where appropriate, they can be referred back to the example in the previous exercises to copy the construction.

Question 1 – Similar to the three illustrations. Give students 5 mins to work in pairs and then discuss their answers, especially parts **d** and **e** (1.1).

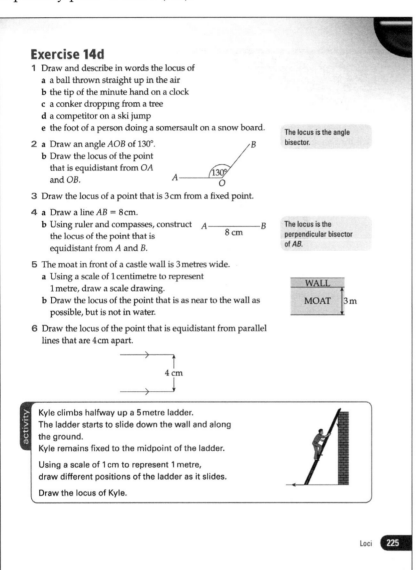

Exercise 14d

1 Draw and describe in words the locus of
   a a ball thrown straight up in the air
   b the tip of the minute hand on a clock
   c a conker dropping from a tree
   d a competitor on a ski jump
   e the foot of a person doing a somersault on a snow board.

> The locus is the angle bisector.

2 a Draw an angle *AOB* of 130°.
   b Draw the locus of the point that is equidistant from *OA* and *OB*.

3 Draw the locus of a point that is 3 cm from a fixed point.

4 a Draw a line *AB* = 8 cm.
   b Using ruler and compasses, construct the locus of the point that is equidistant from *A* and *B*.

> The locus is the perpendicular bisector of *AB*.

5 The moat in front of a castle wall is 3 metres wide.
   a Using a scale of 1 centimetre to represent 1 metre, draw a scale drawing.
   b Draw the locus of the point that is as near to the wall as possible, but is not in water.

| WALL |
| MOAT | 3 m |

6 Draw the locus of the point that is equidistant from parallel lines that are 4 cm apart.

4 cm

activity

Kyle climbs halfway up a 5 metre ladder.
The ladder starts to slide down the wall and along the ground.
Kyle remains fixed to the midpoint of the ladder.

Using a scale of 1 cm to represent 1 metre, draw different positions of the ladder as it slides.

Draw the locus of Kyle.

Loci **225**

Extension

Give students two points 6 cm apart. Can they find all the positions which are within 4 cm of each point? Can they design other questions like this? CT)

Exercise 14d commentary

Question 2 – The locus requires the angle bisector construction from spread **15c** (1.3).

Question 3 – Similar to the second example (1.3).

Question 4 – Similar to the first example. Encourage students to pick a point, P, on the perpendicular bisector and measure the distances AP and PB (1.3)

Question 5 – Centimetre-squared paper will make this easier. Ask students to consider what happens if the wall turns through a right-angle: the moat will include a quarter circle of radius 3 cm (1.1, 1.3)

Question 6 – The locus is a parallel line mid-way between those given. Accurate construction of the locus may prove challenging. It can be drawn by constructing a (mutually) perpendicular transversal and then its perpendicular bisector (1.1, 1.3).

Activity – Ensure that the workings are tidy as the ladder will need to be drawn repeatedly. If the man was not at the centre of the ladder, an ellipse would result (1.1, 1.3).

Assessment Criteria – SSM, level 7: find the locus of a point that moves according to a given rule, both by reasoning and using ICT.

Links

Drinking rocking birds (or dippy ducks) are toys that rock backwards and forwards and appear to drink from a glass of water. The bird consists of two glass bulbs joined by a tube filled with a coloured liquid. There is a video of a drinking bird and an explanation of how it works at http://www.Icefoundry.org/how-the-drinking-bird-works.php

- Use bearings to specify direction   (L6)   *Useful resources*
360° angle indicator
Ruler

Starter – Directions

Ask students to visualise the following shape:

A line 5 cm long towards N, then 4 cm in SE direction, then 4 cm in NE direction, finally 5 cm towards S.

Ask students to sketch what they have visualised. (The letter M)

Students can then make up their own direction puzzle.

Teaching notes

Using a 360° angle indicator is helpful as students are able to start at 0 and rotate the indicator (clockwise) to show the bearing. It is possible to use a protractor but when the bearing is more than 180°, students need to measure from the 180° line and add on the extra angle.

After explaining the conventions for measuring and quoting bearings, ask for a volunteer. Position the volunteer at the entrance to the class room and define the front of the class as North. Go around the class asking students to give a series of bearings and number of paces for the volunteer to follow so that his or her route will take him or her to the students desk.

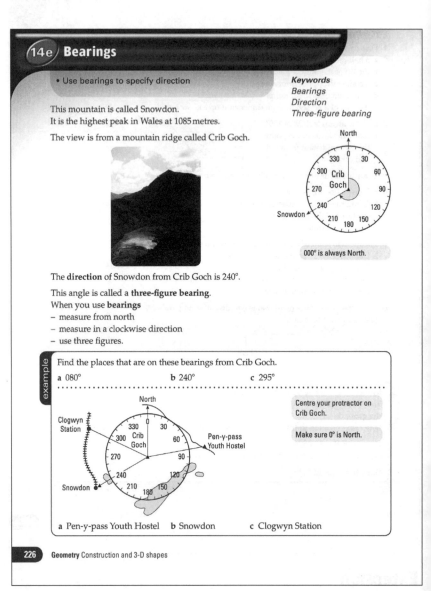

Plenary

In pairs, ask students to, use the map in question **1**, to find the bearings from each of the places to the Middle School. For example, the bearing of College from Middle School is 100°; the bearing of Middle School from the College is 280°. When they have found all the bearings, ask students if they can find a quick way of working out the reverse bearings. Most think that you take away the original bearing from 360° – instead you need to add the original bearing onto 180° (modulo 360º).

Simplification

Use an enlarged version of the compass rose in question **2** so that students can measure the bearings with a protractor. As students gain confidence, add points that do not coincide with the cardinal directions before removing the compass rose altogether.

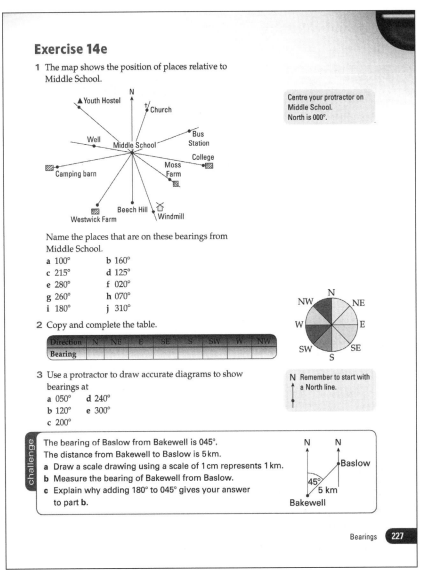

Exercise 14e commentary

Ensure that students measure angles in a clockwise direction, as opposed to the anticlockwise direction used for rotations.

Question 1 – Similar to the example. Ask students to write full answers: 'the bearing of the College is 100° from the Middle School'.

Question 2 – This can be related to geography. It can also be extended to include bearings such as NNE, ENE, *etc*. Check that students start the three figure bearings with a 0 where necessary (1.2, 1.3).

Question 3 – Suggest students look at their answers to question **1** to get an idea of the approximate directions of the bearings (1.3).

Challenge – For part **c**, students may need to be reminded of parallel line theorems. You could challenge students to find Edensor which is on a bearing of 070° from Bakewell and 195° from Baslow (1.3, 1.4).

Links

Live webcams for Mount Snowdon can be viewed at http://www.eryri-npa.co.uk/page/index.php?nav1=enjoying&nav2=12&nav3=1&lang=eng&contrast=1&view=graphic and at http://www.bbc.co.uk/wales/northwest/sites/webcams/pages/snowdon.shtml

Extension

Use the map with question **1**, ask students to find the bearings of Church, College and Westwick farm from Beech Hill. Then set them the task of writing three questions on bearings using the map – with a partner exchange questions, answer them and discuss what makes a good question? (RL)

- Visualise 3D shapes from their nets   (L6)
- Calculate [volumes and] surface areas of cuboids and
  shapes made from cuboids   (L6)

Useful resources

A cuboid and its net

Starter – Area pairs

Write the following three lists on the board.

Triangle bases and heights in cm: 6 and 8, 9 and 12, 7 and 7.
Rectangle lengths and widths in cm: 12 and 8, 14 and 2.5, 9 and 7.
Areas in cm²: 96, 54, 63, 24, 24.5, 35.

Ask students to match up the shapes and areas, and calculate the perimeters of the three rectangles.

Teaching notes

As an introduction to the lesson, show a cuboid and ask students 'what is the surface area of this cuboid?' 'How would they work out this area?' Give students time to discuss with each other and then take feedback. This will lead to a discussion around finding the area of each face and adding them together. This will lead to the introduction of nets and finding the area of each face (CT).

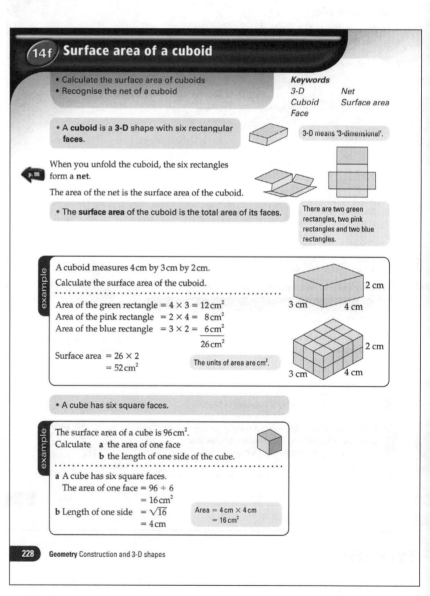

Plenary

If the challenge has been completed by most in the classroom ask students to work in pairs with this problem – if the area of one of the faces of a cuboid is 24 cm² and another is 40 cm² what could the area of the other face if it was not 15 cm²? There are a number of similar questions possible here, for example, area A = 48 cm², area B = 24 cm² find possible values for area face C. Ask students to sketch the cuboids.

Simplification

Where students find question **3** difficult, it may help to draw the net of the cuboid in each case.

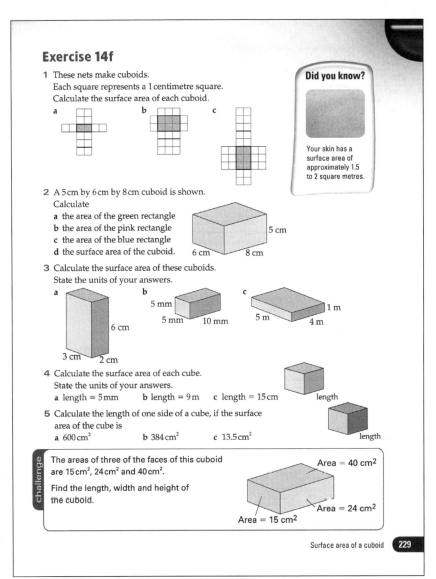

Exercise 14f

1 These nets make cuboids.
   Each square represents a 1 centimetre square.
   Calculate the surface area of each cuboid.
   a   b   c

Did you know?

Your skin has a surface area of approximately 1.5 to 2 square metres.

2 A 5 cm by 6 cm by 8 cm cuboid is shown.
   Calculate
   a the area of the green rectangle
   b the area of the pink rectangle
   c the area of the blue rectangle
   d the surface area of the cuboid.

   5 cm
   6 cm   8 cm

3 Calculate the surface area of these cuboids.
   State the units of your answers.
   a   b   c
            5 mm
            5 mm 10 mm   5 m   4 m   1 m
   6 cm
   3 cm   2 cm

4 Calculate the surface area of each cube.
   State the units of your answers.
   a length = 5 mm   b length = 9 m   c length = 15 cm   length

5 Calculate the length of one side of a cube, if the surface area of the cube is
   a 600 cm^2 b 384 cm^2 c 13.5 cm^2 length

challenge

The areas of three of the faces of this cuboid are 15 cm^2, 24 cm^2 and 40 cm^2.

Find the length, width and height of the cuboid.

Area = 40 cm^2
Area = 24 cm^2
Area = 15 cm^2

Surface area of a cuboid **229**

Extension

Ask questions such as, f the height and width of a cuboid are the same (whole centimetre) length and the area of one face is 12 cm², what could be the dimensions of the cuboid? Ask students to make up similar questions of their own (SM).

Exercise 14f commentary

Check that units are given with all answers (1.3).

Question 1 – Encourage students to multiply the lengths of the sides of the component rectangles rather than count the squares. Highlight the fact that, as the rectangles come in equal pairs, you can double the area of one to find the area of both.

Question 2 – A structured approach to calculating the total surface area, similar to the first example. Students may need to be reminded to double the areas of the three coloured rectangles (1.2, 1.3).

Question 3 – Refer students back to question **2** for scaffolding (1.2, 1.3).

Question 4 – A special case where the total surface area = 6 × area of one face (1.4).

Question 5 – Similar to the second example (1.3).

Challenge – Suggest students work in pairs to find the (prime) factors of the areas. Challenge those who find the question easy to write their own question.

Assessment Criteria – SSM, level 6: calculate surface areas of cuboids.

Links

Human skin accounts for between 15% and 20% of the total weight of the human body and helps to protect the body from the environment. It helps keep body fluids in and water and germs out and constantly renews itself. Over 90% of common house dust is made up of dead skin cells. There is more information about human skin at http://yucky.discovery.com/flash/body/pg000146.html

- Know and use the formula for volume of a cuboid   (L6)   ***Useful resources***
- Calculate volumes [and surface areas] of cuboids and   *Calculator*
  shapes made from cuboids   *Cubes*

Starter – Costly solids

A face cost 5p, an edge cost 8p and a vertex cost 10p.

Ask students to find the cost of different solids, for example, a cuboid, a square based pyramid, a triangular prism, a pentagonal prism.

(Answers: £2.06, £1.39, £1.57 and £2.55)

Can be differentiated by the choice of costs.

Teaching notes

It is important to show students the reasoning why we multiply $L \times W \times H$ to find the volume of cuboids. Use cubes to make the example cuboid, $2 \times 4 \times 3$ units in size. You can show this to be 3 layers of 2×4. That is 3 layers of 8 cubes, 3×8 is 24 cubes.

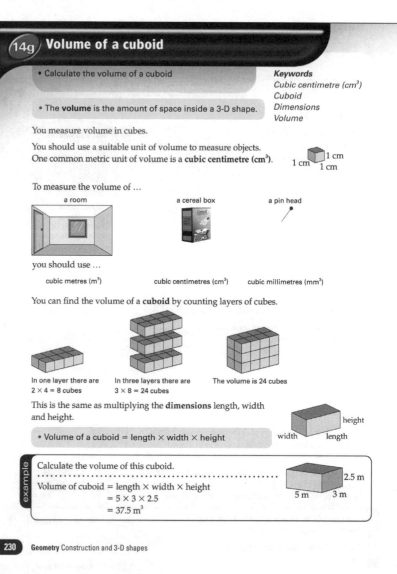

14g Volume of a cuboid

- Calculate the volume of a cuboid

Keywords
Cubic centimetre (cm³)
Cuboid
Dimensions
Volume

- The **volume** is the amount of space inside a 3-D shape.

You measure volume in cubes.

You should use a suitable unit of volume to measure objects. One common metric unit of volume is a **cubic centimetre (cm³)**.

To measure the volume of …
a room   a cereal box   a pin head

you should use …
cubic metres (m³)   cubic centimetres (cm³)   cubic millimetres (mm³)

You can find the volume of a **cuboid** by counting layers of cubes.

In one layer there are   In three layers there are   The volume is 24 cubes
$2 \times 4 = 8$ cubes $3 \times 8 = 24$ cubes

This is the same as multiplying the **dimensions** length, width and height.

- Volume of a cuboid = length × width × height

example
Calculate the volume of this cuboid.
....
Volume of cuboid = length × width × height
$= 5 \times 3 \times 2.5$
$= 37.5$ m³

230   **Geometry** Construction and 3-D shapes

Plenary

Give students the information that the volume of a cube is 24 cm³. What could be the dimensions of the cuboid? Take some answers and then give further information. It looks like a long stick, or the height is half the length. Ask for further answers. Continue with a similar question such as the volume of a cuboid is 60 cm³. The width and length are the same. What are the dimensions? Now ask students, in pairs, to write their own problem. They should share with another pair and exchange questions. You could collect in the questions and put them together for homework! (CT)

Simplification

It will help students find volume difficult to be able to see cuboids. Make the cuboids with small cubes so that the problem can be broken down to: how many layers? How many rows in each layer? How many cubes in each row? This allows students to appreciate $L \times W \times H$. Where appropriate a calculator should be used.

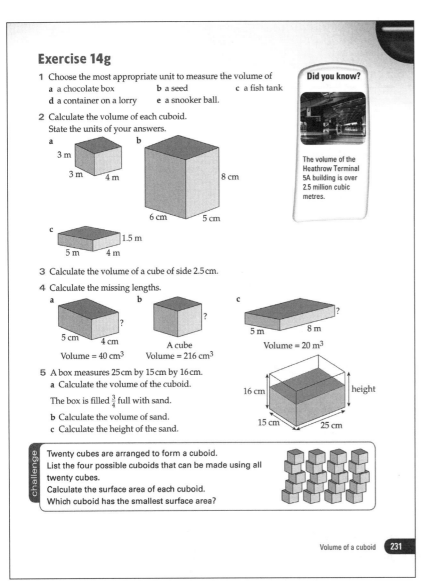

Exercise 14g

1 Choose the most appropriate unit to measure the volume of
 a a chocolate box b a seed c a fish tank
 d a container on a lorry e a snooker ball.

2 Calculate the volume of each cuboid. State the units of your answers.
 a 3 m 3 m 4 m
 b 8 cm 6 cm 5 cm
 c 1.5 m 5 m 4 m

3 Calculate the volume of a cube of side 2.5 cm.

4 Calculate the missing lengths.
 a 5 cm 4 cm ? Volume = 40 cm³
 b ? A cube Volume = 216 cm³
 c ? 5 m 8 m Volume = 20 m³

5 A box measures 25 cm by 15 cm by 16 cm.
 a Calculate the volume of the cuboid.
 The box is filled $\frac{3}{4}$ full with sand.
 b Calculate the volume of sand.
 c Calculate the height of the sand.
 16 cm height 15 cm 25 cm

challenge
Twenty cubes are arranged to form a cuboid.
List the four possible cuboids that can be made using all twenty cubes.
Calculate the surface area of each cuboid.
Which cuboid has the smallest surface area?

Did you know?
The volume of the Heathrow Terminal 5A building is over 2.5 million cubic metres.

Volume of a cuboid **231**

Extension

In question **5,** imagine that the box was placed on its side so that the height was 15 cm. What height would the sand be now? Ask students how else the box could be placed (1.2).

Exercise 14g commentary

Check that all answers have appropriate units (1.3).

Question 1 – Similar to the discussion. Students can work in pairs to agree on the best answers (1.3).

Question 2 – Similar to the example (1.2, 1.3).

Question 3 – A special case where volume = length³. If a calculator is used, ask students to first write down an estimated volume: $8 < 2.5^3 < 27$ (1.3).

Question 4 – Remind students to think which lengths (numbers) they need to multiply to find the volume (1.2, 1.3).

Question 5 – In part **c**, if students used a method similar to that in question **4**, ask them to calculate $\frac{3}{4}$ of the height: why does this give the same answer? (1.3).

Challenge – Make available small cubes for students to experiment with but also encourage students to think about the prime factor decomposition of $20 = 2^2 \times 5$. Refer back to spread **15f** for the surface area calculation (IE).

Assessment Criteria – SSM, level 6: deduce and use the formula for the volume of a cuboid. SSM, level 6: calculate volumes of cuboids.

Links

The largest building in the World by volume is the Boeing aircraft factory at Everett, Washington state in the USA. The volume of the building is 13.3 million m³ and it has a floor area of 398,000 m² or 98 acres. There is more information about the factory at http://www.boeing.com/commercial/facilities/index.html

14a

1 Construct these triangles.
Measure the lengths of the sides and calculate the perimeter of each triangle.

a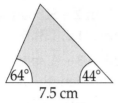
7.5 cm

b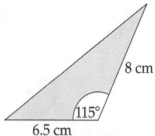
8 cm
115°
6.5 cm

c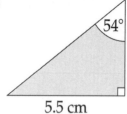
54°
5.5 cm

14b

2 Construct these triangles, using ruler and compasses.
State the mathematical name of each triangle.

a
25 mm   35 mm
30 mm

b
4.5 cm
3 cm
3 cm

c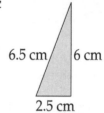
6.5 cm   6 cm
2.5 cm

14c

3 a Draw a large triangle.
   b Construct the perpendicular bisector for each of the three sides.
   c Draw a circle passing through the vertices of the triangle, using the point of intersection as the centre.

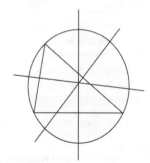

14c²

4 Copy this diagram.
Using compasses, construct a vertical wall through the dot to separate the giraffe from the person.

5 Copy the diagrams on square grid paper.
Draw the locus of the point that is equidistant from P and Q.

a

b

c

> Equidistant means 'equal length'.

6 Measure the three-figure bearings of these places from Manchester.

a Leeds   **b** Liverpool

c Sheffield   **d** Preston

e Bradford   **f** Chester

> You may need to trace the map and extend the lines.

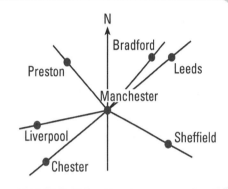

7 Calculate the surface area of these cuboids.

a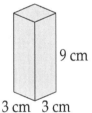
9 cm
3 cm   3 cm

b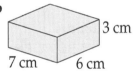
3 cm
7 cm   6 cm

c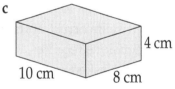
4 cm
10 cm   8 cm

8 Calculate the length of one side of the cube, if the surface area of the cube is

a $294 \, cm^2$ **b** $1536 \, cm^2$ **c** $235.5 \, cm^2$

length

9 Copy and complete the table for the cuboids.

	Length	Width	Height	Volume
a	9 m	8 m	8 m	
b	15 mm	10 mm	20 mm	
c	2.5 cm	2 cm	8 cm	
d	5 m	3.5 m	3.2 m	
e	6 cm	2 cm		$42 \, cm^3$
f	4 m	4 m		$48 \, m^3$
g	7 cm		20 cm	$280 \, cm^3$
h	9 cm		32 cm	$288 \, cm^3$

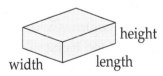
height
width   length

14 Summary

Assessment criteria

- Measure and draw angles to the nearest degree, when constructing models and drawing or using shapes — Level 5
- Use straight edge and compasses to do standard constructions — Level 6
- Deduce and use the formula for the volume of a cuboid — Level 6
- Calculate volumes and surface areas of cuboids — Level 6
- Find the lows of a point that moves according to a given rule, both by reasoning and using ICT — Level 7

Question commentary

Example

The example illustrates a typical problem on constructing the perpendicular bisector of a line segment. Some students may suggest that the bisector could be constructed by drawing a line at right angles without using compasses. Ask why compasses are important when doing constructions. Emphasise that students may lose marks if they rub out the construction lines and that the pencil or compass lead used should be sharp.

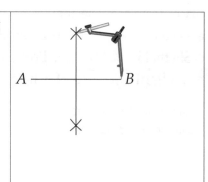

Past question

The question requires students to find the volume of a cuboid. Students may use incorrect units for the volume. Emphasise that 3-dimensions requires a unit containing a '3'. In part **b**, students are asked to find a side length of a cuboid given its area and the other two side dimensions. Either trial and improvement or inverse operations can be used here. Emphasise that a unit for the length should be given. Ask students how they would find the surface area of the two cuboids.

Answer

Level 6

2 **a** $60 \, \text{cm}^3$

  b $x = 6 \, \text{cm}$

Development and links

This topic is developed further in Year 9.

The ability to construct shapes accurately is important in design technology and in some forms of art. Bearings are used in navigation and link to the geography curriculum. Students will use volumes and surface areas in science when dealing with solids, liquids and gases, and in food technology and design technology to calculate required quantities of ingredients or materials.

15 Analysing and interpreting data

Objectives

- Calculate statistics for sets of discrete and continuous data 5
- Recognise when it is appropriate to use the range, mean, median and mode and, for grouped data, the modal class 5
- Interpret tables, graphs and diagrams for both discrete and continuous data, relating summary statistics and findings to the questions being explored 5
- Construct stem-and-leaf diagrams 6
- Construct simple scatter graphs 6

Level

MPA

1.1	15a, b, c, f, CS
1.2	15d, e, CS
1.3	15a, b, c, d, e, f, CS
1.4	15b, c, f, CS
1.5	15b, d, e, CS

PLTS

IE	15b, c, d, CS
CT	15a, b, d, e, f, CS
RL	15c, d, CS
TW	15d, e, CS
SM	15a, b, c, d, e, f, CS
EP	15c, d, e, f, CS

Introduction

The focus of this chapter is on reading and interpreting different types of chart including bar charts, pie charts, frequency diagrams, simple time series line graphs and the construction of stem-and-leaf and scatter diagrams. The chapter begins by calculating statistics for discrete and grouped data from a frequency table. Students begin to consider when it is appropriate to use the mean, mode or median to describe data.

The student book discusses the use of statistics to determine the cost effectiveness of modern medical treatments. In daily life, students are bombarded with statistics in the media. A clear understanding of how statistics are calculated and the features of different types of statistical chart will help prevent students from being mislead by biased reports and will enable them to make informed decisions.

Fast-track

All spreads

Extra Resources

15	Start of chapter presentation
15a	Animation: Mode, median, mean and range
15a	Animation: Median, mean and range
15a	Consolidation sheet
15b	Consolidation sheet
15c	Starter: Statistics multichoice
15c	Worked solution: Q2
15e	Consolidation sheet
15f	Animation: Stem-and-leaf diagrams
15f	Consolidation sheet Maths life: Electricity in the home
	Assessment: chapter 15

- Recognise when it is appropriate to use the range, mean, median and mode   (L5)

Useful resources

● ●

Starter – Today's number is ...111

Ask questions based on today's number', for example,

What is 50% of 111? (55.5)
What is the closest square number to 111? (121)
What is 400 subtract 111? (289)
111 is a multiple of 3.
True or false? (T)
What are the factors of 111?
(1, 3, 37, 111).

Teaching notes

A difficulty is remembering which average is which, ask students how they remember:

Mean: is the hardest one to calculate.

MedIan: is the mIddle term.

Mode: is the mOst common term.

Work through an example, showing how the statistics are calculated. Emphasise the need to order the data; essential to calculate the median and useful for the mode, mean and range. (A table may help with ordering for a large data set.) How to treat no or multiple modes and finding the median for an even number of data values will need special attention.

It will be helpful to discuss when it is appropriate to use the various averages. The mean is useful for continuous data, for example, students' heights. The mode is the only one that is defined for categorical data, for example, students' eye colours. The median is useful when the data has long tails, for example, the typical (average) salary in the UK. (The mean can be distorted by the super-rich) (1.1).

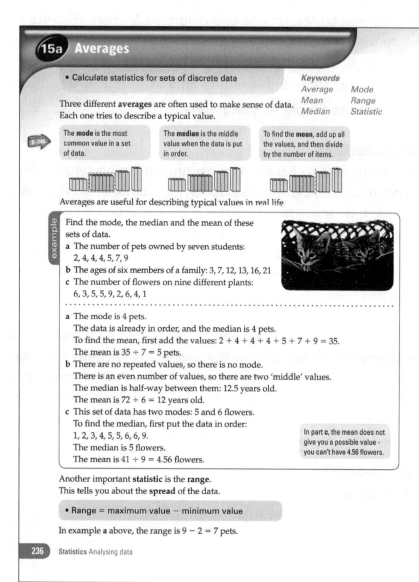

Plenary

Pair up students and give them 30 seconds to agree an answer to each question.

I have four amounts of money:

- the mode is £6   What could the amounts be? ($n, 6, 6, m$)
  Let all groups agree a possible solution.
- the mean is £5   What could they be now? ($n, 6, 6, 8 - n$)
- the median is £6   What could they be now? ($n = 0, 1, 2$)
  Do all groups agree?

You could repeat this with another similar challenge.

Simplification

Give students the numbers or amounts written on cards, for example, a set of five cards with £3, £4, £4, £6 and £8. Ask, what are the mean (5), mode (4) and median? (4)

Using only the last four cards, ask, what is the largest? (8) Smallest? (4) Mean? (5.5) Median? (5)

Can they write questions of their own?

Exercise 15a

1 Four student in class 8A who used the 'MyFace' website recorded the number of messages they received each day.
   Amy   3, 5, 5, 5, 6, 6, 2
   Jen   4, 4, 8, 9, 9, 11
   Selma   5, 6, 1, 2, 5, 7, 8, 4, 6, 2, 6, 6
   Jo   2, 1, 7, 8, 5, 3
   a How many days did each person keep records for?
   b Find the **mode** of each of these sets of data.

2 Mrs Morgan asked student in class 8B to find out about mobile phone use. Some student asked people how many text messages they had sent the previous day.
   Kalid   3, 5, 9
   Dan   2, 6, 7, 9
   James   2, 3, 6, 8, 8, 9, 11
   Emily   4, 5, 4, 7, 1
   Steph   5, 0, 8, 7, 3, 9
   a Who asked most people?
   b Calculate the **median** of each person's data.

3 Five basketball players recorded the number of points they scored in each game that they played in a tournament.
   Sandra   5, 5, 5, 5
   Lizzie   3, 4, 5, 6
   Maya   2, 9, 7, 2
   Abbie   3, 6, 7, 9, 15
   Jill   3, 6, 9, 19, 13, 7
   a How many games did each person play?
   b Calculate the **mean** score for each player.

4 Work out the **range** of each of these sets of data.
   a Number of people living in five houses: 1, 1, 4, 6, 9
   b Number of absences for five student: 5, 4, 3, 9, 6
   c Number of books borrowed from the school library by six student: 9, 5, 0, 6, 4, 8
   d Mass (in kilograms) of 5 cats: 4.8, 6.3, 2.1, 5.8, 2.2

> **discussion**
> Can you make up examples of sets of numbers where:
> a The mean, median and mode are all equal?
> b The mean and median are the same, but the mode is smaller?
> c The mode is bigger than the median, and the median is bigger than the mean?
> d The mode is 1, the median is 2 and the mean is 3?

Averages   **237**

Extension

Following the discussion activity, ask students to make up a question of their own similar to this activity, together with its answer. Then exchange questions with another in the group.

Exercise 15a commentary

Students could be asked to collect their own data to investigate, similar to that in questions **1** and **2** (SM). All questions (1.3)

Question 1 – The case of multiple modes (Jen) and no mode (Jo) may merit discussion.

Question 2 – Students may need to be reminded how to handle an even number of data points and to order the data first. Ask if it makes a difference if the ordered list is ascending or descending.

Question 3 – For Sandra and Lizzie, ask if students can 'see' the answer without doing a calculation – how?

Question 4 – Units should be given for part **d**.

Discussion – Ask students to work in pairs, with one pair's solution being checked by another pair. Further constraints can be introduced, such as, there are only four data values. What is minimum number of data values needed satisfy each constraint? (CT)

Assessment Criteria – HD, level 5: understand and use the mean of discrete data.

Links

The word *average* comes from the French word *averie* which means "damage sustained at sea". Costs of losses at sea were shared between the ship owners and the cargo owners and the calculations used to assess the individual contributions gave rise to the modern sense of the word *average*.

- Calculate statistics for sets of discrete and continuous data (L6)

Useful resources

Starter – DVDs

Ask students to calculate the mean, median and range of the playing times of the following DVDs:

Harry Potter and the Chamber of Secrets 2 hours 34 minutes

Toy Story 2	1 hour 29 minutes
Spiderman	2hours 1 minute
Billy Elliot	1 hour 46 minutes
Batman Begins	2 hours 20 minutes

(Mean = 2hr 2min, median = 2 hr 1min, range = 1hr 5min)

Teaching notes

When using frequency tables, students often get confused over their meaning. To make things more concrete work through an example, simultaneously showing the data in an ordered list and in a table so that they can connect the two (1.1). When finding the total number of entries in the table, encourage students to note down their partial sums, as this will help them identify in which class(es) the 'median value(s)' lie.

When you have a large amount of data, a useful way to find the middle value is to calculate $\frac{(n+1)}{2}$; in the first example $\frac{(27+1)}{2} = 14$. For an even number of data values this gives a fractional value, say $3\frac{1}{2}$, which reminds you to take $\frac{1}{2}$(3rd + 4th) values. When calculating the mean, emphasise the efficiency of performing a single multiplication in the table compared with many repeated additions in the ordered list.

It may help to show a bar chart of the data in the frequency table: the mode is the highest column, the median is the value for which there are as many values to the left as the right and the mean coincides with the centre of gravity or 'balance point'.

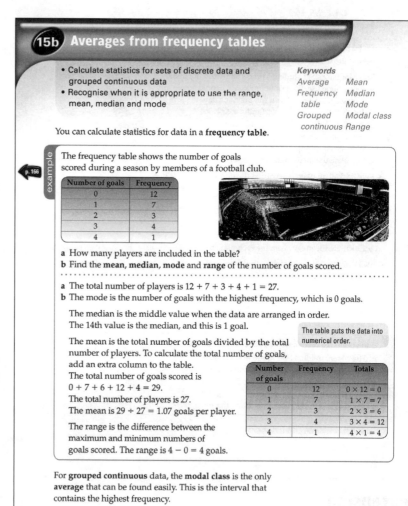

15b Averages from frequency tables

- Calculate statistics for sets of discrete data and grouped continuous data
- Recognise when it is appropriate to use the range, mean, median and mode

Keywords
Average   Mean
Frequency   Median
table   Mode
Grouped   Modal class
continuous   Range

You can calculate statistics for data in a **frequency table**.

The frequency table shows the number of goals scored during a season by members of a football club.

Number of goals	Frequency
0	12
1	7
2	3
3	4
4	1

a How many players are included in the table?
b Find the **mean**, **median**, **mode** and **range** of the number of goals scored.

a The total number of players is 12 + 7 + 3 + 4 + 1 = 27.
b The mode is the number of goals with the highest frequency, which is 0 goals.

The median is the middle value when the data are arranged in order. The 14th value is the median, and this is 1 goal.

> The table puts the data into numerical order.

The mean is the total number of goals divided by the total number of players. To calculate the total number of goals, add an extra column to the table.
The total number of goals scored is
0 + 7 + 6 + 12 + 4 = 29.
The total number of players is 27.
The mean is 29 ÷ 27 = 1.07 goals per player.

The range is the difference between the maximum and minimum numbers of goals scored. The range is 4 − 0 = 4 goals.

Number of goals	Frequency	Totals
0	12	0 × 12 = 0
1	7	1 × 7 = 7
2	3	2 × 3 = 6
3	4	3 × 4 = 12
4	1	4 × 1 = 4

For **grouped continuous** data, the **modal class** is the only **average** that can be found easily. This is the interval that contains the highest frequency.

238   Statistics Analysing data

Plenary

Show the class a number of bar charts for discrete numerical data; include a symmetric distribution and one with a long tail or outlier. Ask them to give the mode and estimate the median and mean for each data set. For each data set, ask which is the best average to quote for a typical value and why. (CT)

Simplification

To help understand the meaning of the frequency table it may be useful to write out in full the list of values, for example in question **1**: 0, 0, 0, 0, 0, 0, 0, 0, 0, 0, 1, 1, 1, 1, 1, 1, 1, 1, 1, 1, 1, 1, 2, 2, 2, 2, 2, 2, 2, 3, 3.

It is also easier (but less efficient) to calculate a mean using the numbers in this format (1.1).

Exercise 15b

1 This frequency table shows the number of pets owned by the students in class 8C5.
   a How many students' data are shown in the table?
   b What is the modal number of pets owned by the students in 8C5?
   c What is the range of the number of pets owned by people in 8C5?

Number of pets	Frequency
0	10
1	12
2	7
3	2

2 Students in class 8P were asked to record the number of portions of fruit and vegetables that they ate one day.
   a How many students' data are shown in the table?
   b What is the modal number of portions eaten?
   c What was the median of the number of portions eaten?

Portions of fruit and vegetables	Frequency
0	1
1	2
2	5
3	4
4	9
5	6
6	4
7	1

3 The table shows the number of absences for the students in class 8K during one term.
   a Find the total number of students whose data are shown in the table.
   b Redraw the table with an extra column to show the total number of absences.
   c Find the overall total number of absences for the whole class.
   d Use your answers to parts **a** and **c** to calculate the mean number of absences per student.

Number of absences	Frequency
0	10
1	5
2	4
3	2
4	3
5	3
6	2
7	0
8	1

discussion

The table shows the heights of students in class 8KT.
a What is the modal class for this set of data?
b Why is the modal class the only average that can be worked out easily?

Height, h cm	Frequency
$170 \leqslant h < 180$	3
$160 \leqslant h < 170$	9
$150 \leqslant h < 160$	14
$140 \leqslant h < 130$	5

Extension

Give students the following data and ask them to calculate the mean (101.4).

102, 101, 101, 100, 103, 104, 102, 104, 99, 98

Is there an easier way to do the calculation? $\left(\frac{100 + 14}{10} \right)$
Will this method work in other cases? (SM) (1.4, 1.5)

Exercise 15b commentary

Again students could be asked to collect their own data and carry out a statistical analysis using the questions below as a guide (IE)

Question 1 – Students could be asked to find the median number of pets (16th data value, 1).

Question 2 – Students may need help with part **c**. Encourage them to note down the partial/running total frequency, 1, 3, 8, 12, 21, 27, 31, 32, so that it is easy to identify the 16th and 17th data values (1.3).

Question 3 – Students need to be methodical and should follow the layout in the example. Check that students do not divide by 9, the number of classes (1.3).

Discussion – Ask students if there are circumstances when it is easy to calculate the median and mean (symmetric distributions).

Assessment Criteria – HD, level 5, understand and use the mean of discrete data.

Links

Bring in some newspaper/web articles on supposed superfoods, for example, http://www.bbc.co.uk/food/food_matters/superfoods.shtml

Do students think these are better than the usual five portions of fruit and vegetables that they eat a day? What do they think about nutraceuticals, like bread with added folic acid (vitamin B9)? In the UK any health claim on a food label must be true and not misleading. How would students decide if a particular food was beneficial?

- Interpret tables, graphs and diagrams for discrete and continuous data, relating summary statistics and findings to the questions being explored (L5)

Useful resources
Protractor or angle indicator
Calculator

Starter – Missing numbers

Ask students to find

Four numbers with a mean of 15 and a range of 11 (11, 12, 15, 22)
Four numbers with a mean of 16 and a range of 14 (9, 13, 19, 23)
Five numbers with a mean of 14, a median of 12 and a range of 9 (10, 12, 12, 17, 19)

(Other solution are possible.)

Teaching notes

Use a classroom discussion to agree what information can be extracted from pie charts and bar charts. For bar charts show how to convert the graph in to a frequency table so that the median and mean can be calculated systematically and more easily. The averages can then be related back to features of the graphs.

Mode: highest bar.
Median: the value for which there are as many values to the left as the right
Mean: coincides with the centre of gravity or 'balance point'

Take the opportunity to highlight the difference between the graphs for discrete data and continuous data. The chart for discrete data has gaps between the bars and the chart for continuous data does not have gaps.

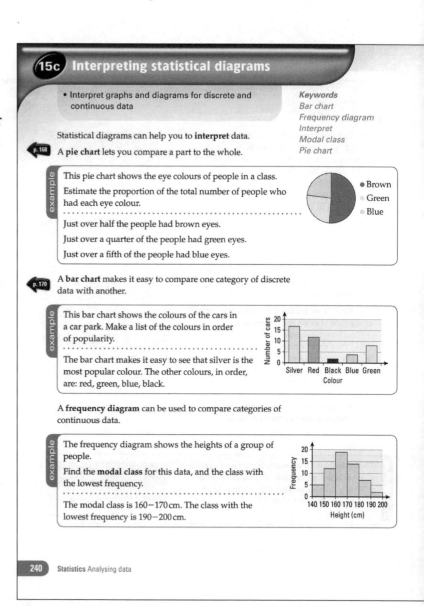

Plenary

Working in pairs, ask students to write four questions to accompany a given bar chart or frequency diagram, for example, whales spotted in the North Atlantic. Students exchange questions with another pair. Each pair should agree which is the best question and explain how one of the questions can be improved. Take whole class feedback (EP, RL).

Simplification

With questions **2** and **3,** it is appropriate to break down the questions. For example, to find the total number of pets owned, ask how many people owned one pet? How many pets is that? How many owned two pets? How many pets is that altogether? *etc.*

Students could be asked to write a short description of the three data sets appearing in the questions.

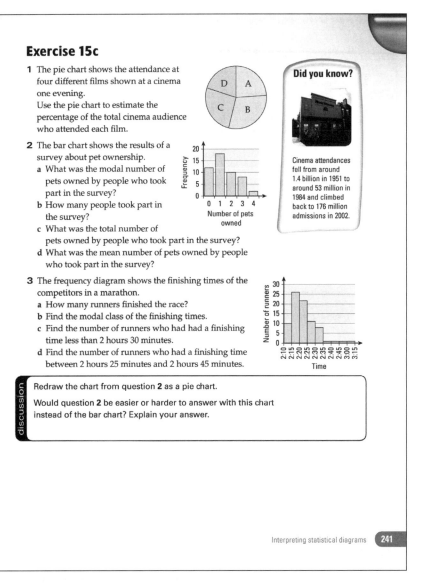

Exercise 15c

1 The pie chart shows the attendance at four different films shown at a cinema one evening.
Use the pie chart to estimate the percentage of the total cinema audience who attended each film.

Did you know?

Cinema attendances fell from around 1.4 billion in 1951 to around 53 million in 1984 and climbed back to 176 million admissions in 2002.

2 The bar chart shows the results of a survey about pet ownership.
a What was the modal number of pets owned by people who took part in the survey?
b How many people took part in the survey?
c What was the total number of pets owned by people who took part in the survey?
d What was the mean number of pets owned by people who took part in the survey?

Frequency
Number of pets owned

3 The frequency diagram shows the finishing times of the competitors in a marathon.
a How many runners finished the race?
b Find the modal class of the finishing times.
c Find the number of runners who had had a finishing time less than 2 hours 30 minutes.
d Find the number of runners who had a finishing time between 2 hours 25 minutes and 2 hours 45 minutes.

Number of runners
Time

discussion

Redraw the chart from question **2** as a pie chart.

Would question **2** be easier or harder to answer with this chart instead of the bar chart? Explain your answer.

Interpreting statistical diagrams **241**

Extension

Give students a frequency diagram to analyse, for example, a diagram showing the heights of boys in year 8. Ask them to bullet point what the diagram shows. Now challenge the students to draw a similar diagram for the heights of girls in year 8. Having drawn this, can they write three or four questions with each question being more challenging than the previous question? (SM)

Exercise 15c commentary

Question 1 – Similar to the first example. Ask students to check their answers – does the biggest percentage correspond to the largest sector? Ask how many saw which film (give them titles) if 400 people went to the cinema (1.3).

Question 2 – Part **a** is similar to the second example. Encourage students to read off the frequencies from the graph and to transfer them into a frequency table, so that they can follow the approach used in the previous spread to calculate the mean.

Question 3 – Part **b** is similar to the third example. Insist that students write down the number of entries in each class so that any mistakes are easier to trace.

Discussion – Students may need reminding how to construct a pie chart, 1 pet $= 9°$. Encourage students to calculate the angles mentally. To answer parts **b–d**, the number of entries in the pie chart, 40, must be supplied separately (IE). (1.1, 1.3, 1.4)

Assessment Criteria – HD, level 5: interpret graphs and diagrams, including pie charts, and draw conclusions.

Links

The largest marathon in the World is the New York City Marathon which had 37 850 finishers in 2006. The course is 42.195 km long and the record time is 2 hours, 7 minutes and 43 seconds for men and 2 hours 22 minutes and 31 seconds for women. Ask the class to give an estimate of the average speed of these runners in kilometres per hour. There is more information about the New York City Marathon at http://www.nycmarathon.org/home/index.php

- Identify key features present in the data (for) line graphs for time series

(L5)

Useful resources
Ruler and pencil
Squared (or graph) paper

Starter – House numbers

Seven consecutive house numbers add up 140.

Ask students: What could the numbers be and what is the mean house number?

(14, 16, 18, 20, 22, 24, 26; mean = 20)

What if the sum of the house numbers is 189?

(21, 23, 25, 27, 29, 31, 33; mean = 27)

Can be differentiated by the choice of total.

Teaching notes

Using the first examples given in the text, ask students to describe what they see. Encourage the use of statistical language and note down the points. Use these as a basis to construct a short paragraph that describes the time series data.

Organise the students into small groups and give them the second time series graph for which they have to write a short description. Ask students to read out their answers and ask the rest of the class to give one good point and one area for improvement in each description (TW, RL) (1.5).

Ask the students why they think the birth rate in Italy will rise steadily until 2050. Do they think this is reliable? Why might it be different? Do they think it is realistic to make such a prediction?

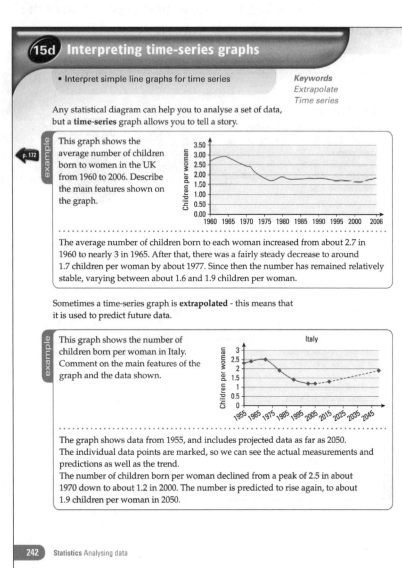

15d Interpreting time-series graphs

- Interpret simple line graphs for time series

Keywords
Extrapolate
Time series

Any statistical diagram can help you to analyse a set of data, but a **time-series** graph allows you to tell a story.

example p. 172 This graph shows the average number of children born to women in the UK from 1960 to 2006. Describe the main features shown on the graph.

The average number of children born to each woman increased from about 2.7 in 1960 to nearly 3 in 1965. After that, there was a fairly steady decrease to around 1.7 children per woman by about 1977. Since then the number has remained relatively stable, varying between about 1.6 and 1.9 children per woman.

Sometimes a time-series graph is **extrapolated** - this means that it is used to predict future data.

example This graph shows the number of children born per woman in Italy. Comment on the main features of the graph and the data shown.

The graph shows data from 1955, and includes projected data as far as 2050. The individual data points are marked, so we can see the actual measurements and predictions as well as the trend.
The number of children born per woman declined from a peak of 2.5 in about 1970 down to about 1.2 in 2000. The number is predicted to rise again, to about 1.9 children per woman in 2050.

242 Statistics Analysing data

Plenary

Show students a time-distance graph for a journey to school. For example, for a bus journey there will be periods of walking, stopping (waiting) and travelling faster. Working in pairs, ask students to describe the journey. Next, ask students to try to draw a time-series graph (time-distance) for one of their journeys to school. Start from when they leave their house to the start of school. There will be a need to discuss and explain a journey and one student will need to describe it carefully to the other (EP).

Simplification

Give students graphs similar to those in the two examples but with modified data values. Then ask them to copy the descriptions given in the text but, where appropriate, to change the details to match the new graphs.

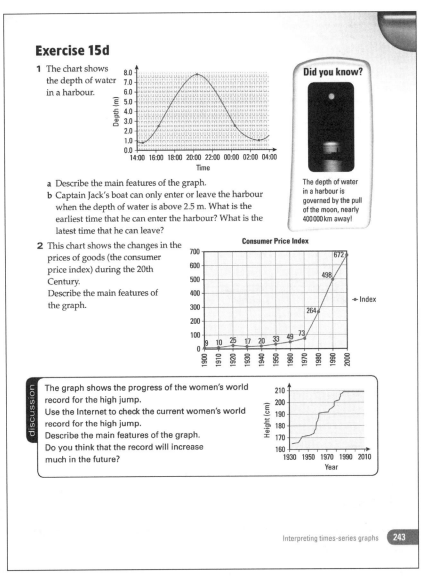

Exercise 15d

1 The chart shows the depth of water in a harbour.

a Describe the main features of the graph.

b Captain Jack's boat can only enter or leave the harbour when the depth of water is above 2.5 m. What is the earliest time that he can enter the harbour? What is the latest time that he can leave?

Did you know?

The depth of water in a harbour is governed by the pull of the moon, nearly 400 000 km away!

2 This chart shows the changes in the prices of goods (the consumer price index) during the 20th Century.
Describe the main features of the graph.

discussion

The graph shows the progress of the women's world record for the high jump.
Use the Internet to check the current women's world record for the high jump.
Describe the main features of the graph.
Do you think that the record will increase much in the future?

Interpreting times-series graphs 243

Extension

Ask students to find data for another Olympic sport and to produce their own time-series graph of the mens' and/or womens' world record against year (SM) (1.3).

Exercise 15d commentary

Encourage a systematic approach to writing the descriptions and the use of bullet points.

Question 1 – Similar to the second example. Ask students how they think the graph would continue for a full 24 hr period; would it be the same in 14 days time? (CT) (1.2).

Question 2 – Explain that the CPI is based on typical prices of the goods an average household would buy. Ask students what they would include and how they would calculate the average cost of their 'shopping basket' (IE).

Discussion – Encourage students to find out who held the world record during the period indicated. Did the same person hold the record during the 'level' periods (SM) (1.2).

Assessment Criteria – HD, level 5: create and interpret line graphs where the intermediate values have meaning.

Links

The highest tides in the United Kingdom are in the Bristol Channel where in extreme cases the water can rise up to 15 metres between low and high tide. Before Bristol's floating harbour was built at the beginning of the 19th century, boats unloading at Bristol were stranded in the mud for a considerable length of time at low tide. Boats had to be in good condition to withstand the stresses and strains this caused and originated the phrase "shipshape and Bristol fashion". A graph showing current tide information for Bristol can be found at http://www.bbc.co.uk/weather/coast/tides/west.shtml

- Construct simple scatter graphs   (L6)
- Construct and identify key features present in the data for scatter graphs to develop further understanding of correlation   (L6)

Useful resources
Ruler
Pencil
Squared paper

Starter – Range bingo

Ask students to draw a 3 × 3 grid and enter nine numbers between 15 and 35 inclusive.

Give pairs of numbers, for example, 17 and 41 and ask students find the range.

Winner is the first student to cross out all their numbers.

Teaching notes

Encourage students to think about the use of appropriate scales before starting to draw a scatter graph. These should start and stop so as to avoid large areas of wasted white space. For example, in question **2**, it would not be appropriate to start the scales from 0.

Students usually find it easy to differentiate between negative, no and positive correlations but often confuse the strength of a correlation with the steepness of the 'line of best fit'. Emphasise that the strength, or weakness, is a measure of how tightly the data cluster about a sloping line. Encourage students to describe correlations in two ways. There is a positive correlation between maths and science test scores. Students who get high (low) scores in maths tests tend to also get high (low) scores in science tests.

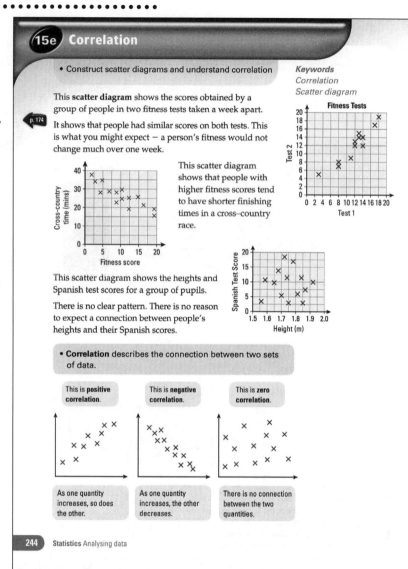

Plenary

Working in pairs, students draw 6 –10 pairs of axes as grids. Give them these axis headings –

- The size of car engine (litres)
- The size of the boot (cubic metres)
- The top speed (mph)
- The price of the car (£)
- The fuel economy rate (miles per litre)
- The age of the car

Label the vertical axis 'size of the engine' and use the other headings on the horizontal axes for different grids. The task is to estimate a scatter graph for each and describe the correlation. This is a discussion topic but, with time, this will allow some students to consider other pairings, for example, top speed against economy rate (TW).

Simplification

Provide assistance in drawing the axes and particularly the scale used. It would be possible to have grids available for students' use, but they need also to be able to draw these sensibly themselves. Extra help may be required.

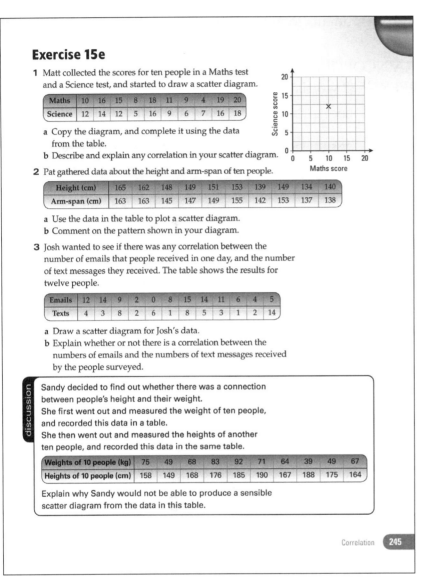

Exercise 15e

1 Matt collected the scores for ten people in a Maths test and a Science test, and started to draw a scatter diagram.

Maths	10	16	15	8	18	11	9	4	19	20
Science	12	14	12	5	16	9	6	7	16	18

a Copy the diagram, and complete it using the data from the table.
b Describe and explain any correlation in your scatter diagram.

2 Pat gathered data about the height and arm-span of ten people.

Height (cm)	165	162	148	149	151	153	139	149	134	140
Arm-span (cm)	163	163	145	147	149	155	142	153	137	138

a Use the data in the table to plot a scatter diagram.
b Comment on the pattern shown in your diagram.

3 Josh wanted to see if there was any correlation between the number of emails that people received in one day, and the number of text messages they received. The table shows the results for twelve people.

Emails	12	14	9	2	0	8	15	14	11	6	4	5
Texts	4	3	8	2	6	1	8	5	3	1	2	14

a Draw a scatter diagram for Josh's data.
b Explain whether or not there is a correlation between the numbers of emails and the numbers of text messages received by the people surveyed.

discussion

Sandy decided to find out whether there was a connection between people's height and their weight.
She first went out and measured the weight of ten people, and recorded this data in a table.
She then went out and measured the heights of another ten people, and recorded this data in the same table.

Weights of 10 people (kg)	75	49	68	83	92	71	64	39	49	67
Heights of 10 people (cm)	158	149	168	176	185	190	167	188	175	164

Explain why Sandy would not be able to produce a sensible scatter diagram from the data in this table.

Correlation **245**

Extension

With questions **1** and **2** the results are quite predictable but do students you agree with question **3**? How would students check that these results were reasonable? Challenge students in small groups to plan a survey that would give accurate information. What are the issues? (can number of messages be counted accurately, should age and/or gender be taken into account.) If there is time before the next lesson (over a day!) then they should carry this out (SM, EP).

Exercise 15e commentary

Question 1 – There is a strong (positive) correlation that suggests a line of best fit. Ask, if someone scored 14 in their maths test could you predict their science score? (CT) (1.3)

Question 2 – Ensure that students use an appropriate scale, 130 –170 for both axes, and that data is plotted at the correct points. The correlation is similar to question **1** (1.3).

Question 3 – Again check that students use appropriate scales and plot the data points correctly (1.2, 1.3).

Discussion – This hints at the key idea of a correlation being evidence for (but not a proof) of a causal link (1.5).

Assessment Criteria – HD, level 6: select, construct and modify scatter graphs.

Links

Bring in some advertisements for the class to use from local car dealers showing prices for second hand cars. Alternatively prices can be found at http://www.autotrader.co.uk/ or http://www.exchangeandmart.co. uk/iad.
Choose a particular brand of car and find prices for models of different ages. Is there any correlation between the age of the car and the price? What other factors affect the price of the car? (mileage, model, condition)

- Construct stem-and-leaf diagrams (L6)

Useful resources
Squared paper

Starter – More jumble

Write a list of anagrams on the board and ask students to unscramble them. Possible anagrams are

AGRAVEE, CQUEENFRY, GAREN, TRAINCOROLE, CASTISITT, EDAMIN, MISTIREESE (2 words), ACIPERTH (2 words)

(average, frequency, range, correlation, statistic, median, time series, pie chart)

Can be extended by asking students to make a data handling word search.

Teaching notes

When drawing a stem-and-leaf diagram it is necessary to follow a process that will not allow data to be missed. After drawing the stem students should plot each result as a leaf (one per square) but not ordered for each ten. They then re-draw the stem-and-leaf diagram but now with the results ordered from smallest to largest along each leaf (RL).

Finding the median can cause problems and a foolproof way (if a little slow) is a tactile method. Starting with the largest and smallest results and count in (and out) one each time until you meet. If you finish on two adjacent results (an even number of data) then the median is the value half-way between them.

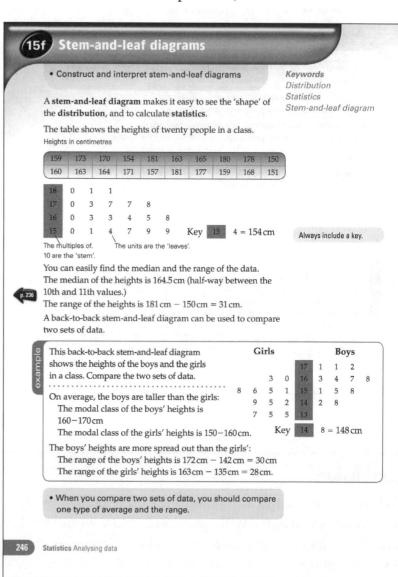

15f Stem-and-leaf diagrams

- Construct and interpret stem-and-leaf diagrams

Keywords
Distribution
Statistics
Stem-and-leaf diagram

A **stem-and-leaf diagram** makes it easy to see the 'shape' of the **distribution**, and to calculate **statistics**.

The table shows the heights of twenty people in a class.

Heights in centimetres

| 159 | 173 | 170 | 154 | 181 | 163 | 165 | 180 | 178 | 150 |
| 160 | 163 | 164 | 171 | 157 | 181 | 177 | 159 | 168 | 151 |

18	0	1	1		
17	0	3	7	7	8
16	0	3	3	4	5	8
15	0	1	4	7	9	9

Key 15 | 4 = 154 cm

Always include a key.

The multiples of 10 are the 'stem'. The units are the 'leaves'.

You can easily find the median and the range of the data.
The median of the heights is 164.5 cm (half-way between the 10th and 11th values.)
The range of the heights is 181 cm − 150 cm = 31 cm.

p. 236

A back-to-back stem-and-leaf diagram can be used to compare two sets of data.

example

This back-to-back stem-and-leaf diagram shows the heights of the boys and the girls in a class. Compare the two sets of data.

Girls		Boys
3 0	17	1 1 2
8 6 5 1	16	3 4 7 8
9 5 2	14	1 5 8
7 5 5	13	2 8

On average, the boys are taller than the girls:
The modal class of the boys' heights is 160–170 cm.
The modal class of the girls' heights is 150–160 cm.

Key 14 | 8 = 148 cm

The boys' heights are more spread out than the girls':
The range of the boys' heights is 172 cm − 142 cm = 30 cm
The range of the girls' heights is 163 cm − 135 cm = 28 cm.

- When you compare two sets of data, you should compare one type of average and the range.

246 **Statistics** Analysing data

Plenary

Give half the class a set of data as a table of numbers, for example, 30 numbers between 0 and 50 mixed in order (the results of a maths test). Give the other half of the class the same numbers but as a stem-and-leaf diagram. Give each student two minutes to bullet point a description of the numbers. Then ask students to compare their descriptions. Those with the diagram should have been able to give far more detail than the groups with the list of numbers (1.1).

Simplification

Making sense of data is a challenge; many students will need additional stem-and-leaf diagrams that they have to describe rather than spending much time drawing the diagrams.

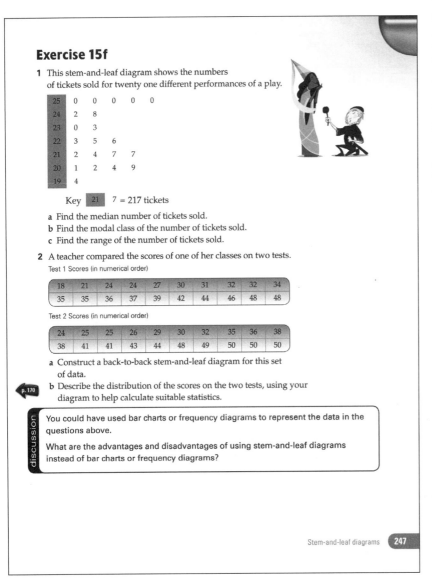

Exercise 15f

1 This stem-and-leaf diagram shows the numbers of tickets sold for twenty one different performances of a play.

25	0	0	0	0	0
24	2	8			
23	0	3			
22	3	5	6		
21	2	4	7	7	
20	1	2	4	9	
19	4				

Key 21 | 7 = 217 tickets

a Find the median number of tickets sold.
b Find the modal class of the number of tickets sold.
c Find the range of the number of tickets sold.

2 A teacher compared the scores of one of her classes on two tests.

Test 1 Scores (in numerical order)

18	21	24	24	27	30	31	32	32	34
35	35	36	37	39	42	44	46	48	48

Test 2 Scores (in numerical order)

24	25	25	26	29	30	32	35	36	38
38	41	41	43	44	48	49	50	50	50

a Construct a back-to-back stem-and-leaf diagram for this set of data.
b Describe the distribution of the scores on the two tests, using your diagram to help calculate suitable statistics.

p. 170

> discussion
>
> You could have used bar charts or frequency diagrams to represent the data in the questions above.
>
> What are the advantages and disadvantages of using stem-and-leaf diagrams instead of bar charts or frequency diagrams?

Stem-and-leaf diagrams   247

Extension

The stems on all the stem-and-leaf diagrams seen have gone up in tens. Challenge students to draw a stem-and-leaf diagram to represent small or large numbers so that the scale is a challenge (involving decimals or thousands.) (SM)

Exercise 15f commentary

Question 1 – This is similar to the discussion. Ask students when they should use mode and modal class? (1.3)

Question 2 – Suggest that students work in two stages: first produce an unordered stem-and-leaf diagram and then produce an ordered one. Part **b** is similar to the example; insist that students back up their statements with appropriate statistics (1.3).

Discussion – Allow students to work in small groups to make a table listing advantages and disadvantages for each type of representation. As a class decide what are the best 'pro' and 'con' points. Ask if there are types of data for which one representation is clearly the best (CT, EP) (1.1, 1.4).

Assessment Criteria – HD, level 6: communicate interpretations and results of a statistical survey using selected tables, graphs and diagrams in support.

Links

The tallest man ever in the world was Robert Pershing Wadlow from the USA who was 2.72 m tall and died in 1940. The World's shortest man is Pingping from China who is 73 cm tall. Add these two pieces of data to the boys' heights in the example. What is the new range? There are pictures of Robert Wadlow and Pingping at http://en.wikipedia.org/wiki/Image:Robert_Wadlow.jpg and at http://www.onlineweblibrary.com/blog/?p=453

15 Consolidation

1 Find the mean, median and range of each of these sets of numbers.
 a 4, 8, 16, 19, 25
 b 34, 67, 92, 108
 c 3.5, 9.2, 7.3, 8.3, 4.1
 d 106.3, 88.9, 71.4, 58.7, 91.9

2 The students in a class were asked how many people live in their house.
 The table shows the result.

Number of occupants	2	3	4	5	6	7	8
Frequency	3	5	11	6	4	0	1

 For this set of data, find
 a the mode   b the median   c the mean   d the range.

3 This bar chart shows the number of house sales in two areas in a town.

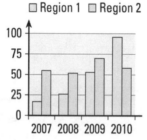

 Describe the main trends shown by this chart.

4 A department store has five floors: ground, 1, 2, 3 and 4.
 There are up and down escalators between each floor.
 Each escalator takes one minute to go up or down one floor, and
 it takes 30 seconds to walk between escalators on each floor.
 Monica starts on the ground floor, and travels up to the top floor
 as quickly as possible.
 Draw a time-series graph for her journey.

5 Sketch a scatter diagram to show the pattern you would expect to see
 when the correlation between variables is
 a positive   b negative   c no correlation.
 In each scatter diagram, suggest what the two sets of data could be.

6 The ages (in years) of 20 customers in a shop were

15	61	42	33	38	29	53	17	44	32
39	45	41	26	22	44	43	49	55	60

Draw a stem-and-leaf diagram for this set of data.

7 The table shows the number of cars issued by two car hire offices in a town, for 20 business days.

Office 1

6	11	21	24	25	9	13	23	13	25
19	7	22	15	17	25	10	8	25	25

Office 2

8	36	36	16	36	34	36	24	36	36
36	27	36	29	36	36	11	36	36	32

Draw a stem-and-leaf diagram for this set of data, and use it to describe the distribution of the data.

15 Case study – Electricity in the home

- Identify the mathematical features of a context or problem
- Give accurate solutions appropriate to the context or problem

(L6)

Useful resources
Data on electricity supply costs
Catalogues for appliances with energy efficiencies

Background

The cost of electricity has recently risen quite rapidly. At the same time, many households have an increasing number of electrical appliances - for example, many houses have multiple computers which would have been unusual until fairly recently. This case study looks at the electrical consumption of household appliances and raises awareness of how the amount of electricity used can be affected by choice of appliance and also by use, especially in terms of leaving appliances on standby rather than turning them off (EP, IE) (1.1).

Teaching notes

Ask the students to look at the pie chart showing electricity usage in a typical household. Read through the list of appliances in the key and ask the students if they have similar items in their homes.

Start to probe the difference between appliances' power rating and their overall consumption by asking questions such as:

Which items do you think will use the most electricity while they are being used? Which items do you think will use less energy?

A kettle uses quite a lot of electricity when it is on in order to boil the water fairly quickly. Why do you think that it only has a fairly small section of the pie chart? (It's not on for very long at a time.)

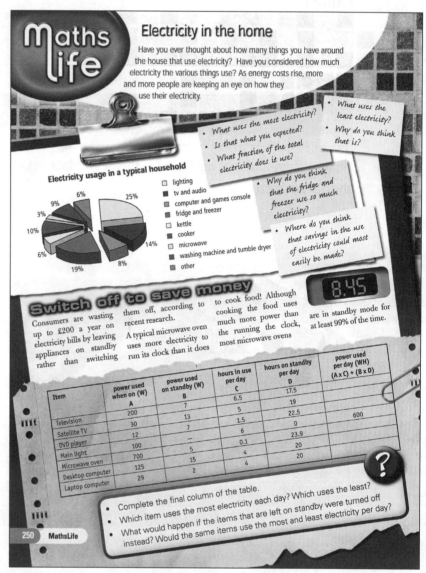

Why do you think the fridge and freezer use the most electricity overall? (They are on 24 hours a day.)

Look at the section titled 'Switch off to save money'. Discuss how the microwave is not just using electricity while it is cooking but also when it is idle but still switched on. Look at the table showing the power usage of a range of appliances when they are switched on and when they are in standby. Make sure that students understand the information and how to work out the total amount of electricity used per day. You might want to work through one row with the students before asking them to work out the daily power usage for each appliance.

Teaching notes continued

When the students have completed the first table that includes power used on standby, ask them to predict what will happen to the power usage if items are switched off when not in use

Do you think that the same item will use the most electricity? And will the same item use the least?

Give them time to work out the results again, this time assuming that all appliances are switched off when not in use (1.2, 1.3, 1.4). These calculations can be done as a paired activity (TW). If ICT is available the problem is naturally suited to being done using a spreadsheet.

Discuss the results, including how much electricity is saved overall per day by switching things off rather than leaving them on standby. Ask the students to convert this saved power into hours of TV viewing to give them something real to measure it by (TW, RL) (1,5).

Look at the information about the fridges, noting that the energy consumption is given in kWh.

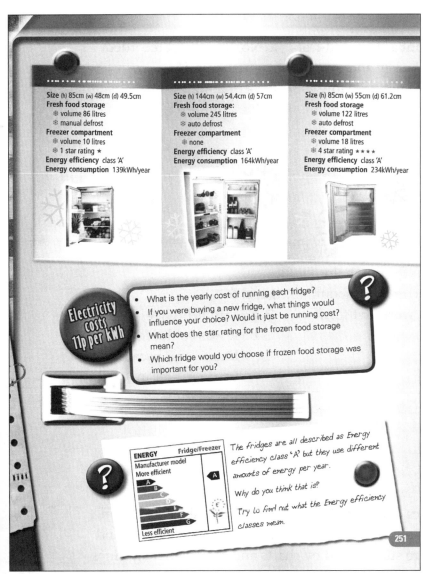

Explain that 1 kWh is equivalent to using 1 kW of energy for 1 hour and that 1 kW is 1000 W. Look together at the first question about the fridges and ask:

If 1 kWh of electricity costs 11p, how can we work out the yearly cost of running each fridge?

Establish that you multiply the cost per kWh by the annual consumption and give the students a few minutes to work out the cost of running each fridge. Then discuss the remaining questions to acknowledge that running cost is very unlikely to ever be the only factor that affects a decision about which appliance to buy, but that it should maybe be one of the factors that are thought about (CT). Discuss how, now that power usage is becoming more topical not only because of cost of energy but also for environmental reasons, labelling is making it easier to compare power usage of different appliances. Look at the energy efficiency label at the bottom of the spread as one example of this.

Extension

Students could research what is meant by the different energy efficiency ratings shown for the fridge, and in doing so answer the questions shown in the case study. They could also find out what other appliances have similar energy efficiency ratings and what they mean (SM).

The students could also research the power consumption and standby consumption of other appliances and add these to their tables/spreadsheet.

Students could produce posters to encourage people to save energy by switching appliances off rather than leaving them on standby or selecting energy efficient appliances (SM, CT).

15 Summary

Assessement criteria

- Interpret graphs and diagrams, including pie charts, and draw conclusions   Level 5
- Understand and use the mean of discrete data   Level 5
- Create and interpret line graphs where the intermediate values have meaning   Level 5
- Select, construct and modify scatter graphs   Level 6
- Communicate interpretations and results of a statistical survey using selected tables, graphs and diagrams in support   Level 6

Question commentary

Example

The example illustrates a problem about a stem-and-leaf diagram. Students may misinterpret the key and give the range as 9, i.e. 9 – 0. Ask questions such as "How many people were in the lift on each journey?" Some students may have difficulty in finding the median. Encourage students to check that the value found is the median by checking that the number of values higher than the median is the same as the number lower than the median.

a No, the largest recorded number is 24
b 24 – 2 = 22
c 8th value = 11

Past question

The question requires students to interpret a time-series graph. Part **a** can be read directly from the graph. In part **b**, emphasise that students must comment about the weather and not just that the eggs were first seen later than in other years.

Answer

Level 5
2 a   15th February
   b   It was colder for longer or it became warmer later in the year

Development and links

This topic is developed further in Year 9.

Students will read and interpret statistics and statistical diagrams across the curriculum, particularly in geography but also in science, physical education, history, economics and citizenship. In everyday life, students are surrounded by charts and graphs in magazines, newspapers and on television.

Objectives

Level

MPA

1.1	16a, b, c, d, e
1.2	
1.3	16a, b, c, d, e
1.4	16e
1.5	16d, e

PLTS

IE	16a, b, c, d, e
CT	16b, d, e
RL	16a, b, c, d, e
TW	16d, e
SM	16a, b, d, e
EP	16a, c, e

Introduction

The focus of this chapter is on using efficient methods and strategies to build confidence when solving problems involving calculations. Students identify the information necessary to solve a problem, break the problem down into smaller steps, consolidate mental, written and calculator methods for addition, subtraction, multiplication and division, check working and check answers using inverse operations and estimations.

The student page discusses the use of approximations to simplify problems. The shape of the Earth is described as a geoid, although scientists approximate it to the shape of a perfect sphere. The diameter of the Earth at the equator is 43 km larger than the diameter through the North and South Pole. Some approximations are forced upon us, for example, measurements are never exact. The accuracy of a measurement is limited by the accuracy of the measuring equipment, so all measurements are in fact approximations.

Extra Resources

16   Start of chapter presentation

16c   Starter: Multiplying with negatives T or F

16e   Starter: Rounded decimal matching

       Assessment: chapter 16

⚡Fast-track

No spreads

- Strengthen and extend mental methods of calculation (L5)
- Make and justify estimates and approximations of calculations (L5)

Useful resources
Mini whitboards

Starter – Estimate

Ask students
> to estimate the number of seconds in July
> to calculate the number of seconds in July (2 678 400)

How close were their estimates?

Can be extended by comparing the accuracy of boys versus girls or students with names beginning A to M versus names beginning N to Z.

Teaching notes

Students will have a lot of ideas on how to solve mental maths problems. Give them six quick questions to answer and then ask students to explain how they did the calculations. Collect a selection of these methods on the board and then ask the students who offered explanations to set a question that can be calculated using their method.

Give students another six questions, with too many digits to do mentally, and challenge them to provide estimated answers as fast as they can. Again, ask students to explain how they approximated to simplify the calculation. Once the methods have been shared, check understanding with a few more questions (EP).

Finally, pose a more complex word problem and work through it with the class. Ask students how they identify the important information and decide what to calculate: ask them to write down this calculation. How should they estimate the answer and perform the exact calculation? Does the estimate agree with the exact answer? (IE)

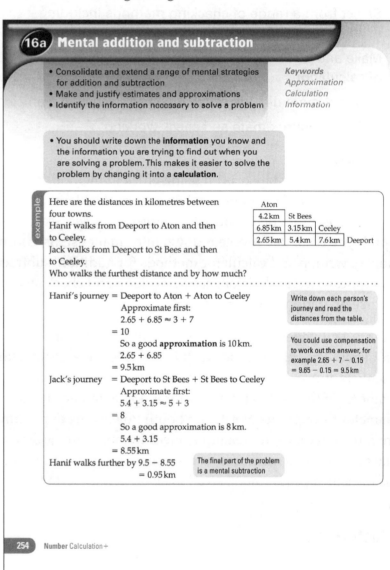

Plenary

Show students a map with a number of routes between towns and their lengths. Ask students quick-fire questions on estimating or mentally calculating the distances between destinations. Ask students to say which methods they prefer to use to do the calculations and explain why.

Simplification

Give students simple addition and subtraction questions to solve mentally. Then, in pairs, ask them to explain to one another how they did the calculations; if they disagree on the answer, ask them to work out what mistake was made. They can then attempt the exercises, working in pairs to identify what calculations to carry out (RL).

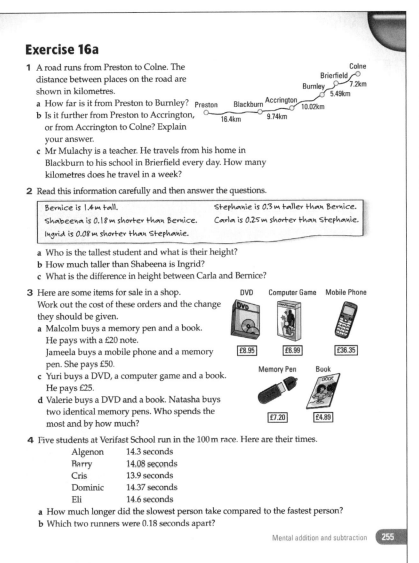

Exercise 16a

1 A road runs from Preston to Colne. The distance between places on the road are shown in kilometres.
  a How far is it from Preston to Burnley?
  b Is it further from Preston to Accrington, or from Accrington to Colne? Explain your answer.
  c Mr Mulachy is a teacher. He travels from his home in Blackburn to his school in Brierfield every day. How many kilometres does he travel in a week?

2 Read this information carefully and then answer the questions.

  Bernice is 1.4 m tall.
  Shabeena is 0.18 m shorter than Bernice.
  Ingrid is 0.08 m shorter than Stephanie.
  Stephanie is 0.3 m taller than Bernice.
  Carla is 0.25 m shorter than Stephanie.

  a Who is the tallest student and what is their height?
  b How much taller than Shabeena is Ingrid?
  c What is the difference in height between Carla and Bernice?

3 Here are some items for sale in a shop. Work out the cost of these orders and the change they should be given.
  a Malcolm buys a memory pen and a book. He pays with a £20 note.
    Jameela buys a mobile phone and a memory pen. She pays £50.
  c Yuri buys a DVD, a computer game and a book. He pays £25.
  d Valerie buys a DVD and a book. Natasha buys two identical memory pens. Who spends the most and by how much?

  DVD £8.95   Computer Game £6.99   Mobile Phone £36.35
  Memory Pen £7.20   Book £4.89

4 Five students at Verifast School run in the 100 m race. Here are their times.

Algenon	14.3 seconds
Barry	14.08 seconds
Cris	13.9 seconds
Dominic	14.37 seconds
Eli	14.6 seconds

  a How much longer did the slowest person take compared to the fastest person?
  b Which two runners were 0.18 seconds apart?

Mental addition and subtraction 255

Extension

Give students a train or coach schedule. Set questions on finding the times taken to travel between various places and the total time spent in stations. Remind students that there are 60 minutes in an hour (SM).

Exercise 16a commentary

Insist that students show some workings for each answer.

All questions (1.1, 1.3)

Question 1 – Similar to the example.

Question 2 – It may help to use a number line to organise the information in the question.

Question 3 – Encourage students to first find an approximate answer and use this to check their calculation.

Question 4 – Remind students to add an extra 0 to help do the calculations, for example, 14.60 – 14.37. In part **b**, finding approximate differences will eliminate several possibilities.

Assessment Criteria – Calculating, level 4: use a range of mental methods of computation with all operations. UAM, level 5: identify and obtain necessary information to carry through a task and solve mathematical problems. UAM, level 5: check results considering whether these are reasonable.

Links

The world record for adding 100 single digit numbers randomly generated by a computer is 19.23 seconds and is held by Alberto Coto from Spain. Details of other mental calculation world records can be found at http://www. recordholders.org/en/list/memory. html#adding10digits

- Use efficient written methods to add and subtract integers and decimals of any size, including numbers with differing numbers of decimal places (L5)
- Select from a range of checking methods, including estimating in context and using inverse operations (L5)

Useful resources

Starter – How many sweets?

Five students estimated the number of sweets in a jar.

Their estimates were: 41, 45, 49, 58, and 62.

The errors in their estimates, not necessarily in the same order, were:
2, 6, 7, 11 and 15.

Ask students how many sweets were in the jar. (56)

Can be extended by asking students to explain their strategies and making their own puzzles.

Teaching notes

Written addition and subtraction methods are consolidated here in a variety of real-life settings to support the development of functional skills in mathematics. Through the worked examples, students are shown the importance of lining up the digits in calculations and the outcomes of failing to do this correctly are illustrated.

Discussion of the examples should consider not only the processes for correct addition and subtraction but also the importance of using inverse operations for checking and, indeed, common sense (RL).

Throughout, the mathematical problems are word-based and students are offered the opportunity to develop their skills in identifying key points from text and to ascertain the mathematical operation required to find a solution. These skills are equally as important as the actual mathematics itself and students build on the key processes in number, particularly representing, as they identify the type of problem and the operations needed to reach a solution (IE).

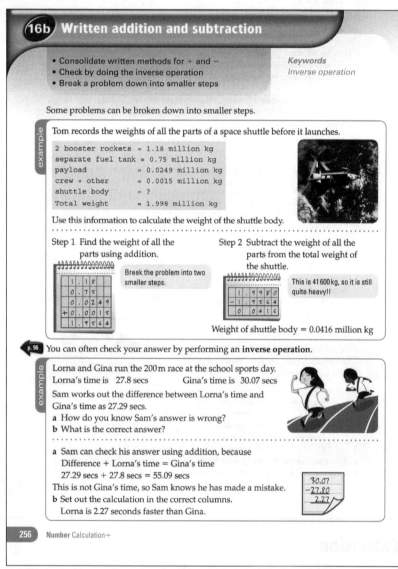

16b Written addition and subtraction

- Consolidate written methods for + and −
- Check by doing the inverse operation
- Break a problem down into smaller steps

Keywords
Inverse operation

Some problems can be broken down into smaller steps.

Tom records the weights of all the parts of a space shuttle before it launches.

```
2 booster rockets = 1.18 million kg
separate fuel tank = 0.75 million kg
payload           = 0.0249 million kg
crew + other      = 0.0015 million kg
shuttle body      = ?
Total weight      = 1.998 million kg
```

Use this information to calculate the weight of the shuttle body.

Step 1 Find the weight of all the parts using addition.

Break the problem into two smaller steps.

Step 2 Subtract the weight of all the parts from the total weight of the shuttle.

This is 41 600 kg, so it is still quite heavy!!

Weight of shuttle body = 0.0416 million kg

You can often check your answer by performing an **inverse operation**.

Lorna and Gina run the 200 m race at the school sports day.
Lorna's time is 27.8 secs   Gina's time is 30.07 secs
Sam works out the difference between Lorna's time and Gina's time as 27.29 secs.
a How do you know Sam's answer is wrong?
b What is the correct answer?

a Sam can check his answer using addition, because
   Difference + Lorna's time = Gina's time
   27.29 secs + 27.8 secs = 55.09 secs
This is not Gina's time, so Sam knows he has made a mistake.
b Set out the calculation in the correct columns.
   Lorna is 2.27 seconds faster than Gina.

```
  30.07
− 27.80
   2.27
```

Plenary

Ask students to revisit the puzzle from the exercise and to write a different set of questions using the same map. Swap with a friend to complete and then swap back to mark.

Simplification

If students experience difficulty here, it is likely to be due to the need to support their literacy skills. Scaffold their attempts to highlight and extract key information from word-based problems.

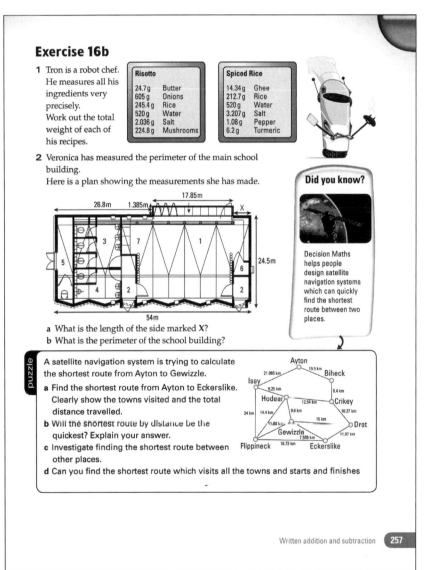

Exercise 16b

1 Tron is a robot chef. He measures all his ingredients very precisely.
Work out the total weight of each of his recipes.

Risotto

24.7 g	Butter
605 g	Onions
245.4 g	Rice
520 g	Water
2.036 g	Salt
224.8 g	Mushrooms

Spiced Rice

14.34 g	Ghee
212.7 g	Rice
520 g	Water
3.207 g	Salt
1.08 g	Pepper
6.2 g	Turmeric

2 Veronica has measured the perimeter of the main school building.
Here is a plan showing the measurements she has made.

17.85m
26.8m   1.385m   X
24.5m
54m

a What is the length of the side marked X?
b What is the perimeter of the school building?

Did you know?

Decision Maths helps people design satellite navigation systems which can quickly find the shortest route between two places.

puzzle

A satellite navigation system is trying to calculate the shortest route from Ayton to Gewizzle.

a Find the shortest route from Ayton to Eckerslike. Clearly show the towns visited and the total distance travelled.
b Will the shortest route by distance be the quickest? Explain your answer.
c Investigate finding the shortest route between other places.
d Can you find the shortest route which visits all the towns and starts and finishes

Ayton
Isay   21.065 km   19.5 km   Biheck
9.25 km   8.4 km
Hodear   12.54 km   Crikey
24 km   14.4 km   9.6 km   10.37 km
11.08 km   15 km   Drat
Gewizzle   11.97 km
7.538 km
Flippineck   16.73 km   Eckerslike

Extension

Invite more able students to research an area of interest from which they can construct their own written addition and/or subtraction problems (CT, SM).

Exercise 16b commentary

All questions (1.1, 1.3)

Question 1 – This question helps to consolidate the importance of lining up the digits in the correct columns when using written methods. Students could break the additions down into sub-sums if this helps.

Question 2 – Students may find it helps to make their own, simplified, drawing and write in the various dimensions.

Puzzle – Encourage students to perform approximate calculations first to eliminate a number of possibilities before doing full calculations. Part **d**, is related to the famous 'travelling salesman' problem.

Assessment Criteria – Calculating, level 4, use efficient written methods of addition and subtraction. Calculating, level 5: apply inverse operations and approximate to check answers to problems are of the right magnitude.

Links

To date over 120 US space shuttle flights have been made since the space shuttle Columbia made its first test flight into space in 1981. There is more information about the US space shuttle at http://www.nasa.gov/mission_pages/shuttle/main/index.html

- Strengthen and extend mental methods of calculation (L5) **Useful resources**
- Make and justify estimates and approximations of calculations (L5)

•••

Starter – Sums and products

Ask students for

Three numbers where the sum is the same as the product.
$(1 + 2 + 3 = 1 \times 2 \times 3)$
Four numbers where the sum is the same as the product.
$(1 + 1 + 2 + 4 = 1 \times 1 \times 2 \times 4)$
Five numbers where the sum is the same as the product.
$(1 + 1 + 2 + 2 + 2 = 1 \times 1 \times 2 \times 2 \times 2)$

Hint: the same number may be used more than once.

Teaching notes

In this spread, all problems are word-based and involve real-life scenarios. The focus is on students becoming functional mathematicians and using their mental skills to not only solve problems but, equally importantly, to be able to estimate sensibly to know whether they are going to be right.

Question **3** illustrates how to answer a question without actually doing the mathematics and this is very important for students to understand. The task is about seeing the bigger picture of the question and not getting drowned in the detail. For many students, this is a difficult concept to grasp (IE).

Again, the value of the use of approximation as a checking mechanism cannot be over-estimated (RL).

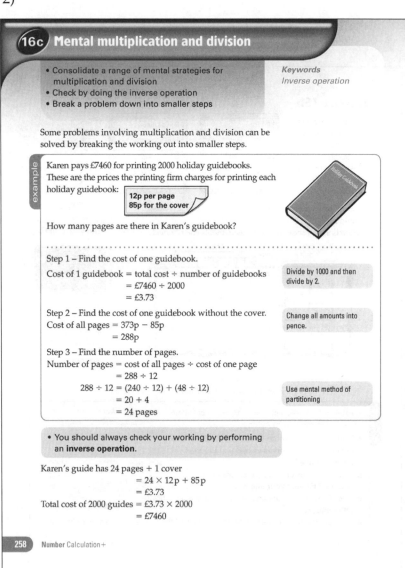

16c Mental multiplication and division

- Consolidate a range of mental strategies for multiplication and division
- Check by doing the inverse operation
- Break a problem down into smaller steps

Keywords
Inverse operation

Some problems involving multiplication and division can be solved by breaking the working out into smaller steps.

example

Karen pays £7460 for printing 2000 holiday guidebooks. These are the prices the printing firm charges for printing each holiday guidebook:

12p per page
85p for the cover

How many pages are there in Karen's guidebook?

Step 1 – Find the cost of one guidebook.
Cost of 1 guidebook = total cost ÷ number of guidebooks
= £7460 ÷ 2000
= £3.73

Divide by 1000 and then divide by 2.

Step 2 – Find the cost of one guidebook without the cover.
Cost of all pages = 373p − 85p
= 288p

Change all amounts into pence.

Step 3 – Find the number of pages.
Number of pages = cost of all pages ÷ cost of one page
= 288 ÷ 12
$288 ÷ 12 = (240 ÷ 12) + (48 ÷ 12)$
= 20 + 4
= 24 pages

Use mental method of partitioning

- You should always check your working by performing an **inverse operation**.

Karen's guide has 24 pages + 1 cover
= 24 × 12p + 85p
= £3.73
Total cost of 2000 guides = £3.73 × 2000
= £7460

258 Number Calculation +

Plenary

Ask students to choose one of the examples from the exercise, or something similar, and turn this into an annotated example as if they were writing a revision guide. Use the worked examples as a model.

Simplification

Weaker students find it difficult to know when and how to approximate when working towards a solution. Development of these skills will require particular support.

Exercise 16c

1 Jody is trying to improve her fitness levels.
   When she is cycling, Jody's heartbeat is 84 beats per minute.
   When she is running, Jody's heartbeat is 92 beats per minute.
   Jody cycles for 21 minutes per day and runs for 19 minutes per day.
   Does Jody's heart beat more in total when she cycles or when she runs?
   Explain and justify your answer.

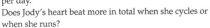

2 Debbie pays £56 300 for printing 10 000 holiday guidebooks.
   These are the prices the printing firm charges for printing each holiday guidebook.
   How many pages are there in Debbie's guidebook?

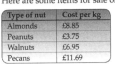

13p per page
95p for the cover

3 Here are some items for sale on the 'Nut-e-nuts' website.

Type of nut	Cost per kg
Almonds	£8.85
Peanuts	£3.75
Walnuts	£6.95
Pecans	£11.69

   Use an approximation to decide if each person has enough money for the cost of their orders. Explain and justify your answers.
   a Maurice orders 5 kg of almonds and 4 kg of peanuts. He has £65.
   b Jenna orders 11 kg of walnuts and 12 kg of pecans. She has £215.
   c Boris orders 2 kg of almonds, 3 kg of peanuts and 5 kg of pecans. He has £95.

4 Every person is recommended to consume five portions of fruit and vegetables every day. A 150 ml glass of orange juice counts as one daily portion.
   A carton of orange juice contains 1000 ml (= 1 litre).
   a How many recommended daily portions of orange juice are there in one carton?
   b A family of four decide to each drink 150 ml of orange juice every day.
   How much orange juice will they drink in one week?
   How many cartons of orange juice will they need to buy?

Orange Juice

1 litre

Extension

More able students could spend additional time discussing the methods they have used with a friend. As they follow the logic of someone else's calculation, this will further develop the process skill of interpreting and evaluation, particularly the effects of rounding.

Exercise 16c commentary

All questions (1.1, 1.3)

Question 1 – Remind students that $21 = 20 + 1$ and $19 = 20 - 1$. The explanation and justification element is very important.

Question 2 – Similar to the example which can serve as a model for how to breakdown the calculation into smaller steps – developing the process skill of representing.

Question 3 – Insist that students write down the calculation that they need to do and also the approximation/rounding before doing any mental calculations.

Question 4 – Encourage students to write down the calculations that they need to perform before calculating them mentally. Ask students also to supply an approximate calculation that they can use to check their calculation (EP).

Assessment Criteria – Calculating, level 4: use a range of mental methods of computation with all operations.

UAM, level 6: solve problems and carry through substantial tasks by breaking them into smaller, more manageable tasks, using a range of efficient techniques..

Links

There is a suggestion that the total number of heart beats that an animal has over a lifetime is a fixed number. Information on heart rates and lifetimes for various animals can be found at http://www.sjsu.edu/faculty/watkins/longevity.htm

Ask students to work out the total number of heart beats for an animal or, using the heart rate, estimate a lifetime. Do they think fitter people live longer?

- Use efficient written methods for multiplication and division of integers and decimals   (L5)
- Select from a range of checking methods, including estimating in context and using inverse operations   (L5)

Useful resources

Starter – Last minute?

Write the following list of times on the board:

3 days, 192 hours, 47 hours, 24 600 seconds, 8 days, 17 hours, 12 300 minutes, 1020 minutes, 169 200 seconds, 4320 minutes, 410 minutes, 205 hours.

Challenge students to match up six pairs in the shortest possible time.

Teaching notes

The main aim of the spread is to pose real-life problems to solve. These are presented in a fashion that invites the development of skill transferability, as students extract the mathematics from a situation. There is, however, thorough discussion of the worked examples, a review of methods of calculation and a reminder of the value of checking with inverse operations (RL).

The key process of representing is central to the work here, as students identify the type of problem and the mathematics needed to reach a solution and break down more complex problems into a sequence of steps.

Where the opportunity is taken to use the questions in the exercise to develop purposeful mathematical discussion, students will take responsibility, showing confidence in themselves and their contribution. (TW)

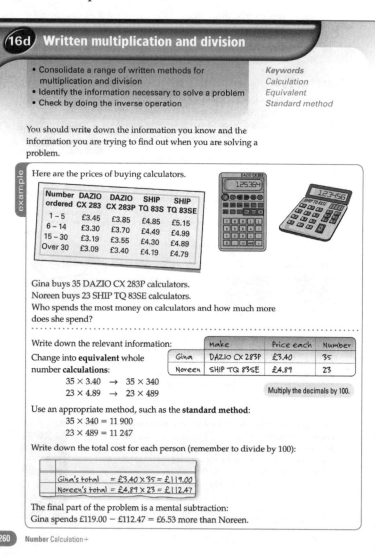

Plenary

While students are working, identify some of the best interaction and high quality discussion. Invite a selection of students to present their working methods to the class (TW).

Simplification

Where students experience difficulty, support them in the analysis of the questions. Offer highlighters to help pick out key points from (copies of) individual questions and encourage dialogue with other students.

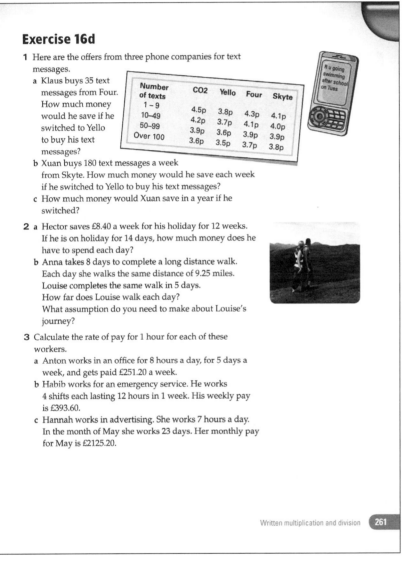

Exercise 16d

1 Here are the offers from three phone companies for text messages.

a Klaus buys 35 text messages from Four. How much money would he save if he switched to Yello to buy his text messages?

Number of texts	CO2	Yello	Four	Skyte
1 – 9	4.5p	3.8p	4.3p	4.1p
10–49	4.2p	3.7p	4.1p	4.0p
50–99	3.9p	3.6p	3.9p	3.9p
Over 100	3.6p	3.5p	3.7p	3.8p

b Xuan buys 180 text messages a week from Skyte. How much money would he save each week if he switched to Yello to buy his text messages?

c How much money would Xuan save in a year if he switched?

2 a Hector saves £8.40 a week for his holiday for 12 weeks. If he is on holiday for 14 days, how much money does he have to spend each day?

b Anna takes 8 days to complete a long distance walk. Each day she walks the same distance of 9.25 miles. Louise completes the same walk in 5 days. How far does Louise walk each day? What assumption do you need to make about Louise's journey?

3 Calculate the rate of pay for 1 hour for each of these workers.

a Anton works in an office for 8 hours a day, for 5 days a week, and gets paid £251.20 a week.

b Habib works for an emergency service. He works 4 shifts each lasting 12 hours in 1 week. His weekly pay is £393.60.

c Hannah works in advertising. She works 7 hours a day. In the month of May she works 23 days. Her monthly pay for May is £2125.20.

Written multiplication and division **261**

Extension

More able students can be challenged to produce high quality explanations and justifications of their thinking that could be used as model strategies (CT, SM).

Exercise 16d commentary

Consider pairing students so that they can work together to extract the relevant information and discuss their methods for finding the solution (IE) (1.5).

Question 1 – Similar to the example (1.1, 1.3).

Question 2 – In pairs, students could answer one part each and then explain to their partner how they arrived at their solution (1.1, 1.3, 1.5).

Question 3 – Again, an effective learning strategy is to share out the parts of the question, asking students to explain to each other and to account for their methods (RL) (1.1, 1.3, 1.5).

Assessment Criteria – Calculating, level 4: use efficient written methods of short multiplication and division. Calculating, level 5: understand and use appropriate non-calculator methods for solving problems that involve multiplying and dividing any three digit number by any two digit number.

MPA, Level 6 – students interpret, discuss and synthesise information presented in a variety of mathematical forms.

Links

The division symbol ÷ is called an obelus. The symbol was originally used in manuscripts to mark passages containing errors but first appeared as a division symbol in a book called *Teutsche Algebra* by Johann Rahn in 1659. In Denmark, the obelus was used to represent subtraction. There is more information about mathematical symbols at http://members.aol.com/jeff570/operation.html

- Calculate accurately, selecting mental methods or calculating devices as appropriate   (L5)
- Carry out more difficult calculations effectively and efficiently   (L5)
- Estimate, approximate and check working   (L5)

Useful resources

• •

Starter – Same digit

Ask students to find pairs of numbers that will give the same digit when one of the numbers is divided by the other, for example,

$385 \div 50 = 7.7$

$33 \div 6 = 5.5$

$3000 \div 9 = 333.33333\ldots$

Can be extended by asking students to explain any methods they have used.

Teaching notes

The work here supports process skill development throughout. In terms of representing, students will be involved in choosing between mental, written and calculator methods; they will strengthen analysing skills as they develop routines for estimating, approximating and checking; and they will communicate and reflect as they discuss different approaches to solving numerical problems and compare the efficiency of calculation procedures.

Through collaborating and sharing thinking, students will question their own and others' assumptions and try out alternatives or new solutions and follow ideas through (TW, CT).

Students will be encouraged to feel successful here as there is no necessarily correct approach. The key focus is on the ability to explain and justify the method chosen and to be able to use a checking strategy to know whether a solution is likely to be right.

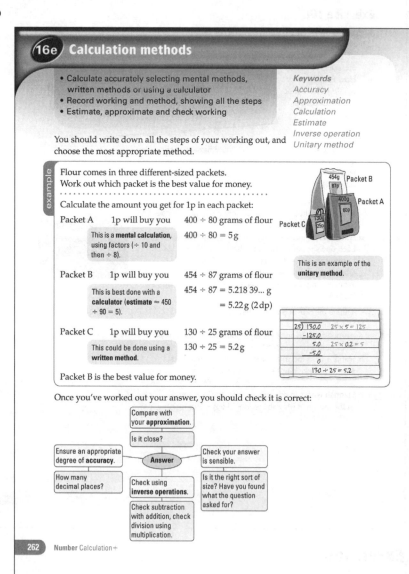

Plenary

Invite students to list good checking strategies and when they would use them. Take feedback and discuss as a class (RL).

Simplification

Students who are not confident here will learn most effectively through discussion and sharing ideas with others. Also, the opportunity to solve the same problem using more than one method will build confidence.

Exercise 16e

1 ChocChoc biscuits come in three different-sized packets. Work out which packet is the best value for money by calculating the cost of one biscuit for each packet.

Packet B

Packet C

Packet A

2 Choose the most appropriate method (mental, written or calculator) for solving each of these calculations. In each case, explain your choice and then use that method to solve the problem.
Oliver goes on a sponsored charity walk from Maryport to Liverpool, dressed as a giant leek.
He takes 17 days to walk from Maryport to Liverpool.
Each day he walks the same distance.
He returns by car along the same route in 3 hours at an average speed of 47.6 mph.
He raises £28.63 for charity for each mile he walks.
i How far did Oliver travel each day on his charity walk?
ii How much money did Oliver raise for charity by completing his walk?

3 Jayne lives in Halifax and works in Manchester. Her journey to work each day is 55 km. She is trying to decide whether it is better to travel to work by car or by train. Here are some ideas she has written down.

HALIFAX

MANCHESTER

Travelling by car...
Car insurance = £280.45 (each year)
Road tax = £180 (yearly)
Servicing = £195 (twice a year)
MOT = £50.63 (each year)
Fuel costs
Petrol = 113.9 p per litre
Consumption = 12.3 km per litre
Travelling time
Each journey lasts about 55 mins

Travelling by train...
Monthly season ticket = £285.29
Daily return ticket = £13.75
Travelling time
Home to station = 19 mins walk
Journey on train = 38 mins
Station to work = 13 mins walk

Jayne works for 46 weeks a year. She has a 4-week holiday in August.
Write a short report recommending which form of transport Jayne should take. Explain and justify your answer.

Calculation methods 263

Extension

Challenge more able students with further opportunities to use mathematics to substantiate a point of view or a recommendation as in question 3 (SM).

Exercise 16e commentary

Question 1 – Similar to the example. Encourage students to explain which method of calculation they used and to say why they chose this method (IE) (1.1, 1.3).

Question 2 – After completing this question, ask students to work with a friend to compare methods and solutions. Where they have used different strategies they should discuss the relative merits of each (EP) (1.1, 1.3, 1.4, 1.5).

Question 3 – This also supports developing skills in persuasive writing (1.1, 1.3).

Assessment Criteria – Calculating, level 5: use known facts, place value, knowledge of operations and brackets to calculate including using all four operations with decimals to two places.

Links

Before electronic calculators became widely available in around 1974, slide rules were used to perform multiplication and division calculations at school and in science and engineering. A slide rule is a mechanical calculator shaped like a large ruler with 2 or more scales that can slide against each other. Using a slide rule converted a multiplication or division into an addition or subtraction. There is more information about slide rules at http://en.wikipedia.org/wiki/Slide_rule and at http://www.sliderulemuseum.com/

16 Summary

Assessment criteria

- Identify and obtain necessary information to carry through a task and solve mathematical problems — Level 5
- Check results, considering whether these are reasonable — Level 5
- Solve problems and carry through substantial tasks by breaking them into smaller, more manageable tasks, using a range of efficient techniques, methods and resources — Level 6
- Use a range of mental methods of computation with all operations — Level 4
- Use efficient written methods of addition and subtraction and of short multiplication and division — Level 4
- Apply inverse operations and approximate to check answers to problems are of the right magnitude — Level 5
- Use known facts, place value, knowledge of operations and brackets to calculate including using all four operations with decimals to two places — Level 5

Question commentary

Example

The example illustrates a typical calculation problem which needs to be broken down into smaller steps. Encourage a logical approach. Some students may forget to subtract the weight of the empty box and give an answer for the weight of each drawing pin as 0.33 g. Emphasise the importance of checking the final answer and of giving units.

$$50 - 12.5 = 37.5$$
$$37.5 \div 150 = 0.25 \text{ g}$$

Past question

The question requires students to calculate the number of cans of drink sold in a day from the coins inserted into a vending machine. Ask questions such as "What information do you have?" and "Is there anything else you need to know?" Some students may use the total number of coins inserted instead of the total value. It may help to add an extra column to the table and calculate the total value inserted for each denomination of coin. Emphasise that students will lose marks if they do not show their working.

Answer

Level 5

2 $50 \times 31 = 1550$
$20 \times 22 = 440$
$10 \times 41 = 410$
$5 \times 59 = \underline{295}$
2695

$2695 \div 55 =$
49 cans of drink

Development and links

Students will continue to practise their calculation and problem-solving skills in Year 9.

Students need to perform calculations in many areas of the curriculum, particularly in science, technology and geography. The skills and techniques practised in this chapter can be applied to a variety of problems and puzzles in daily life and it is often quicker and easier to use a mental or written method than to try to find a calculator.

17 Functional Maths

Objectives

Objectives	Level
Calculate accurately, selecting mental methods or calculating devices as appropriate	5
Identify exceptional cases or counter examples	5
Use logical argument to interpret the mathematics in a given context or to establish the truth of a statement	5
Relate the current problem and structure to previous situations	5
Identify the mathematical features of a context or problem	5
Evaluate the efficiency of alternative strategies and approaches	5
Select appropriate procedures and tools	5
Give accurate solutions appropriate to the context or problem	5
Make accurate mathematical diagrams, graphs and constructions on paper and on screen	5
Refine own findings and approaches on the basis of discussions with others	6
Give accurate solutions appropriate to the context or problem	6

MPA

1.1	17a, b, c, d, e
1.2	17a, b, c, d, e
1.3	17a, b, c, d, e
1.4	17a, b, c, d, e
1.5	17a, b, c, d, e

PLTS

IE	17a, b, c, d, e
CT	17a, b, c, d, e
RL	17a, b, c, d, e
TW	17a, c, d, e
SM	17a, b, c, d, e
EP	17a, b, c, d, e

Introduction

The chapter consists of a sequence of five spreads based on the theme of a year 8 school camping trip to France. This allows questions to cover a wide range of topics taken from algebra, statistics, geometry and number. The questions are word-based and often do not directly indicate what type of mathematics is involved. Therefore students will need to work to identify the relevant mathematics and in several instances which of a variety of methods to apply before commencing. This approach is rather different from the previous topic based spreads and students may require additional support in this aspect of functional maths.

Extra Resources

17 Start of chapter presentation

Assessment: chapter 17

Fast-track

All spreads

17a Planning the trip to France

- Calculate accurately, selecting mental methods or calculating devices as appropriate (L5)
- Identify exceptional cases or counter examples (L5)

Useful resources
Calculators
Travel brochures

Background

This spreads focuses on the logistics and finances of the trip and largely exercises number skills. Issues surrounding costs, deposits and exchange rates may be familiar to some students from family holidays and this knowledge can be used to both enliven discussion and provide a source of illustrative examples (EP).

As the first spread in the chapter it is important to establish how the students should approach the work, whether as individuals, as pairs or small groups, *etc.*, should they work at their own pace, will they be expected to start straightaway or will an introduction be given, *etc.*

Teaching notes

Invite pairs of students to imagine that they are a teacher planning a school trip and ask them to suggest what they need to consider. Focus on the costs involved: how should they calculate deposits, deal with exchange rates, *etc.* The student book can be used as a prompt. The subsequent discussion should concentrate on generic approaches, what is required and a suitable method, rather than specific details (TW).

Supply students with some example calculations and ask them to explain how they would complete them. Total cost £3782, 20 students: cost per student, 15% deposit, £435 to be paid in Euros at £1= €1.28. Ask how they decide whether to use mental, written or calculator methods: can they give two pros and two cons for each method? (EP) (1.5) Also ask how they would go about checking their answers (against an approximation, using an inverse operation, is it reasonable, is it to an appropriate degree of accuracy?) (RL).

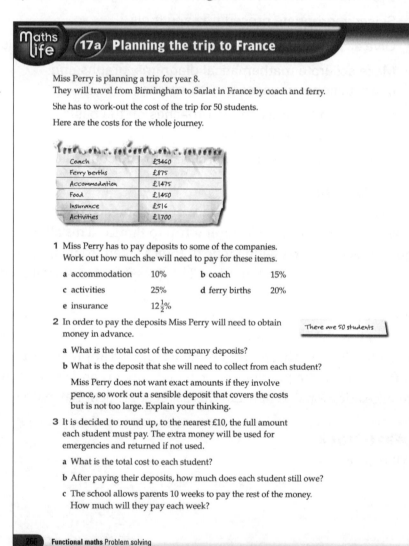

Ask students if they can supply some handy hints for doing calculations, especial using mental methods. These can be collected on the board as a reminder for students as they work through the spread (1.5).

.

Simplification

The questions in this spread involve straightforward arithmetic (discourage falling back on a calculator), the difficulties are likely to arise from the language and understanding what is required. Pair weaker students and encourage breaking down a problem using a checklist (SM):

1) Understand the problem/question
2) Underline/copy important information
3) Decide which operation is necessary
4) Make an approximation
5) Work out the answer
6) Check that the answer is sensible

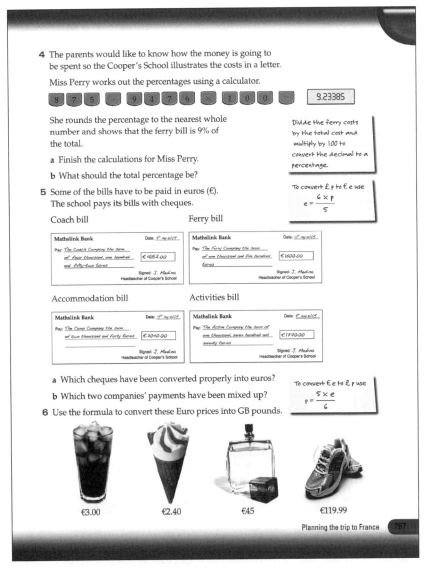

4 The parents would like to know how the money is going to be spent so the Cooper's School illustrates the costs in a letter.

Miss Perry works out the percentages using a calculator.

`8 7 5 ÷ 9 4 7 6 × 1 0 0 =` `9.23385`

She rounds the percentage to the nearest whole number and shows that the ferry bill is 9% of the total.

Divide the ferry costs by the total cost and multiply by 100 to convert the decimal to a percentage.

a Finish the calculations for Miss Perry.

b What should the total percentage be?

5 Some of the bills have to be paid in euros (€). The school pays its bills with cheques.

To convert £ p to € e use
$$e = \frac{6 \times p}{5}$$

Coach bill

Mathslink Bank   Date: 4th May 2009
Pay: *The Coach Company the sum of four thousand, one hundred and fifty-two euros*   €4152.00
Signed: J. Medina
Headteacher of Cooper's School

Ferry bill

Mathslink Bank   Date: 10th May 2009
Pay: *The Ferry Company the sum of one thousand and five hundred euros*   €1600.00
Signed: J. Medina
Headteacher of Cooper's School

Accommodation bill

Mathslink Bank   Date: 19th May 2009
Pay: *The Camp Company the sum of two thousand and forty euros*   €2040.00
Signed: J. Medina
Headteacher of Cooper's School

Activities bill

Mathslink Bank   Date: 1st June 2009
Pay: *The Active Company the sum of one thousand, seven hundred and seventy euros*   €1770.00
Signed: J. Medina
Headteacher of Cooper's School

a Which cheques have been converted properly into euros?

b Which two companies' payments have been mixed up?

6 Use the formula to convert these Euro prices into GB pounds.

To convert € e to £ p use
$$p = \frac{5 \times e}{6}$$

€3.00   €2.40   €45   €119.99

Planning the trip to France   267

Extension

Provide students with a holiday brochure and ask them to calculate an approximate cost for two adults and two children to go on holiday. Ask them to include approximate costs for food, activities, *etc*. If access to the internet is available, encourage students to investigate total costs of flights including insurance and all surcharges *etc*.

Exercise 17a commentary

All questions (IE) (1.1, 1.3, 1.4)

Question 1 – (spread 4e) Initially avoid using a calculator and encourage the use of 'tricks' supported by workings; can students see an equivalent decimal multiplication? For example, 15% of $= \frac{1}{10} + \frac{1}{2} \times \frac{1}{10} = 0.15 \times$ (CT).

Question 2 – (spreads 8a-c, e, g, 16a-d) In part **a**, check that students line their calculation up on the decimal point. In part **b**, ask, is their an easy way to do the division? ($\div 50 = \times 2 \div 100$)

Question 3 – (spreads 8a-d, 16a-d) In part **a**, ask, how does 'rounding up' differ from 'rounding'? In part **b**, check that the rounded-up deposit is subtracted.

Question 4 – (spreads 4f, 8a, i, 16e) Encourage students to check that the total cost is £9476. In part **a**, ask if it is possible to speed up the calculations using the memory function? (store $100 \div 9476$)

Question 5 – (spreads 5e, 13b, f, g) A slightly simpler formula to use is $e = 12 \times p / 10$. Can students spot any errors without doing a calculation? (€2040 > €1770 but £1475 < £1770) Ask students to try and explain how the errors happened and how they might try to avoid making such mistakes themselves (RL) (1.2).

Question 6 – (spreads 5e, 13b, f, g) It should not be necessary to use a calculator. Does it make sense to round the price of the trainers?

Assessment Criteria – UAM, level 5: check results, considering whether these are reasonable. UAM, level 5: solve word problems and investigations from a range of contexts.

- Use logical argument to interpret the mathematics in a given context or to establish the truth of a statement **(L6)**
- Relate the current problem and structure to previous situations **(L5)**

Useful resources
Local area map
Small name cards

Background

This spread takes up the theme of the school trip and arriving at the camp where they have to organise the accommodation and familiarise themselves with the campsite. The mathematics involves areas of rectangles, arithmetic with decimals, the use of coordinates (with direct cross-curricular links to geography) and logical reasoning (EP).

Teaching notes

The first question involves the multiplication and division of decimals. It will be useful revision to ask students to explain how to do this and how to check their answer. Test their understanding by asking them to calculate the area of a rectangular tent. Ask how they think this is related to how many people the tent will comfortably sleep Is the ground area the only thing that needs to be considered? What about the tent's shape? (CT) (1.3)

Question **2** is likely to be new to the students in the context of mathematics. It may help to provide a similar example and ask students to provide a 'method' for solving the puzzle and for verifying any solution (IE, RL) (1.4).

Questions **3** to **6** involve interpreting a map and finding locations. Using a local area map, ask students to specify the positions of local landmarks. Can they do this is such a way that they don't refer to other locations on the map? This may be familiar from geography and the method used in the questions is easily tied in with the use of coordinates in mathematics. The map is also a scale drawing and students could be asked to think about how they could calculate real-life distances based on either the local area or campsite maps. One way to approach this is by asking students to say how they would go about creating an accurate map of the school (SM). (1.2, 1.5)

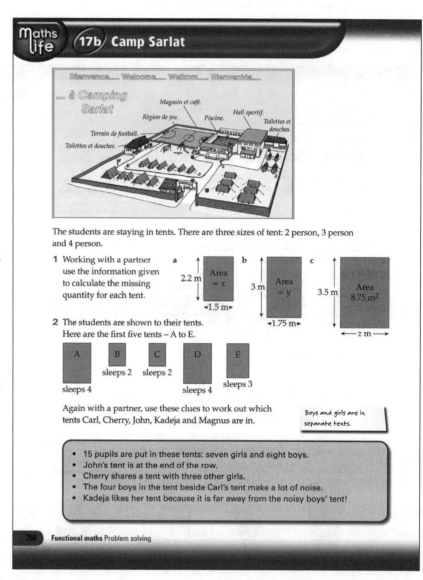

Simplification

In question **1**, the decimals may cause difficulty. Suggest, for example, that students think about the area of a 2 × 2 tent and a 22 × 15 tent. Encourage students to think this through rather than simply use a calculator.

For question **2**, provide cards with the five names on to allow students to experiment with their order

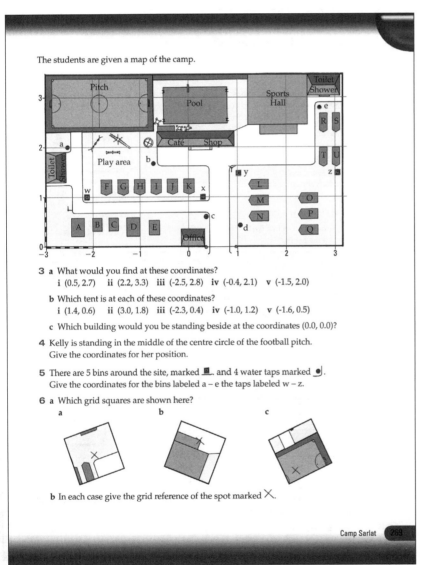

The students are given a map of the camp.

3 a What would you find at these coordinates?
  i (0.5, 2.7) **ii** (2.2, 3.3) **iii** (-2.5, 2.8) **iv** (-0.4, 2.1) **v** (-1.5, 2.0)

  b Which tent is at each of these coordinates?
  i (1.4, 0.6) **ii** (3.0, 1.8) **iii** (-2.3, 0.4) **iv** (-1.0, 1.2) **v** (-1.6, 0.5)

  c Which building would you be standing beside at the coordinates (0.0, 0.0)?

4 Kelly is standing in the middle of the centre circle of the football pitch. Give the coordinates for her position.

5 There are 5 bins around the site, marked 🗑 and 4 water taps marked ⬤. Give the coordinates for the bins labeled a – e the taps labeled w – z.

6 a Which grid squares are shown here?

  a   **b**   **c**

  b In each case give the grid reference of the spot marked ✕.

Camp Sarlat 269

Extension

Supply students with a scale for the map, the large squares are 10 m × 10 m, and ask them to determine the sizes of various features. How long is the football pitch? What is its area? If the swimming pool is 1.5 m deep, what is its volume? (IE)

Exercise 17b commentary

All questions (CT) (1.1)

Question 1 – (spread 2c) Calculators are not necessary. It will be important to check the size of the answers using approximate integer calculations; inverse operations can be used for part **c**. Check that units are given.

What is the area per person for the three tents? (1.3)

Question 2 – This is a new type of question; students will need to be systematic in their reasoning and be careful to use each piece of information given (IE) (1.4).

Questions 3 – (spread 7e) Check that students can read the scales and find coordinates in the first two quadrants. The postcard on the previous page may help with interpreting the map (1.2, 1.3).

Questions 4 and 5 – (spread 7e) Check that students are reading the scales to sufficient accuracy. It may help to accurately copy the sub-scale accurately onto a piece of paper which students can use instead of a ruler to measure distances on the map. In question **5**, some help might be needed to find all the bins and taps (1.2, 1.3).

Question 6 – (spread 7e) As a first step, check that students have correctly identified and oriented the grid squares. Ask them to make a sketch and label the coordinates of the bottom-left corner and show the directions of the coordinate axes. For example, part **a** shows the grid square (2, 1).containing tent T rotated through 160° (1.2, 1.3).

Assessment Criteria – UAM, level 6: use logical argument to establish the truth of a statement. UAM, level 5: identify and obtain necessary information to carry through a task and solve mathematical problems.

- Identify the mathematical features of a context or problem (L5)
- Evaluate the efficiency of alternative strategies and approaches (L5)
- Refine own findings and approaches on the basis of discussions with others (L6)

Useful resources
Large copies of question 2 table

Background

The sports day theme can be made even more real for the students if data from sports competitions in which they are involved can be used as illustrations or to replace numeric values in the questions (EP).

A large range of mathematics is encountered in this spread broadly on the theme of statistics, including: interpreting pie charts, collating data and finding summary statistics, solving 'algebraic' problems, rounding and making and justifying decisions.

Teaching notes

Given the breadth of knowledge being tested here it will be most useful to focus attention on those areas which are likely to cause the students most difficulty, rather than try to address all potential issues.

A majority of the class is likely to be familiar with scoring in football. Using results from the school or an international competition will allow several of the issues associated with question **2** to be discussed. In particular, cover how to systematically collate the raw results into the summary table.

Put students into groups and pose a question similar to **3**. Ask students for their ideas on how to go about solving it; did they get it right? How do they know? Several approaches are possible and it will be instructive to get students to compare their relative merits (TW, EP) (1.4, 1.5).

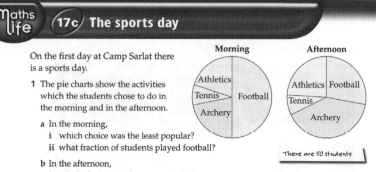

Maths life — 17c The sports day

On the first day at Camp Sarlat there is a sports day.

1 The pie charts show the activities which the students chose to do in the morning and in the afternoon.

Morning — Athletics, Tennis, Football, Archery
Afternoon — Athletics, Football, Tennis, Archery

There are 50 students

  a In the morning,
   i which choice was the least popular?
   ii what fraction of students played football?

  b In the afternoon,
   i which choice was the most popular?
   ii approximately how many students did athletics?

  c Overall, what was the most popular choice?

2 Five teams take part in a five-a-side soccer competition. These are the results.

Round 1				
High 5	2	v	Superstars	4
Champions	3	v	Cheetahs	1
High 5	2	v	All Stars	2
Cheetahs	0	v	Superstars	2
Champions	2	v	All Stars	1

Round 1				
High 5	3	v	Cheetahs	2
All Stars	1	v	Superstars	1
High 5	0	v	Champions	0
All Stars	2	v	Cheetahs	5
Champions	2	v	Superstars	2

  a How many goals were scored in total in the competition?
  b What is the modal average number of goals scored in all the matches?
  c Copy this table of results and complete it. You can work with a partner.

	Games				Goals		Points
	played	won	drawn	lost	for	against	
All Stars	4						
Champions	4						
Cheetahs	4						
High 5	4						
Superstars	4						

Points scoring system
win = 3pt
draw = 1pt
lose = 0pt

  d Calculate the mean average for the number of goals scored in all matches.

  e There is a tie for first place. Without playing another match, explain how you would decide who was the winning team.

Simplification

The wide variety of topics in this spread may prove discouraging to a student. At the same time it presents an opportunity to find questions with which they feel more comfortable, allowing them to build confidence. It will also help to organise students into small groups to share ideas and work together; this will be most helpful for questions **2** and **3** (TW). For question **2**, provide students with a large version of the table and, working one team at a time, enter tallies of games' results and goals scored. In a second stage these can be added up and entered into the final table.

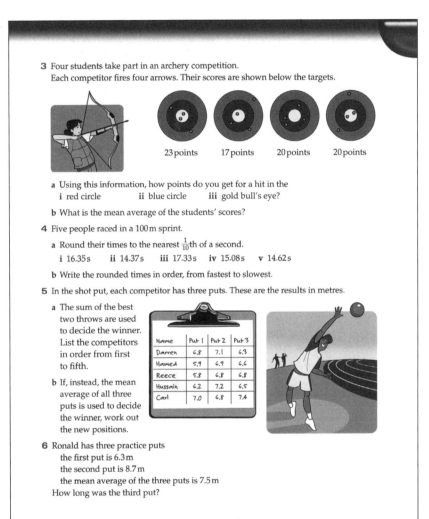

3 Four students take part in an archery competition.
Each competitor fires four arrows. Their scores are shown below the targets.

23 points   17 points   20 points   20 points

a Using this information, how points do you get for a hit in the
  i red circle   ii blue circle   iii gold bull's eye?

b What is the mean average of the students' scores?

4 Five people raced in a 100 m sprint.

a Round their times to the nearest $\frac{1}{10}$th of a second.
  i 16.35 s   ii 14.37 s   iii 17.33 s   iv 15.08 s   v 14.62 s

b Write the rounded times in order, from fastest to slowest.

5 In the shot put, each competitor has three puts. These are the results in metres.

a The sum of the best two throws are used to decide the winner. List the competitors in order from first to fifth.

Name	Put 1	Put 2	Put 3
Darren	6.8	7.1	6.3
Hamed	5.9	6.9	6.6
Reece	5.8	6.8	6.8
Hussain	6.2	7.2	6.5
Carl	7.0	6.8	7.4

b If, instead, the mean average of all three puts is used to decide the winner, work out the new positions.

6 Ronald has three practice puts
  the first put is 6.3 m
  the second put is 8.7 m
  the mean average of the three puts is 7.5 m
  How long was the third put?

The sports day 271

Extension

A number of checks are available for question **2c**. Ask students to investigate the following questions. Does the total number of won games have to equal the total number of lost games? Can the total number of drawn games be odd? How does the total number of won, drawn and lost games relate to the total number played? What is the difference between the total number of goals for and total number of goals against? How does this relate to the answer in part **a**? (RL)

Exercise 17c commentary

All questions (1.1, 1.3)

Question 1 – (spread 11c) One student = 7.2° so angles are only needed to this accuracy; can they be estimated? Ask students to measure the angles and calculate the numbers to check their answers (RL) (1.4).

Question 2 – (spreads 11b, 15a, b) In part **b**, encourage students to write down an ordered list of the number of goals scored per match and also to calculate the median. In part **c** students need to be very careful, what checks can be made? In part **d**, ask which is the more representative average, the mode, the median or the mean? (SM, IE)

Question 3 – (spreads 5f, 7a-c, 13c, 15a) Encourage abstracting four equations: R + B + 2G = 23, 2R + B + G = 17, 4B = 20, 2R + 2G =20. The third target gives B = 5 and hence R + 2G = 18, 2R + G = 12 and G + R = 10. Possible approaches are, trial-and-improvement, algebraic, graphical – encourage trying out more than one (CT) (1.2, 1.4, 1.5).

Question 4 – (spreads 4a, 8a) In part **b**, check times go from smallest (fastest) to largest (slowest).

Question 5 – (spreads 4a, 8b, c, 16a, b) In part **b**, how can the answers from part **a** help? Do they need to divide by 3 to rank the competitors?

Question 6 – (spreads 15a) Encourage students to set up an equation rather than use an *ad hoc* method or trial-and-improvement.

Assessment Criteria – UAM, level 6: interpret, discuss and synthesise information presented in a variety of mathematical forms. UAM, level 5: show understanding of situations by describing them mathematically using symbols, words and diagrams.

- Select appropriate procedures and tools (L5)
- Give accurate solutions appropriate to the context or problem (L6)

Useful resources

Protractor

Ruler

Enlargement of map and cliff face

OS maps

Background

Students who are involved in the Duke of Edinburgh award scheme, Boy Scouts, Girl Guides, Woodcraft Folk, Combined Cadet Force, *etc.* may have direct experience of going on expeditions. Sailors and orienteers may also have knowledge of navigation. These students' experiences of how mathematics can be applied should be used to enliven and inform classroom discussion (EP) (1.2).

The mathematics in this spread is broadly on the theme of geometry and includes giving compass bearings, measuring angles and measuring distances on scale drawings, as well as averages, time scales and finding proportions. There are direct links to the geography syllabus.

Teaching notes

In question **1c**, the mean can be thought of as a 'balance point' for the distribution of students' weights. This provides a means of checking the answer: The sum of the differences between individual students' weights and the mean should be zero. This provides a more formal definition of 'it should be in the middle' (RL).

Question **2** has obvious links to geography with directions being specified by compass points, whilst in question **4**, angles are measured in degrees. The approaches can be combined to give directions as three-figure bearings (EP).

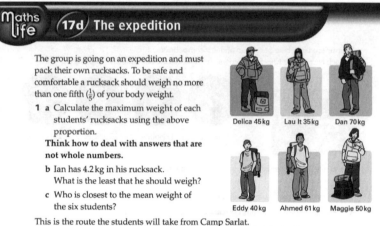

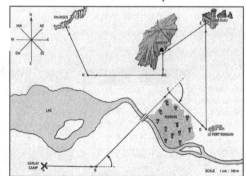

A further option is to show how locations can be 'triangulated': what place is on a bearing 045° as seen from point A and 030° as seen from point B? (cave/grotte) Can students provide their own examples, perhaps using a different base-line that requires larger angles to be measured? This could even be used as a challenge: can students produce an accurate scale drawing given the line AB and pairs of bearings for other locations? Distance can then be measured with a ruler and converted into a real-life distance using a scale; this skill is required for question **5** (CT).

The first part of question **3** is likely to cause trouble due to the lack of a year zero – which some students might not appreciate. This is most easily clarified using small values and a number line.

Simplification

Measuring some of the angles may prove awkward. Provide students with an enlarged copy of the map/cliff face or advise laying a ruler over the top of the protractor to make reading the scale easier. Using a 360° protractor should make measuring the reflex angles in question **5c** easier. (Enlarged diagrams should not be used for measuring distances.)

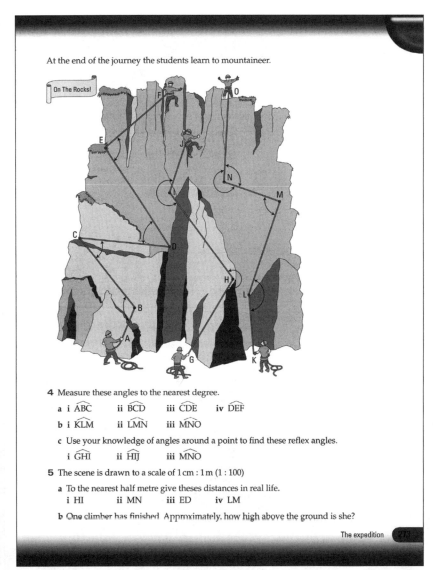

At the end of the journey the students learn to mountaineer.

On The Rocks!

4 Measure these angles to the nearest degree.

    a i $\widehat{ABC}$ ii $\widehat{BCD}$ iii $\widehat{CDE}$ iv $\widehat{DEF}$

    b i $\widehat{KLM}$ ii $\widehat{LMN}$ iii $\widehat{MNO}$

    c Use your knowledge of angles around a point to find these reflex angles.

      i $\widehat{GHI}$ ii $\widehat{HIJ}$ iii $\widehat{MNO}$

5 The scene is drawn to a scale of 1 cm : 1 m (1 : 100)

    a To the nearest half metre give theses distances in real life.

      i HI   ii MN   iii ED   iv LM

    b One climber has finished. Approximately, how high above the ground is she?

The expedition 273

Extension

Supply students with an OS map and ask them to plan a walk between given points. To begin they should specify the route as a list of directions/bearings and distances read off the map. They should then give timings for the journey assuming a typical walking pace of 15mins/km on even, level ground. This could be refined to take account of the type of terrain, any altitude changes (allow 10mins/100m climbed) and any rest stops (SM, IE).

Exercise 17d commentary

All questions (1.1, 1.3)

Question 1 – (spreads 4d, 15a) A calculator is not necessary (1/5 = 2/10) especially if rounded answers are given. In part **c**, how does the mean compare to the median? Which is more representative?

Question 2 – (spreads 9f, 14e) Aim for ±1 mm accuracies in measurements. Students could be asked to give three-figure bearings as well as compass directions (1.2, 1.5).

Question 3 – (spreads 8b, c, 16a, b) In part **a**, link BC dates to negative numbers. Explain the numbering 2 BC, 1 BC, 1 AD, 2 AD (no year zero). Encourage students to test their calculation with smaller numbers that can be shown on a 'number/time line'. In part **b**, what is a sensible accuracy for the answer? (TW, 1.2)

Question 4 – (spreads 6a, 9f, 14e) Aim for ±1° accuracy. Encourage students to make estimates of the angles first (at least to the level acute, obtuse, reflex) to help to choose the correct scale to read. In parts **a** and **b**, can the obtuse angles be calculated by subtraction? How does this relate to the reflex angles in part **c**?

Question 5 – (spread 9f) If 1cm represents 1m and you want answers accurate to 0.5 m, how accurately do you have to measure the lines with your ruler? (1.2, 1.4)

Assessment Criteria – UAM, level 5: solve word problems, and investigations from a range of contexts. UAM, level 6: give solutions to an appropriate degree of accuracy

- Make accurate mathematical diagrams, graphs and constructions on paper and on screen   (L5)
- Relate the current problem and structure to previous situations   (L5)

Useful resources
Graph paper
Tracing paper

Background

The spread has a loose focus on incidents that occur in the life of Miss Perry and the students. It allows a breadth of mathematics to be covered including: finding areas, applying algebra, rotations, using systematic approaches to problem solving and the speed-distance-time relationship.

An aspect of camp life is giving awards for various types of achievement. This could be mirrored in this final spread with, for example, bronze, silver and gold awards being given to students in recognition of their 'effort', 'achievement' and 'support to others'. This ties in with a suggestion for an **extension** activity.

Teaching notes

Question **1** involves finding areas. It may be instructive to ask students to explain where the formula for the area of a triangle comes from. Can they use this argument to simplify calculating the area of the two triangles?

Question **2** should be tackled using algebra. Supply a similar question, for example $12a + 2 = 6a + 20$, and ask students to explain how they would solve this equation. Also ask how they could check that their answer is correct (RL).

Question **3** may prove confusing to students given the apparent diagonal axes. It will be useful to get students to explain their methods for how to rotate a shape, drawn on a grid,

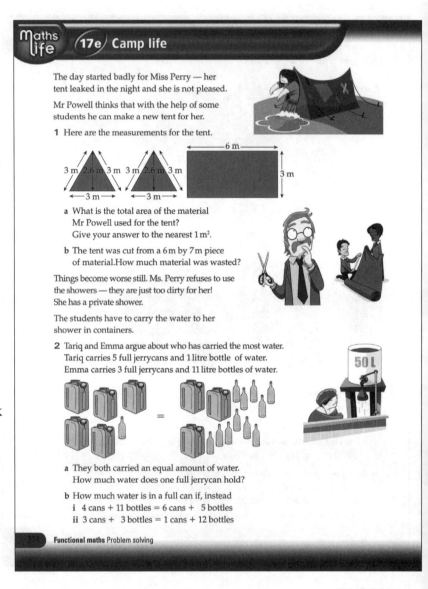

through a right-angle. Do they get the same result if the same problem were posed but with the axes in a different orientation? In fact, are axes required at all? (CT)

Question **4**, requires students to work systematically through the possible combination of weights. Ask students to explain their methods for listing and testing the various possibilities

Question **5** involves the relationship between speed, distance and time, which students have not previously encountered. This can be left for students to reason through what is required using common sense and experience or a simple example could be discussed to demonstrate how they should proceed (CT).

Simplification

Each question in this spread focuses on a different aspect of mathematics. In the first instance, direct students to questions on topics that they feel more comfortable with and are most likely to succeed. Allow students to work in small groups and do the questions collectively; as well as providing mutual support it is important that they discuss and share their approaches (TW, EP) (1.5)

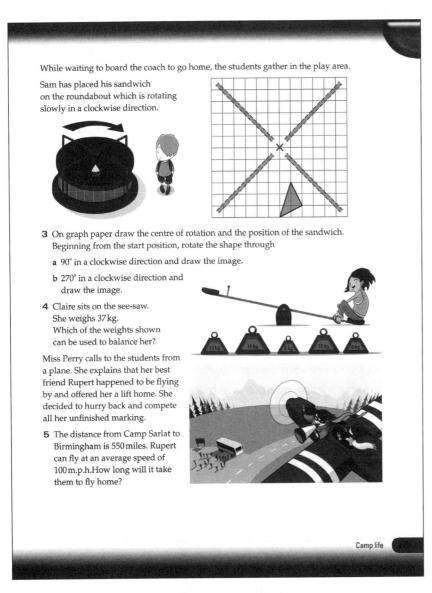

While waiting to board the coach to go home, the students gather in the play area.

Sam has placed his sandwich on the roundabout which is rotating slowly in a clockwise direction.

3 On graph paper draw the centre of rotation and the position of the sandwich. Beginning from the start position, rotate the shape through

   a 90° in a clockwise direction and draw the image.

   b 270° in a clockwise direction and draw the image.

4 Claire sits on the see-saw. She weighs 37 kg. Which of the weights shown can be used to balance her?

Miss Perry calls to the students from a plane. She explains that her best friend Rupert happened to be flying by and offered her a lift home. She decided to hurry back and compete all her unfinished marking.

5 The distance from Camp Sarlat to Birmingham is 550 miles. Rupert can fly at an average speed of 100 m.p.h. How long will it take them to fly home?

Camp life

Extension

If it was decided to run an awards scheme students could be charged with collating class members' marks in the various categories, ranking them and deciding appropriate boundaries for gold, silver and bronze awards.

This can be developed further with students producing pie charts and other graphs showing, for example, the differences between boys' and girls' achievements (SM, RL).

Exercise 17e commentary

All questions (CT, IE) (1.1, 1.3)

Question 1 – (spreads 2c, d, 8a) In part **a**, check that the slope height of the triangles is not used. Could the two triangles be made into a rectangle? (1.4, 1.5)

Question 2 – (spreads 7a-c, 13c) Whilst part **a** can be done using a pictoral balancing approach suggest setting up equations, $5C + B = 3C + 11B$, and collecting like terms to give, $2C = 10B$ or $C = 5B$. To simplify further you could replace $11B$ by 11 *etc*. Part **b ii** gives a fractional answer.

Question 3 – (spread 9a) In part **a**, a counting squares approach could be used instead of tracing paper. Ask, are the 45° lines relevant? Are you allowed to first turn the paper through 45° and then do the rotations? In part **b**, can students see the relationship between a +270° rotation and a -90° rotation? (1.2, 1.4)

Question 4 – Ask students to explain their strategies. Are they sure their answer is unique? (1.2)

Question 5 – (spreads 13f, g) Challenge students to give their answer in hours and minutes without using a calculator: $550 \div 100 = 55 \div 10 = 5\frac{1}{2} = 5\ 30/60 = 5$ hrs 30 mins. An intuitive approach based on flying 100 miles in an hour may prove helpful.

Assessment Criteria – UAM, level 6: present concise, reasoned argument, using symbols, diagrams, graphs and related explanatory texts. UAM, level 5: draw simple conclusions of their own and give an explanation of their reasoning.

17 Summary

Assessment criteria

• Check results, considering whether these are reasonable	Level 5
• Solve word problems and investigations from a range of contexts	Level 5
• Identify and obtain necessary information to carry through a task and solve mathematical problems	Level 5
• Show understanding of situations by describing them mathematically using symbols, words and diagrams.	Level 5
• Solve word problems, and investigations from a range of contexts	Level 5
• Draw simple conclusions of their own and give an explanation of their reasoning.	Level 5
• Use logical argument to establish the truth of a statement	Level 6
• Interpret, discuss and synthesise information presented in a variety of mathematical forms.	Level 6
• Give solutions to an appropriate degree of accuracy	Level 6
• Present concise, reasoned argument, using symbols, diagrams, graphs and related explanatory texts	Level 6

Development and links

Students will continue to practise functional mathematics throughout Year 9.

The student book start of chapter suggests five areas of everyday life where aspects of the ability to apply mathematical ideas prove highly valuable.

Using mathematical reasoning: the most efficient way to lay out a pattern on a piece of cloth is an example of an optimisation problem. Similar problems have to be solved when arranging files on a hard drive so as to minimize wasted space or packing oranges in a box so as reduce the need for packaging.

Representing: scale drawings based on careful measurements and 3-D reasoning are used throughout building and engineering to plan projects. Many of the same ideas are used to produce the graphics in computer games.

Using mathematical procedures: spreadsheet type programs are often used to help compile business accounts. However, to check and understand what they are doing you often need to be able to do simpler versions of the calculations yourself

Interpreting and evaluating: distribution centres face a difficult task in planning the best routes, loads and timetables. To guide them, they collect data on fuel cost, journey times, stock shortages, etc, and by carefully looking at this data can monitor and improve their performance.

Communicating: it isn't always enough to just find the correct answer to a problem, often you have to be able to convince other people using appropriately chosen graphs and mathematical explanations

Student book answers

Chapter 1

Check in   page 1

1 a 0.47, 0.5, 0.512, 0.52, 0.55

b -10, -6, -4, 3, 5, 9

2 a -5 **b** -3 **c** -9

3 a 35 **b** 3 **c** 54 **d** 9

4 a 7, 14, 21 **b** 1, 2, 3, 4, 6, 12

5 2, 3, 5, 7, 11

Exercise 1a   page 3

1 a -8 < 6 **b** -7 < -5

c -5 < -4.5 **d** -3.2 < -3

e -1.5 < -1.49 **f** -2.7 > -2.8

g -0.37 > -0.39 **h** -0.0235 > -0.024

2 a -12, -8, -6, 3, 5 **b** -8, -3.5, -1.5, 0.5, 1.4

c -2.9, -1.6, -1.4, 3.2, 4.7 **d** -3, -2.9, -2.5, -2.3, 1.35

3 a 15 **b** 8 **c** -4

d -9 **e** -20 **f** -1

g 4 **h** -4 **i** 6

4 a 0 **b** 0 **c** 3 **d** -11

e -14 **f** 11 **g** 14 **h** 0

i 3 **j** 4 **k** -5 **l** -25

5 a £21.50 **b** 6 m

Puzzle

a

b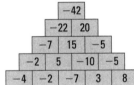

Exercise 1b   page 5

1 a 21 **b** -15 **c** -27

14 -10 -18

7 -5 -9

0 0 0

-7 5 9

-14 10 18

-21 15 27

-28 20 36

2

	-4	-3	-2	-1	0	1	2	3	4
4	-16	-12	-8	-4	0	4	8	12	16
3	-12	-9	-6	-3	0	3	6	9	12
2	-8	-6	-4	-2	0	2	4	6	8
1	-4	-3	-2	-1	0	1	2	3	4
0	0	0	0	0	0	0	0	0	0
-1	4	3	2	1	0	-1	-2	-3	-4
-2	8	6	4	2	0	-2	-4	-6	-8
-3	12	9	6	3	0	-3	-6	-9	-12
-4	16	12	8	4	0	-4	-8	-12	-16

3 a -6 **b** 6 **c** 3 **d** -3

e What is $4 \times$ -3 or $3 \times$ -4 or -3×4 or -4×3?

4 a -12 **b** -10 **c** 12 **d** 35

e -40 **f** 24 **g** -25 **h** 81

i -121 **j** 150 **k** 4 **l** -3

m 13 **n** 11 **o** -3 **p** -44

q -60 **r** 90 **s** 6 **t** -16

5 a -6 **b** -5 **c** 5

d -8 **e** -28 **f** -4

6 a -4 **b** 5 **c** -7 **d** -5

e 7 **f** 9 **g** -36 **h** 36

Puzzle

×	4	5	-6	8
-2	-8	-10	12	-16
3	12	15	-18	24
-7	-28	-35	42	-56
-9	-36	-45	54	-72

Exercise 1c   page 7

1 a 5, 10, 15 **b** 14, 28, 42

c 21, 42, 63 **d** 35, 70, 105

e 48, 96, 144 **f** 115, 230, 345

2 a 1, 2, 4, 5, 10, 20

b 1, 2, 4, 7, 14, 28

c 1, 3, 5, 9, 15, 45

d 1, 2, 4, 13, 26, 52

e 1, 2, 3, 6, 11, 22, 33, 66

f 1, 2, 3, 4, 6, 7, 12, 14, 21, 28, 42, 84

3 a 120 **b** 135

4 a yes **b** yes **c** no **d** yes

e yes **f** yes **g** yes **h** no

i yes **j** yes

5 a 5 **b** 8

6 a No because the sum of the digits is not divisible by 9

b Yes because 540 is divisible by both 3 and 4

c He can have

2 groups of 12

3 groups of 8

4 groups of 6

6 groups of 4

8 groups of 3

12 groups of 2

i.e. 6 ways

7 a   **b**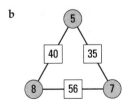

277

Investigation

a Yes

b ÷9 and ÷2, ÷6 and ÷3

Exercise 1d   page 9

1   3, 5, 7, 11, 13 because they all have exactly 2 factors, themselves and 1

2   a   1, 2, 3, 5, 6, 10, 15, 30

   b   1, 2, 3, 4, 6, 8, 12, 16, 24, 48

   c   1, 67

   d   1, 2, 29, 58

   e   1, 53

   67 and 53 are prime

3   a   yes as number ends in 5

   b   yes as digits add up to a multiple of 3

   c   yes as number is even

   d   yes as $1 + 3 = 4$

4   a   59 is prime   b   67 is prime

5   a   not   b   prime   c   prime   d   not

   e   not   f   prime   g   not   h   prime

   i   not   j   not

6   a   211   b   yes

7   a 3×5 b $2 \times 2 \times 2 \times 3$

   c $2 \times 2 \times 2 \times 5$ d $3 \times 3 \times 3$

   e $2 \times 2 \times 2 \times 2 \times 2 \times 2$ f $2 \times 2 \times 2 \times 7$

   g $2 \times 2 \times 2 \times 2 \times 3$ h $2 \times 2 \times 2 \times 3 \times 3$

   i $2 \times 2 \times 2 \times 2 \times 5$ j $2 \times 2 \times 2 \times 2 \times 2 \times 3$

Investigation

a   multiple answers   b   102 $(= 2 \times 3 \times 17)$

Exercise 1e   page 11

1

Numbers	Factors	HCF	First five multiples	LCM
4	1, 2, 4	2	4, 8, 12 16, 20	12
6	1, 2, 3, 6		6, 12, 18, 24, 30	

2   $4 = 2 \times 2, 6 = 2 \times 3$ so HCF = 2 and

   LCM = $2 \times 2 \times 3 = 12$

3   a   4   b   7   c   8

   d   13   e   8   f   7

4   a   24   b   42   c   160

   d   156   e   120   f   196

5   a   14, 84   b   7, 728   c   36, 216

   d   24, 720   e   45, 1350   f   64, 384

6

Problem

a   378 seconds   b   16 cm by 16 cm

Exercise 1f   page 13

1   a   36   b   81   c   144

   d   225   e   216   f   1000

2   a   343   b   289   c   1, 64, 729

---

3   a   196   b   324   c   576   d   2197

   e   4913   f   49   g   2744   h   6.25

   i   1000   j   3.375   k   421.875   l   -125

4   a   A-4   B-7   C-1   D-2

      E-8   F-3   G-6   H-5

   b   A = 12.25   B = 60.84   C = 123.21   D = 28.09

      E = 193.21   F = 42.88   G = 185.19   H = 24.93

5   a   11 cm   b   56   c   32 m

6   $\sqrt{2070} = 45.5$

   $45 \times 46 = 2070$

Investigation

Only 1 and 121 cannot be written as the sum of 2 primes.

Consolidation   page 14

1   a   -5, -3, -1, 2, 4   b   -3.1, -2.4, -2, -0.9, 1.5

   c   -2.6, -1.4, -1.1, 2.7, 3.2   d   -2.1, -1.6, -1.5, -0.5, 0.15

2   a   4   b   -3   c   -4   d   -10

   e   -13   f   11   g   15   h   6

   i   -4   j   5   k   -6   l   6

   m   -34   n   -3   o   29   p   -10

   q   37   r   5   s   56   t   -48

3   a   -10   b   -12   c   24   d   40

   e   -54   f   63   g   -81   h   36

   i   -70   j   90   k   5   l   -5

   m   4   n   7   o   -5   p   -65

   q   -100   r   165   s   12   t   -12

4   a   -9   b   9   c   -7   d   -13

   e   4   f   12   g   -256   h   126

5   a   1, 2, 3, 5, 6, 10, 15, 30

   b   1, 2, 3, 4, 6, 8, 12, 16, 24, 48

   c   1, 5, 13, 65

   d   1, 2, 3, 4, 6, 8, 9, 12, 18, 24, 36, 72

   e   1, 2, 3, 4, 6, 8, 12, 16, 24, 32, 48, 96

   f   1, 2, 4, 5, 10, 20, 25, 50, 100

   g   1, 2, 5, 10, 13, 26, 65, 130

   h   1, 2, 3, 4, 6, 9, 12, 18, 27, 36, 54, 108

   i   1, 2, 3, 4, 5, 6, 8, 10, 12, 15, 20, 24, 30, 40, 60, 120

   j   1, 2, 3, 4, 6, 11, 12, 22, 33, 44, 66, 132

   k   1, 2, 3, 4, 6, 8, 9, 12, 16, 18, 24, 36, 48, 72, 144

   l   1, 2, 3, 5, 6, 10, 15, 25, 30, 50, 75, 150

6   a   yes   b   yes   c   yes   d   no

   e   no   f   no   g   no   h   no

   i   no   j   yes

7   a   not   b   not   c   prime   d   prime

   e   not   f   prime   g   not   h   prime

   i   not   j   not   k   not   l   not

8   a   45   b   30   c   18   d   225

   e   54   f   75   g   343   h   105

   i   175   j   1155

9   a $2 \times 3 \times 3$ b $2 \times 2 \times 7$

   c $3 \times 3 \times 5$ d 3×19

   e $3 \times 3 \times 7$ f $2 \times 2 \times 19$

   g $2 \times 2 \times 2 \times 11$ h $2 \times 2 \times 23$

   i $2 \times 2 \times 3 \times 3 \times 3$ j 5×23

   k $2 \times 5 \times 13$ l $2 \times 2 \times 3 \times 11$

   m $2 \times 2 \times 2 \times 2 \times 3 \times 3$ n $2^5 \times 5$

   o $2 \times 5 \times 17$ p $5 \times 5 \times 7$

   q $2 \times 2 \times 47$ r $2^4 \times 3 \times 5$

10 a 2   b 4   c 9   d 5
    e 8   f 5   g 4   h 8

11 a 30   b 48   c 54   d 60
    e 48   f 120   g 224   h 1400

12

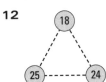

13 a 343   b 169   c 441   d 729
    e 3375   f 9   g 13824   h 0.25
    i 0.01   j 0.001   k 4   l -8

Summary   page 16

2 a 'My number must be even'
     All multiples of 4 are divisible by 2
   b 'My number could be even or odd'
     4 and 5 are both factors

Chapter 2

Check in   page 17

1 a 3000   b 56   c 46   d 8.5
2 a 132   b 36   c 8   d 9
3 a 20   b 8   c 100   d 4
4 a 6cm²   Perimeter = 12cm
   b 5cm²   Perimeter = 12cm
   c 8cm²   Perimeter = 12cm

Exercise 2a   page 19

1 a cl   b m   c litres
    d g   e mm
2 6m, 10litres, 2tonnes, 3min, 15kg, 3m.
3 50,000m
4 a 7.5kg   b 65cm
    c 8.5litres   d 0.5km
    e 2.5tonnes   f 0.085m
    g 0.007litres   h 19,500g
    i 400,000cm
5 5 type A and 1 type B
6 The 250ml bottle as this costs £4 × 0.99 per litre,
    i.e. £3.96 per litre

Problem

a 1 million paper clips weighs 500kg and is therefore heavier
b Depends on how strong they are and the way
     the paper clips are packaged, but should be able to carry about
     10kg, i.e. 20000 paper clips

Exercise 2b   page 21

1 a inch   b kg   c litre
    d mile   e metre
2 a 11lb   b 4.8litres   c 30cm
    d 154lb   e 90cm   f 176lb
    g 84litres   h 1.1lb   i 60cm
3 a 6inches   b 45kg   c 20pints
    d 0.5pints   e 16inches   f 60kg
    g 20inches   h 40pints   i 50inches

4 Manchester   48km
    Stoke   104km
    Birmingham   176km
    Bath   320km
5 £4.95
6 a 100miles, 350miles   b 4°C, 18°C
    c 50ml, 175ml   d 60m, 120m
    e 40cm, 80cm

Problem

a i 14.6kg
    ii after 1250 days (3 years 155 days)
b 76 pints

Exercise 2c   page 23

1 a 42cm, 90cm²   b 40cm, 75cm²
    c 23m, 33m²   d 18.8mm, 22mm²
2 a 30cm   b 46m   c 64m   d 21cm
3 a 28cm, 23cm²   b 32cm, 28cm²
    c 60cm, 125cm²

Challenge

a 500cm² and 900cm²   b 1200cm²

Exercise 2d   page 25

1 a 6cm²   b 4cm²   c 8cm²   d 4.5cm²
2 Multiple answers possible
3 a 42m²   b 22cm²   c 64cm²   d 17.5mm²
4 a 5cm   b 12m   c 10cm   d 7.5mm
5 a 18units²   b 8units²   c 15units²   d 6units²

Challenge

a 200cm²   b 100cm²

Exercise 2e   page 27

1 a 171cm²   b 360cm²   c 34m²
2 a 63cm²   b 50mm²   c 10.5m²
3 a 6cm   b 12.5m   c 30mm

Activity

a 36cm²
b i    ii 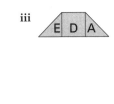   iii

c i 18cm²   ii 18cm²   iii 18cm²
d 9cm²

Consolidation   page 28

1 a g   b mile   c g
    d cm   e cl
2 a 38cm   b 4500g   c 650cl
    d 3.5m   e 2.5tonnes
3 a 20m   b 60kg   c 25miles
    d 3litres   e 45litres
4 a 5cm, 25cm   b 1.5gallons, 3.5gallons
    c 200ml, 800ml
5 a 50cm, 150cm²   b 90cm, 500cm²
    c 7cm, 32cm   d 4cm, 17cm

e 4.5 cm, 24.75 cm² **f** 5.5 cm, 41.25 cm²
g 8 cm, 6 cm
6 a 28 cm, 38 cm² **b** 54 cm, 100 cm²
c 34 cm, 60 cm²
7 a 140 cm² **b** 68 cm²
8 a, b, c multiple answers possible
9 a 216 m² **b** 65 cm² **c** 28 m²
10 a 266 cm² **b** 165 m² **c** 100 cm²

Summary page 30

2 a 12 cm **b** 1.2 m **c** 0.12 km

Chapter 3

Check in page 31

1 For this question, any suitable choice of word such as likely, unlikely and so on is fine.
2 a 0.5 **b** 0.4 **c** 0.375
3 a 0.35 **b** 0.635 **c** 0.07

Exercise 3a page 33

1 a 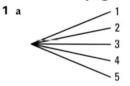 **b**

2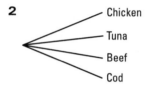

3 a i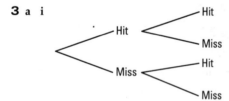

ii

		Second throw	
		Hit	Miss
First throw	Hit	Hit Hit	Hit Miss
	Miss	Miss Hit	Miss Miss

b i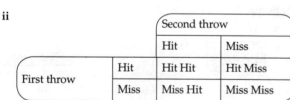

ii

		Second test	
		Pass	Fail
First test	Pass	PP	PF
	Fail	FP	FF

c i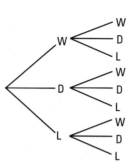

ii

		2nd match		
		W	D	L
1st match	W	WW	WD	WL
	D	DW	DD	DL
	L	LW	LD	LL

4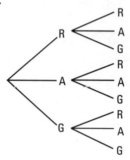

5

		2nd day		
		P	L	A
1st day	P	PP	PL	PA
	L	LP	LL	LA
	A	AP	AL	AA

Discussion

A tree diagram

Exercise 3b page 35

1 very unlikely ⟶ show at $\frac{1}{10}$
likely ⟶ show at $\frac{7}{10}$
almost certain ⟶ show at $\frac{9}{10}$
certain ⟶ show at 1

2 Various answers possible but approximately:
10% almost impossible
50% even chance
20% fairly unlikely
60% better than evens
100% certain
0% impossible

3 Various answers possible e.g.
a unlikely 0.1 (depends on time of year and location)
b very likely 0.9 (some will be 11)
c very likely 0.9 (could be half-term!)
d evens 0.5 (provided dice is unbiased)

4 a $\frac{1}{6}$ **b** $\frac{1}{3}$ **c** $\frac{1}{2}$ **d** 0

5 a 0.7 **b** 55% **c** $\frac{9}{16}$

Discussion

No. The 7 days of the week do not have equal chances of being chosen, as the teenages will probably be at work or school from Monday to Friday.

Exercise 3c   page 37

1 $\frac{5}{28}$

2 $\frac{18}{69} = \frac{6}{23}$

3 $\frac{13}{20}$

4 a $\frac{32}{50} = \frac{16}{25}$ or 0.64

   b $\frac{58}{100} = \frac{29}{50}$ or 0.58

   c Ben's as he did double the number of trials

Discussion

One way would be to look at the Christmas Day weather records for London for the past 100 years and see on what proportion of these days snow was recorded.

Exercise 3d   page 39

1 a 80   **b** $P(H) = \frac{31}{80}$ $P(T) = \frac{49}{80}$

   c It does not appear to be fair as $P(H)$ is less than $P(T)$

   d Increase the number of trials

2 a $\frac{62}{77}$ **b** more likely since $\frac{23}{28} > \frac{62}{77}$

3 a green $= \frac{3}{10} = 0.3$
   red $= \frac{4}{10} = 0.4$
   blue $= \frac{2}{10} = 0.2$
   yellow $= \frac{1}{10} = 0.1$

   b $P(G) = \frac{48}{130}$ $P(R) = \frac{51}{130}$ $P(B) = \frac{19}{130}$ $P(Y) = \frac{12}{130}$
       ≈ 0.37 ≈ 0.39 ≈ 0.15 ≈ 0.09

   c The experimental results are only estimates

Discussion

a No

b This would be a good idea as it increases the number of trials

Consolidation   page 40

1 a   **b**

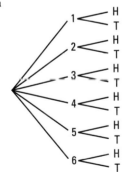

c

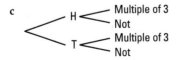

2 a

1st card		2nd card			
		H	C	D	S
	H	HH	HC	HD	HS
	C	CH	CC	CD	CS
	D	DH	DC	DD	DS
	S	SH	SC	SD	SS

b

1st letter		2nd letter	
		V	C
	V	VV	VC
	C	CV	CC

c

1st can		2nd can		
		C	B	F
	C	CC	CB	CF
	B	BC	BB	BF
	F	FC	FB	FF

3 a $\frac{13}{100}$ **b** $\frac{1}{2}$ **c** $\frac{1}{10}$ **d** $\frac{7}{50}$

4 a $\frac{13}{20}$ **b** 325 $(= \frac{13}{20} \times 500)$

5 a $P(B) = \frac{25}{100} = 0.25$, $P(R) = \frac{50}{100} = 0.5$,
   $P(Y) = \frac{25}{100} = 0.25$

   b $P(B) = \frac{61}{200} = 0.305$, $P(R) = \frac{108}{200} = 0.54$,
   $P(Y) = \frac{31}{200} = 0.155$

   c Various answers possible e.g. he may not have shaken the bag or the counters may not all be the same size (anything that prevents each counter having an equal chance of being chosen on each occasion)

Case Study   page 42

Cartoon answers

1 Students' own reasons. The new opponent might play paper to beat rock as they guess that an experienced player will play stone as it sounds 'strong'.

2 Students' own reasons. On the assumption that no one plays the same move twice in succession, then if they play the move that would have lost this time, they have a 50% chance of winning and a 50% chance of drawing.

3 Students' own reasons. They might assume that no one is likely to play the same move three times, thereby narrowing down the choices that they have to beat. Similar to the previous strategy.

Notepaper answers

4 Students should use their own systematic way of recording.

Player 1	paper	paper	paper	scissors	scissors	scissors	stone	stone	stone
Player 2	paper	scissors	stone	paper	scissors	stone	paper	scissors	stone

5 a There are 3 winning moves: paper against rock, rock against scissors and scissors against paper.

   b There are 3 drawing moves: paper against paper, rock against rock and scissors against scissors.

   c There are 3 losing moves: rock against paper, scissors against rock and paper against scissors.

6 a win: 1 in 3 or 1/3,   lose: 1 in 3 or 1/3,   draw: 1 in 3 or 1/3
Playing randomly over a number of games is likely to result in an overall draw, so it is not a good winning strategy.

Clipboard answers

7 $P(7) = \frac{6}{36} = \frac{1}{6}$,

$P(6) = P(8) = \frac{5}{36}$,

$P(5) = P(9) = \frac{4}{36} = \frac{1}{9}$

$P(4) = P(10) = \frac{3}{36} = \frac{1}{12}$

$P(3) = P(11) = \frac{2}{36} = \frac{1}{18}$

$P(2) = P(12) = \frac{1}{36}$

$P(1) = 0$

7 easiest to get, 2 hardest, 1 impossible.

8 All numbers 1 to 6, for a single dice $P = \frac{1}{6}$

9 a Both dice need for 7 to 9

b One die needed for 1

10 a (6), (1, 5), (2, 4), (1, 2, 3)

b (8), (1, 7), (2, 6), (3, 5), (1, 2, 5), (1, 3, 4)

c (2, 9), (3, 8), (4, 7), (5, 6), (1, 2, 8), (1, 3, 7), (1, 4, 6), (2, 3, 6), (2, 4, 5), (1, 2, 3, 5)

11 Students' answers
The game is hard to analyse. On the first play with two dice, scores allowing low numbers to be covered are more likely: 1 can be covered for all scores except 2 whilst 9 can only be covered for scores 9 to 12. Furthermore low scores occur in many more combinations than high scores. These suggest covering the higher numbers if possible. In the second play, all probabilities remain the same but the number of allowed combinations change according to the first play.

Number to cover	Probability of being able to cover	Number of combinations containing number	Expected number of combinations
1	$\frac{35}{36}$	29	$2\frac{5}{12}$
2	$\frac{33}{36} = \frac{11}{12}$	26	$2\frac{1}{6}$
3	$\frac{35}{36}$	23	$1\frac{31}{36}$
4	$\frac{33}{36} = \frac{11}{12}$	19	$1\frac{7}{12}$
5	$\frac{30}{36} = \frac{5}{6}$	16	$1\frac{13}{36}$
6	$\frac{26}{36} = \frac{13}{18}$	13	$1\frac{1}{12}$
7	$\frac{21}{36} = \frac{7}{12}$	10	$\frac{7}{9}$
8	$\frac{15}{36} = \frac{5}{12}$	7	$\frac{1}{2}$
9	$\frac{10}{36} = \frac{5}{18}$	5	$\frac{11}{36}$

Summary   page 44

2 a $\frac{1}{3}$ or 0.333…   **b** 3

Chapter 4

Check in   page 45

1 a 0.3   **b** 7.4   **c** 0.05

2 0.23, 0.3, 0.35, 0.39, 0.4

3 a £30   **b** 3.5 kg   **c** 18 m

Exercise 4a   page 47

1 a 700   **b** 7000   **c** 70   **d** 700 000
e 700 000   **f** 7   **g** $\frac{7}{10}$ **h** $\frac{7}{1000}$

2 a <   **b** >   **c** <
d >   **e** >   **f** <

3 a 3, 3.23, 3.3, 3.39, 3.4
b 3.72, 3.74, 3.757, 3.8, 3.88
c 0.03, 0.033, 0.035, 0.0351, 0.0362

4 a <   **b** >   **c** <   **d** >

5

Height h (metres)	Frequency
$1.3 \le h < 1.4$	3
$1.4 \le h < 1.5$	4
$1.5 \le h < 1.6$	5
$1.6 \le h < 1.7$	5
$1.7 \le h < 1.8$	3

Numbers most difficult to place are those on the boundaries

Group work

Multiple answers possible

Exercise 4b   page 49

1 a, d, e

2 a $\frac{3}{10}$ **b** $\frac{3}{5}$ **c** $\frac{3}{4}$ **d** $\frac{7}{25}$
e $\frac{33}{50}$ **f** $\frac{1}{20}$ **g** $\frac{3}{8}$ **h** $\frac{37}{200}$
i $\frac{19}{200}$ **j** $\frac{1}{125}$

3 a 0.1   **b** 0.65   **c** 0.28   **d** 0.66
e 0.6   **f** 0.6   **g** 0.95   **h** 0.75
i 0.7   **j** 0.22

4 a >   **b** >   **c** >   **d** >
e <   **f** <   **g** >   **h** >

5 a $\frac{3}{7}, \frac{3}{4}, \frac{4}{5}, \frac{7}{8}$
b $\frac{4}{19}, \frac{2}{9}, \frac{3}{13}, \frac{1}{3}$
c $\frac{2}{5}, \frac{7}{16}, \frac{4}{9}, \frac{9}{20}$

6 a $1\frac{1}{2}$ **b** $2\frac{3}{4}$ **c** $3\frac{2}{5}$
d $1\frac{7}{20}$ **e** $1\frac{19}{40}$

7 a 1.7   **b** 1.75   **c** 1.35
d 2.3125   **e** 3.44

Investigation

He is not correct as denominators of 8, 16 will also give terminating decimals

Exercise 4c   page 51

1 a 3   **b** 10   **c** 12   **d** 28
e 72   **f** 48   **g** 21   **h** 80

2 a $\frac{5}{7}$ **b** $\frac{3}{4}$ **c** $\frac{3}{5}$ **d** $\frac{1}{4}$
 e $\frac{8}{11}$ **f** $\frac{3}{13}$ **g** $\frac{4}{3}$ or $1\frac{1}{3}$ **h** $\frac{4}{5}$

3 a $\frac{7}{10}$ **b** $\frac{11}{12}$ **c** $\frac{11}{15}$

4 a $\frac{7}{12}$ **b** $\frac{13}{15}$ **c** $\frac{11}{30}$ **d** $\frac{11}{15}$
 e $\frac{23}{24}$ **f** $\frac{19}{30}$ **g** $\frac{13}{45}$ **h** $\frac{5}{33}$

5 $\frac{7}{20}$

Challenge

a She has not used a common denominator and just added the tops and the bottoms
b Because $\frac{4}{7}$ is more than $\frac{1}{2}$
c $\frac{50}{63}$

Exercise 4d   page 53

1 a £5 **b** 5 MB **c** 4 DVDs
 d 5 pupils **e** 20 shops **f** 80 g

2 a $\frac{4}{9}$ **b** $\frac{1}{2}$ **c** $\frac{2}{3}$
 d $1\frac{1}{3}$ **e** $1\frac{3}{4}$ **f** $3\frac{1}{3}$

3 a $5\frac{1}{4}$ feet **b** $9\frac{1}{3}$ million **c** $18\frac{3}{4}$ km
 d 120 kg **e** $17\frac{6}{7}$ m **f** $11\frac{1}{5}$ mm

4 a 88.8 kg **b** £122.92 **c** 6.82 km
 d 2.86 kg **e** 88.89 litres **f** 54°

5 a £9.07 **b** $\frac{9}{14}$

6 a $\frac{2}{5}$ **b** $\frac{11}{12}$ **c** $\frac{2}{3}$ **d** $\frac{7}{31}$

Challenge

a 66 feet **b** 72.6 feet **c** 6 years

Exercise 4e   page 55

1 a £35 **b** 4.5 kg **c** 15 m **d** 64 MB

2 a £8 **b** 3 DVDs **c** 3 MB **d** £30
 e £420 **f** 27° **g** £605 **h** 75 N
 i 19.2 ml **j** £7000 **k** 286 goods **l** 380 kJ

3 a £7.20 **b** 4.97 kg **c** 6.38 km
 d 13.6 euros **e** 2.25 mm **f** 13.2 kB
 g 5.22 litres **h** 32.2 mph **i** 21.16 m
 j 44.1 MB **k** 9.45 cm **l** 69.3°

4 a £17.76 **b** 33.6 kg **c** 266.24 MB
 d 305.3 km **e** 7.4 mm **f** 231 ml
 g £20 **h** 434.75 g **i** 6 N
 j £3280 **k** 613.2 kJ **l** 5.5 million

5 a 54 **b** 5.2 GB **c** 9 g

Investigation

a sugar 12 g, fat 0.48 g, protein 11.76 g, carbohydrates 31.2 g
b varies with each student

Exercise 4f   page 57

1 a 35% E, 0.8 C, $1\frac{1}{4}$ J, 60% G, 0.45 F, 0.1 A,
 110% I, $\frac{19}{20}$ D, $\frac{3}{4}$ H, $\frac{1}{5}$ B

 b A = 10%, $\frac{1}{10}$, 0.1
   B = 20%, $\frac{1}{5}$, 0.2
   C = 80%, $\frac{4}{5}$, 0.8

 D = 95%, $\frac{19}{20}$, 0.95
 E = 35%, $\frac{7}{20}$, 0.35
 F = 45%, $\frac{9}{20}$, 0.45
 G = 60%, $\frac{3}{5}$, 0.6
 H = 75%, $\frac{3}{4}$, 0.75
 I = 110%, $1\frac{1}{10}$, 1.1
 J = 125%, $1\frac{1}{4}$, 1.25

2 a $\frac{2}{5}$ **b** $\frac{3}{4}$ **c** $\frac{17}{20}$ **d** $\frac{9}{20}$
 e $\frac{8}{25}$ **f** $\frac{1}{20}$ **g** $\frac{1}{100}$ **h** $\frac{5}{4}$ or $1\frac{1}{4}$
 i $\frac{21}{20}$ or $1\frac{1}{20}$ **j** $\frac{1}{40}$

3 a 0.8 **b** 0.25 **c** 0.08 **d** 0.35
 e 0.99 **f** 1.3 **g** 0.235 **h** 0.072
 i 0.0475 **j** 1.45

4 a 30% **b** 58% **c** 56% **d** 175%
 e 32.5% **f** 160% **g** 144% **h** 115%
 i 117.5% **j** 287.5%

5 a 56.3% **b** 67.5% **c** 68% **d** 125%
 e 52% **f** 87.5% **g** 77.8% **h** 135%
 i 83.3% **j** 66.7%

6 a 58% **b** 8% **c** 80% **d** 108%
 e 180% **f** 3.5% **g** 41.5% **h** 105%
 i 155.5…% **j** 99.9%

7 a 60% **b** 60%

Challenge

a Science as he got 77.5%
b History as he got 68.6% (1 dp)
c History, Geography, RE, English, French, Maths, Science

Consolidation   page 58

1 a 4, 4.34, 4.4, 4.48, 4.5
 b 5.9, 5.94, 5.96, 5.979, 6
 c 0.06, 0.066, 0.068, 0.0684, 0.0695
 d 2.16, 2.77, 2.771, 2.776, 2.8

2 a < **b** > **c** < **d** >
 e < **f** < **g** > **h** <

3 a 0.875 **b** 0.4375 **c** 0.35
 d $0.1\dot{6}$ **e** $0.\dot{5}$

4 a $\frac{2}{9}$ $\frac{1}{4}$ $\frac{3}{10}$ $\frac{4}{13}$
 b $\frac{2}{3}$ $\frac{13}{18}$ $\frac{11}{15}$ $\frac{3}{4}$
 c $\frac{5}{9}$ $\frac{9}{16}$ $\frac{3}{5}$ $\frac{13}{20}$

5 a 10 **b** 9 **c** 16 **d** 35
 e 121 **f** 9 **g** 30 **h** 35

6 a $\frac{13}{20}$ **b** $\frac{22}{35}$ **c** $\frac{8}{15}$ **d** $\frac{31}{36}$
 e $\frac{13}{20}$ **f** $\frac{7}{18}$ **g** $1\frac{5}{24}$ **h** $\frac{16}{21}$

7 a 67.43 g **b** £218.75 **c** 10.29 km **d** 75.2 miles
 e 20 hours **f** $15 **g** 19.44 tonnes **h** 144°

8 a $\frac{2}{5}$ **b** $\frac{1}{6}$

9 a £3.50 **b** 5.4 kg **c** 5.58 km
 d 22.23 euros **e** £25.23 **f** 25.83 kg

10

Fraction	Decimal	Percentage
$\frac{17}{25}$	0.68	68%
$\frac{61}{100}$	0.61	61%
$\frac{21}{50}$	0.42	42%
$\frac{9}{40}$	0.225	22.5%

Summary   page 60

2 a 7 out of 10 is the same as 70%

   10 out of 20 is the same as 50%

  b You could put many different numbers in this sentence. One example is '1 out of 20 is the same as 5%'

Chapter 5

Check in   page 61

1 $(4 \times 12) + (3 \times 6) = 66$

2 a 16   **b** 100   **c** 180

3 a 5   **b** 4   **c** £4.50

4 a 800 km   **b** 3

Exercise 5a   page 63

1 a $5x$ **b** $3y$ **c** $4z$

  d $3z$ **e** $\frac{8}{x}$ **f** $5y$

  g $\frac{x}{2}$ **h** $\frac{y}{5}$ **i** $\frac{z}{3}$

  a 50   **b** 15   **c** 24

  d 18   **e** 0.8   **f** 25

  g 5   **h** 1   **i** 2

2 a $x + 4$ **b** 16

3 a $y - 10$ **b** 40

4 a $4n$ **b** 24

5 a $4y$ metres   **b** 48 metres

6 a $2r + 2s$ metres   **b** 170 metres

Challenge

$3x + 2y$ cm, 10

31 cm

Exercise 5b   page 65

1 a 2^4 **b** 7^5 **c** 9^6

  d n^3 **e** y^7 **f** z^2

2 a $4^3 \times 5^4$ **b** $8^2 \times 6^5$ **c** $2^3 \times 3^4$

  d $5^3 \times 9^2$ **e** $6^4 \times 4^2$ **f** $3^4 \times n^2$

  g $r^2 \times s^5$ **h** $a^2 \times b^2 \times c^3$

3 a $2 \times 2 \times 2 \times 2 = 16$ **b** $3 \times 3 \times 3 = 27$

  c $5 \times 5 = 25$ **d** $2 \times 2 \times 2 \times 2 \times 2 = 32$

  e $1 \times 1 \times 1 \times 1 \times 1 \times 1 = 1$

  f $2 \times 2 \times 3 \times 3 = 36$

  g $2 \times 2 \times 2 \times 5 \times 5 = 200$

  h $1 \times 1 \times 1 \times 1 \times 6 \times 6 = 36$

  i $3 \times 3 \times 10 \times 10 \times 10 \times 10 = 90000$

  j $0 \times 0 \times 0 \times 0 \times 0 \times 7 \times 7 \times 7 = 343$

4 a 6^7 **b** 8^6 **c** 2^9 **d** 3^9

  e 5^8 **f** 7^{12} **g** 10^9 **h** 3^9

  i 5^7 **j** 6^8 **k** 4^{10} **l** 10^8

5 a 3   **b** 2   **c** 3

  d 3   **e** 3   **f** 5

6 a 10^3 cm³   **b** 1000   **c** x^2 cm² x^3 cm³

  d Discuss students' answers.

Research

Billion: 10^9 in USA; 10^{12} in UK

Trillion: 10^{12} in USA; 10^{18} in UK

Quadrillion: 10^{15} in USA; 10^{24} in UK

although the American definitions are increasingly common in Britain.

Googol: 10^{100} introduced by U.S. mathematician Edward Kasner (1878–1955), whose 9 year-old nephew invented it.

Googolplex: $10^{10^{100}}$ is 10 to the power of a googol.

Exercise 5c   page 67

1 a $10x$ **b** $7y$ **c** $7z$

2 a $6n$ **b** $12m$ **c** $5p$ **d** $3q$

  e $12t$ **f** $3r$ **g** 0   **h** x

  i $2x$ **j** $7y$ **k** $2z$ **l** 0

3 a $6x + 2y$ cm   **b** $4x + 6y$ cm

4 a $5x + 2y$ **b** $2x + 8y$ **c** $2x + 5y$

  d $6s + 6t$ **e** $2u + 7v$ **f** $7r + 3s$

  g $5x + 2y$ **h** $5a + 3b$ **i** $2a + 2b$

5 a $6x + 8x^2$ **b** $9y + 3y^2$ **c** $7z + 4z^2$

  d $5u + 3u^2$ **e** $2v + 3v^2$ **f** $6x + 5x^2$

  g $2z$ **h** $9h^3 + 2h$ **i** $2j^2 + j$

6 a $3x^2 + 3xy$ **b** $8x + 2y$

Puzzle

4

Exercise 5d   page 69

1 a $3x + 6$ **b** $2x + 8$ **c** $4x + 4$

2 a $5x + 15$ **b** $10x + 25$ **c** $5x + 15$

  d $10x - 25$ **e** $18u + 12$ **f** $3v - 12$

  g $10a - 15$ **h** $12b - 4$ **i** $15 - 10c$

3 a $x^2 + 3x$ **b** $5x^2 + 3x$ **c** $2x^2 + 4x$

  d $3x^2 - 2x$ **e** $4x^2 - 5x$ **f** $2x^2 - 7x$

  g $4x - 5x^2$ **h** $2x + x^2$ **i** $7x + 3x^2$

4 a $4(x + 8)$ kg   **b** $4x + 32$ kg

5 a $3(20y + 50)$ g   **b** $60y + 150$ g

6 a $14x + 9$ **b** $27x + 4$ **c** $17x + 2$

  d $10x + 3$ **e** $16x - 9$ **f** $15x - 8$

Challenge

4 ways: $3(4x + 8)$, $4(3x + 6)$, $6(2x + 4)$, $12(x + 2)$ wing only integers

Exercise 5e   page 71

1 a 50   **b** 80   **c** 170

2 a 50 mph   **b** 35 mph   **c** 23 mph

3 a 34   **b** 88   **c** 60

4 a 50   **b** 30   **c** 50

5 a 19   **b** 14   **c** 18

6 a 32   **b** 24   **c** 30

Research

a $212°F = 100°C$; $32°F = 0°C$

  These are the boiling points and freezing points of water.

b All 5 surnames are used for temperature scales.

Exercise 5f page 73

1 a i $P = 13 + x$ **ii** $P = 18 + x$ **iii** $P = 4x + 12$
  b i 17 cm   **ii** 22 cm   **iii** 28 cm

2 a $y = 180° - x$ **b** 120°

3 a $A = 190° - x$ **b** 90°

4 a 48 cm², 3x cm²   **b** $A = 48 + 3x$
  c 63 cm²

5 $A = 110° - Q; 50°$

6 a $C = 4p + 6$
  b i 126   **ii** 146

Challenge

a $A = 80 - x^2$ cm²   **b** 44 cm²

Consolidation page 74

1 a 3x   **b** 4y   **c** 6z

2 a 5x   **b** $5x - 6$ **c** 44

3 a 20   **b** 20   **c** 3   **d** 2

4 a 27   **b** 16   **c** 100   **d** 25 000

5 a $a^3b^2c^3$ **b** 3^6 **c** 6^8

6 a 4   **b** 4   **c** 2

7 $5x + 5y$

8 $6x + 4y$

9 a $5p + 3q$ **b** $5m - n$ **c** $9x^2 - x$ **d** 0

10 a $5x + 10$ cm² or $5(x + 2)$ cm²
   b $5x + 10$

11 a i $4x + 12$ **ii** $15x + 12$
    b i $22y + 14$ **ii** $8z + 2$

12 a $20 + 10x$ grams
    b $5(20 + 10x)$ grams
    c $100 + 50x$

13 a 5   **b** 7   **c** $7\frac{2}{3}$

14 a 60   **b** 225   **c** 0

15 a 10   **b** 0   **c** 9

16 a $P = 2x + 13$ cm   **b** 21 cm

17 a $A = 180° - x - y$ **b** 75°

18 a i $5x + 10$ **ii** $15x + 12$
    b i $22y + 14$ **ii** $8z + 2$

Summary page 76

2

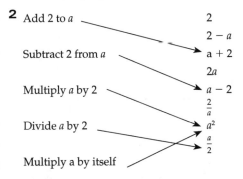

Add 2 to a 2
2 − a
Subtract 2 from a a + 2
2a
Multiply a by 2   a − 2
$\frac{2}{a}$
Divide a by 2   a²
$\frac{a}{2}$
Multiply a by itself

Chapter 6

Check in page 77

1 a acute   **b** obtuse   **c** straight
  d reflex   **e** right

---

2 a 105°   **b** 125°

3 a 90°   **b** 50°   **c** 72°

Exercise 6a page 79

1 a r   **b** s   **c** p
  d q   **e** u

2 a 42°   **b** 58°   **c** 38°

3 a 64°   **b** $c = 95°, b = d = 85°$
  c $c = 28°, d = 81°, e = f = 71°$

4 a 30°   **b** 43°   **c** 120°
  d 144°   **e** 119°   **f** 45°

Challenge

a 90°   **b** 30°   **c** 105°

Exercise 6b page 81

1 a 55°   **b** 53°   **c** 107°

2 a 60° equilateral   **b** 71° isosceles
  c 90° right-angled   **d** 78° scalene
  e 42° isosceles

3 a 37°   **b** $b = 48°, c = 84°$
  c $c = 57°, d = 66°$

4 $a = b = c = d = 45°, e = 90°$

5 a 115°   **b** 69°   **c** 42°

Activity

Check student explanations

Exercise 6c page 83

1 a Student's own diagram measurements.
  b corresponding

2 a    **b**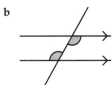

  c

3 a 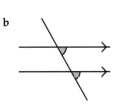   **b**

  c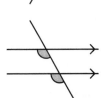

4 a 115° (alternate)   **b** 108° (corresponding)
  c 135° (alternate)

5 a $a = 35°, b = 15°, c = 130°$
  b $b = 45°$ $c = 135°$ $d = e = 55°$ $f = 125°$

Activity
Check students' work

Exercise 6d   page 85

1 a 90° rectangle or square

b 60° rhombus, parallelogram or isosceles trapezium

c 70° trapezium or kite

d 30° arrowhead

e 100° quadrilateral (no parallel sides or equal angles)

2

	Equal in length	Bisect each other	Perpendicular
Parallelogram	No	Yes	No
Kite	No	No	Yes
Rhombus	No	Yes	Yes
Square	Yes	Yes	Yes
Rectangle	Yes	Yes	No

3 isosceles trapezium, rhombus

4 a Square, rectangle

b Kite, arrowhead and trapezium with two 90° angles

c Square, rhombus

Activity
Check students' work

Exercise 6d² page 87

1 a equilateral triangle **b** square

2 a 57° **b** 104°

c $c = 125°, d = 70°, e = 55°, f = 55°$

3 b 140° **c** 1260°

4 Check students' answers

5 135°

Activity
Check students' work

Exercise 6e   page 89

1 a ↕ **b** ◯ i.e. the large circle

c ▢ **d** (L-shape)

e △ i.e. the small triangle

2 $A = B = 68°, C = D = 112°$

3 a 5 cm **b** 12 cm **c** 13 cm

Activity
Check students' work

Exercise 6f   page 91

1 a A cuboid B cube C square-based pyramid
D tetrahedron (triangular-based pyramid)
E tetrahedron F triangular prism G hexagonal prism
H hexagonal pyramid

b A **i** 6 **ii** 8 **iii** 12

B **i** 6 **ii** 8 **iii** 12

C **i** 5 **ii** 5 **iii** 8

D **i** 4 **ii** 4 **iii** 6

E **i** 4 **ii** 4 **iii** 6

F **i** 5 **ii** 6 **iii** 9

G **i** 8 **ii** 12 **iii** 18

H **i** 7 **ii** 7 **iii** 12

2 a 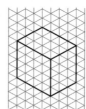 **b**
Other nets possible

c 80 cm²

3 a, b

4 a 4 **b** 8 **c** 0

5 b Depends on shape drawn e.g. shape C and F in question 1.

Activity

C × is opposite ◯
diamond is opposite spade
heart is opposite club

Exercise 6g   page 93

1 a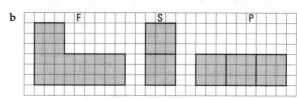

b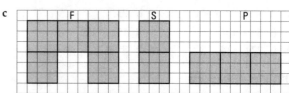

c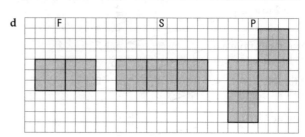

d

2 a i   **ii** Plan view
Front elevation
Side elevation

b i   **ii** Plan view

Front elevation

Side elevation

c i   **ii** Plan view

Front elevation

Side elevation

d i   **ii** Plan view

Front elevation

Side elevation

3 a   **b** 4

4 a A ↔ H, B ↔ G, C ↔ E, D ↔ F
   b A triangular prism   B cube   C square pyramid   D cylinder

Challenge
a 2 by 2 by 2 cube
b   **c**

c Discuss students answers.

Consolidation   page 92
1 a 138°   **b** 37°   **c** c 108°, $d = e = 72°$
2 14°, 24°, 142°
   214°, 22°, 124°
3 a 75° isosceles   **b** 90° right-angled
   c 36° isosceles   **d** 81° scalene
   e 45° right-angled isosceles
4 70°, 40°   or   55°, 55°
5 a $a = 115°, b = c = 65°$
   b $b = c = 43°, d = 137°$
   c $c = d = 55°, e = f = 125°$
6 a 95°   **b** $b = 115°, c = d = 65°$ **c** 15°
7 A, F, K, H equilateral triangles
   B, J isosceles trapeziums
   C, I equilateral triangles
   D, E, G rhombuses
8 $a = 108°$ b 144°
9 A, B, C, D, E, G, H
10 a i 8   **ii** 12   **iii** 18
   b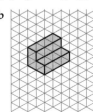

11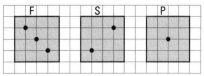

Summary   page 94
2 a 8 faces
   b You should have a drawing on a cube, cuboid or pentagonal pyramid on isometric paper.

Chapter 7

Check in   page 95
1 a 9   **b** 6   **c** 7   **d** 20
2 a 7   **b** 8
3 8
4 a 56   **b** 56 ÷ 8 = 7   56 ÷ 7 = 8

Exercise 7a   page 97
1 a 5   **b** 5   **c** 2
2 a i 2   **ii** 9   **iii** 1
     iv 2   **v** 4   **vi** 9
     vii 1   **viii** 2
   b i 9   **ii** 9   **iii** 7
     iv 10   **v** 14   **vi** $3\frac{1}{2}$
     vii 11   **viii** 14
   c i 5   **ii** 4   **iii** 2
     iv 9   **v** 4   **vi** 2
     vii 11   **viii** 5
   d i 10   **ii** 12   **iii** 8
     iv 18   **v** 14   **vi** 20
     vii 60   **viii** 18
3 a 1   **b** 15   **c** 4   **d** 6
   e 10   **f** 4   **g** 18   **h** 18
   i 11   **j** 6   **k** 60   **l** 5

Challenge
a $x = 3$ **b** $x = 4$ **c** $x = 5$

Exercise 7b   page 99
1 a 3   **b** 5   **c** 4   **d** 2
   e 4   **f** 5   **g** 6   **h** $4\frac{1}{2}$
   i 4   **j** 3   **k** 4   **l** $3\frac{1}{2}$
2 a 5   **b** 5   **c** 3
3 a 3   **b** 2   **c** 4   **d** 9
   e 3   **f** 2   **g** 1   **h** $2\frac{1}{2}$
   i 2   **j** 4   **k** 2   **l** 3
4 a 6   **b** 6   **c** 3   **d** 4
   e 4   **f** 2   **g** 2   **h** 3
   i 2   **j** 2   **k** 7   **l** 2
5 a 4   **b** 7   **c** 4   **d** 2
   e 3   **f** 7   **g** 2   **h** 0
   i 4   **j** 6   **k** 3   **l** 3
6 a $7n + 4 = 5n + 28$ **b** $n = 12$

Puzzle
   a 6
   b Triangle 120 g, Hexagon 150 g, Circle 75 g, star 165 g

Exercise 7c   page 101

1
a 5 b 2 c 2 d 3
e 4 f $4\frac{1}{2}$ g $4\frac{1}{2}$ h 2
i $2\frac{1}{2}$ j 2 k 5 l 2
m 7 n $6\frac{1}{2}$ o 11 p 9
q 10 r 3

2
a 1 b 3 c 4 d 1
e 2 f 0 g 5 h 1
i $\frac{1}{2}$ j 1 k 2 l 2

3 10

4
a 0 b -1 c -3 d -6
e -2 f -3 g -1 h -2
i -2 j 0 k -2 l -2
m 5 n 0 o -1 p 2

5
a $3\frac{1}{2}$ b $2\frac{1}{2}$ c $1\frac{1}{4}$ d $3\frac{1}{8}$
e $4\frac{1}{2}$ f $4\frac{1}{4}$ g $2\frac{1}{2}$ h $1\frac{1}{8}$

Challenge

-40°

Somewhere very cold, e.g. Siberia

Exercise 7d   page 103

1 a x y

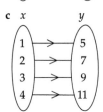

b x y

c x y

2 a add 6   or $y = x + 6$

b multiply by 3   or $y = 3x$

3 a x y

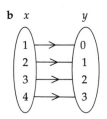

x	1	2	3	5
y	3	5	7	11

b x y

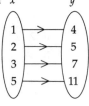

x	1	2	3	5
y	4	6	8	12

4 a 2 miles

b
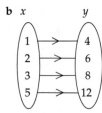

seconds x	5	10	15	20	25	30
miles y	1	2	3	4	5	6

Task

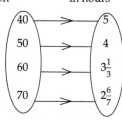

Speed (x) in mph   Time taken (t) in hours

40 → 5
50 → 4
60 → $3\frac{1}{3}$
70 → $2\frac{6}{7}$

Rule is divide 200 by the average speed

Exercise 7e   page 105

1 a

x	0	1	2	3	4	5
y	4	5	6	7	8	9

b, c
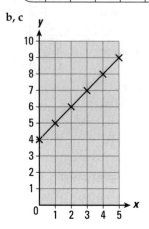

2 a

x	0	1	2	3	4	5
y	2	3	4	5	6	7

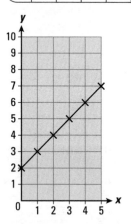

b

x	0	1	2	3	4	5
y	-1	0	1	2	3	4

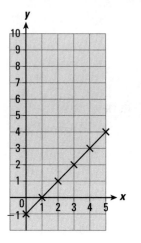

c

x	0	1	2	3	4	5
y	1	3	5	7	9	11

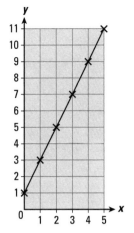

d

x	0	1	2	3	4	5
y	-3	-1	1	3	5	7

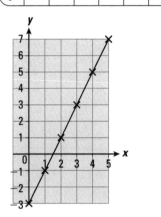

e

x	0	1	2	3	4	5
y	10	9	8	7	6	5

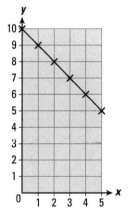

f

x	0	1	2	3	4	5
y	5	4	3	2	1	0

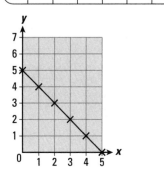

3 a

x	0	2	4	6	8	10
y	10	8	6	4	2	0

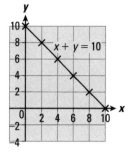

$x + y = 10$

b

x	0	2	4	6	8	10
y	6	4	2	0	-2	-4

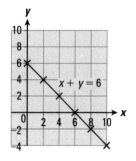

$x + y = 6$

c

x	0	2	4	6	8	10
y	8	6	4	2	0	-2

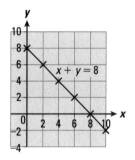

$x + y = 8$

d

x	0	2	4	6	8	10
y	9	7	5	3	1	-1

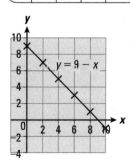

$y = 9 - x$

4 a

x	1	2	3	4	5	6
c	6	10	14	18	22	26

b

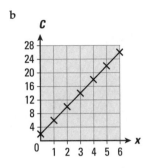

c 2.5 metres

Research

Function can mean:

a an activity or purpose which something is designed for (*The function of a car is to carry people about.*)

b a public occasion or social gathering (*The hotel has a function room where you can have a wedding reception.*)

c A factor which is dependant on other factors. (*The price of petrol is a function of supply and demand.*)

d In mathematics, a relationship where every member of one set maps onto just one member of another set.

The Latin word *functio* means 'to perform'.

Exercise 7f   page 107

1 a ∥ to x-axis   **b** ∥ to y-axis   **c** sloping   **d** sloping

e ∥ to y-axis   **f** sloping   **g** ∥ to x-axis   **h** sloping

2 A $x = 2$ **B** $x = 3$ **C** $x = 5$

D $y = 3$ **E** $y = 1$

3 a

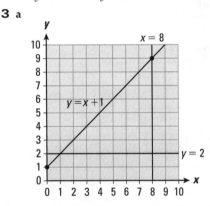

b

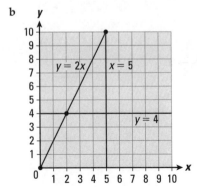

c

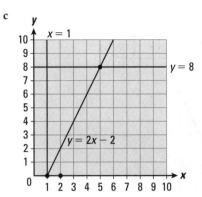

d

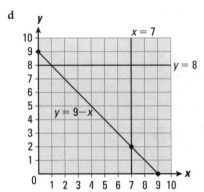

4 a yes   **b** yes   **c** yes   **d** no

e no   **f** yes   **g** yes   **h** no

5

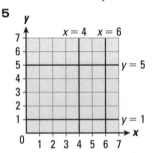

Area = 8 units^2

6

x	20	30	40	50	60
y	60	50	40	30	20

Investigation

The line cross at (2, 5)

Consolidation   page 108

1 a 3   **b** 6

2 a 4   **b** 10   **c** 4   **d** 10

e 2   **f** 5   **g** 6   **h** 8

3 a 5   **b** 7   **c** 6

d 3   **e** 10   **f** 18

4 a 5   **b** 4

5 a 5   **b** 12   **c** 2

d 2   **e** 3   **f** 5

6 a 5   **b** 4   **c** 2

d 3   **e** -2   **f** -3

7 $6x + 2 = 3x + 17$

$x = 5$

8 a 2   **b** 4   **c** 6

d 10   **e** 4   **f** 2

9 a 2   **b** 2   **c** 1

d $\frac{2}{3}$ **e** $1\frac{1}{3}$ **f** $1\frac{1}{3}$

10 a x y

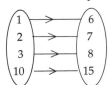

x	1	2	3	10
y	6	7	8	15

b x y

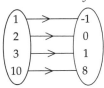

x	1	2	3	10
y	-1	0	1	8

c x y

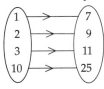

x	1	2	3	10
y	7	9	11	25

11 a x y

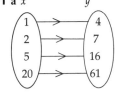

x	1	2	5	20
y	4	7	16	61

b x y

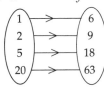

x	1	2	5	20
y	6	9	18	63

12 a

x	-1	0	1	2	3	4
y	4	5	6	7	8	9

Check graph is a straight line passing through $(0, 5)$ and $(4, 9)$

b i

x	-1	0	1	2	3	4
y	1	2	3	4	5	6

Check graph is a straight line passing through $(0, 2)$ and $(4, 6)$

ii

x	-1	0	1	2	3	4
y	-3	-1	1	3	5	7

Check graph is a straight line through $(0, -1)$ and $(4, 7)$

iii

x	-1	0	1	2	3	4
y	10	9	8	7	6	5

Check graph is a straight line through $(0, 9)$ and $(4, 5)$

13 A $x = 4$ **B** $y = 3$ **C** $x = 2$ **D** $y = 1$

14 a ‖ to x-axis   **b** ‖ to y-axis
  c sloping   **d** sloping
  e ‖ to y-axis   **f** ‖ to x-axis

Summary   page 110

2

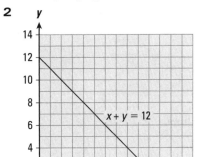

$x + y = 12$

Chapter 8

Check in   page 111

1 a i 3000   ii 3500   iii 3460
  b i 5000   ii 5300   iii 5280

2 a 95   **b** 56   **c** 12.1   **d** 6.3

3 a 86.1   **b** 67.8

4 a 390   **b** 480   **c** 0.58   **d** 48.5

5 a 153   **b** 195   **c** 228   **d** 198

6 a 480   **b** 16   **c** 4770   **d** 13

Exercise 8a   page 113

1 a i 4000   ii 4100   iii 4070
  b i 7000   ii 7200   iii 7190
  c i 4000   ii 3700   iii 3650
  d i 8000   ii 7500   iii 7530
  e i 6000   ii 5600   iii 5590
  f i 7000   ii 6600   iii 6570
  g i 5000   ii 4900   iii 4940
  h i 13000   ii 13400   iii 13390
  i i 28000   ii 27600   iii 27590
  j i 32000   ii 31700   iii 31690
  k i 66000   ii 66000   iii 65960
  l i 75000   ii 75000   iii 75000

2 a i 4   ii 3.7   iii 3.74
  b i 4   ii 4.2   iii 4.22
  c i 7   ii 7.3   iii 7.29
  d i 9   ii 9.3   iii 9.35
  e i 14   ii 13.9   iii 13.86
  f i 13   ii 13.0   iii 13.04
  g i 4   ii 4.3   iii 4.31
  h i 8   ii 7.9   iii 7.94
  i i 2   ii 2.0   iii 2.04
  j i 3   ii 2.6   iii 2.64
  k i 1   ii 1.3   iii 1.32
  l i 4   ii 3.6   iii 3.58

3 Yes, 17.999 is nearer to 18 than to 17. Also 17.999 is nearer to 18 than to 17.9. A better way of showing the answer to the nearest tenth would have been to write $17.999 \approx 18.0$.

4 a £3.25   **b** £2.29 (nearest penny)
  c £10.93   **d** 16p (nearest penny)

Challenge

Check students' newspaper articles.

Appropriate approximations would be

6 hours 43 mins ≈ $6\frac{3}{4}$ hours

115.0779 mph ≈ 115 mph

86.3085 mph ≈ 86 mph

63 mins ≈ 1 hour

14 983 000 trees ≈ 15 million trees.

Exercise 8b   page 115

1 **a** 15.1   **b** 20.6   **c** 7.8   **d** 25.4

2 **a** 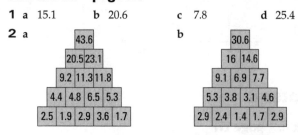   **b**

3 **a** 24.1   **b** 8.47   **c** 10.53   **d** 10.18

4 **a** 3.57   **b** 3.15   **c** 5.95   **d** 13.01

5 **a** 47 seconds

   b 0.75 MB

   c 12 mins 47 seconds

   d Total storage needed is 12.03 MB. No.

Problem

a Depends on year when question is attempted

b 2476, 2327, 2228, 1954, 1900, 1727

c 3905, 4054, 4153, 4427, 4481, 4654

Exercise 8c   page 117

1 **a** 96.1   **b** 44.3   **c** 83.7   **d** 123.3

   e 54.2   **f** 104.1   **g** 11.5   **h** 6.5

2 **a** 14.85   **b** 25.94   **c** 44.12   **d** 58.14

   e 43.66   **f** 73.29   **g** 4.71   **h** 28.02

3 **a** 483.7   **b** 706.66   **c** 897.58   **d** 1112.34

   e 578.31   **f** 226.83   **g** 682.58   **h** 1267.47

4 **a** 413.79   **b** 290.87   **c** 20.1   **d** 40.37

   e 497.8   **f** 504.65

5 **a** 52.24 km   **b** 100.233 kg   **c** 43.485 m

Problem solving

a 8.72 kg   **b** 4.69 km

c 60.78 litres   **d** 4.39 tonnes

Exercise 8d   page 119

1 **a** 70   **b** 4   **c** 490   **d** 0.78

   e 300   **f** 0.47   **g** 0.94   **h** 0.0593

2 **a** £30   **b** 4.5 kg   **c** $40   **d** 3.85 km

3 **a** 30   **b** 200

4 **a** 0.3   **b** 0.5   **c** 0.09   **d** 0.07

   e 60   **f** 90   **g** 500   **h** 300

5 **a** 2.5   **b** 0.29   **c** 360   **d** 4500

   e 29   **f** 3.7   **g** 4100   **h** 20 000

6 **a** 3.9   **b** 2470   **c** 0.29   **d** 41

   e 1740   **f** 93   **g** 3.45   **h** 0.027

   i 0.548   **j** 0.037   **k** 2700   **l** 0.8

7 **a** 10   **b** 100   **c** 10   **d** 10

   e 0.1   **f** 0.01   **g** 0.1   **h** 10

Exercise 8e   page 121

1 **a** 240   **b** 200   **c** 1800   **d** 3500

   e 500   **f** 510   **g** 600   **h** 6600

   i 4.8   **j** 0.15   **k** 9   **l** 0.6

   m 9.6   **n** 3.06   **o** 17.6   **p** 0.52

   q 26   **r** 35   **s** 22   **t** 24

   u 27   **v** 7   **w** 9   **x** 9

2 **a** 40.8   **b** 31.5   **c** 18.2   **d** 56

   e 30.8   **f** 64.5   **g** 46.8   **h** 86.8

   i 21 r3   **j** 23 r6   **k** 18 r4   **l** 21 r7

   m 22 r2   **n** 25 r10   **o** 25 r10   **p** 21 r18

3 **a** 60.9   **b** 83.6   **c** 170.1   **d** 111.6

   e 66.5   **f** 121.8   **g** 36.1   **h** 123.9

   i 72.2   **j** 52.8   **k** 28.8   **l** 122.5

4 **a** 94.5 g   **b** 71.3 litres

Problem solving

Each student will come up with different methods − check they are correct by checking answers obtained, i.e.

152, 147, 195, 39, 304, 408, 186, 12 r6, 270, 165

Exercise 8f   page 123

1 **a** 180   **b** 456   **c** 468   **d** 714

   e 1075   **f** 488   **g** 408   **h** 1986

2 **a** 32.2   **b** 17.5   **c** 63.2   **d** 54.4

   e 47.7   **f** 36.8   **g** 27.9   **h** 140.4

3 **a** 10.8   **b** 9.8   **c** 48.93   **d** 74.24

   e 48.87   **f** 30.72   **g** 25.02   **h** 45.45

4 **a** 70.2   **b** 38.4   **c** 39.9   **d** 281.2

   e 214.6   **f** 264   **g** 436.1   **h** 112

5 **a** 30.48   **b** 53.76   **c** 61.23   **d** 104.22

   e 262.86   **f** 304.5   **g** 373.32   **h** 759.05

6 **a** £106   **b** £33.17

   c Megan, as Megan has run 374 m whereas Hayden has run 331.1 m

7 **a** 57.72 m   **b** 166.17 cm

Investigation

a 705   **b** 7.05   **c** 7050

d Missing boxes are 15 × 4700, 15 × 47 000 and 15 × 4.7 (on top line), but many others are possible, e.g. 15 × 0.047

e Several, for example 70.5 ÷ 15, 70.5 ÷ 1.5, and 70.5 ÷ 4.7, 70.5 ÷ 47 etc.

Exercise 8g   page 125

1 **a** 23   **b** 26   **c** 31   **d** 34

   e 15   **f** 16 r1   **g** 24   **h** 21 r3

2 **a** 27   **b** 21   **c** 18

3 **a** 7.6   **b** 7.9   **c** 7.2   **d** 5.6

   e 5.4   **f** 6.3   **g** 7.1   **h** 6.6

4 **a** 6.9   **b** 2.9   **c** 5.8   **d** 5.1

   e 10.9   **f** 11.4   **g** 9.4   **h** 12.4

5 **a** 3.2   **b** 4.6   **c** 2.4   **d** 2.7

   e 2.5   **f** 2.5   **g** 2.6   **h** 2.6

6 **a** Devvon's, because Devvon's travels 14.8 km per litre and Alec's travels 14.6 km per litre

   b 1 month = 31 × 24 hours = 744 hours, so she has been alive 1 month and 21 hours (assuming 31 days in a month). So, she is more than a month old.

   c 3.4 seconds

Puzzle

a The blue packet.

b The blue packet costs the least per gram at 0.14 p per gram.

Exercise 8h   page 127

1 a 29   b 9   c 22   d 20

   e 9   f 46   g 5   h 20

2 a X   b Y   c Y   d X

   e Y   f X   g Y

3 a 4   b 1.8   c 25   d 3.3

4 a $(8 + 5) \times 4 - 3 = 49$ b $8 + 5 \times (4 - 3) = 13$

   c $8 + 5 \times 4 - 3 = 25$

5 a 1.11   b 22.44   c 21.66   d 45.5

   e 5.33   f 1.54   g 1   h 1

Investigation

$\frac{60}{3} \times 4 = \frac{60}{12} = 60 \div 12 = 5$

but $\frac{60}{3} \times 4 = (60 \div 3)/4 = (60 \div 3) \div 4$

Exercise 8i   page 129

1 a £22.22   b 8 cakes r2   c $2\frac{3}{4}$ hours   d 6.43 kg

2 a 3 hours 20 mins   b 33 days 8 hours

   c 107 mins 30 secs   d 103 hours 40 mins

3 a 27 boxes with 2 eggs left unpacked so 28 boxes

   b £76923.08

   c 1 and $\frac{9}{10}$ of a pizza

   d 3.57 m

Investigation

a 591780821 years, 328 days and 12 hours

b 17077 years, 301 days, 10 hours, 35 minutes and 18.5 seconds

c open ended

Consolidation   page 130

1 a i 3000   ii 3200   iii 3180

   b i 6000   ii 6300   iii 6270

   c i 5000   ii 4800   iii 4770

   d i 9000   ii 8600   iii 8630

   e i 7000   ii 6700   iii 6710

   f i 8000   ii 7700   iii 7680

   g i 5000   ii 5000   iii 5050

   h i 25000   ii 24500   iii 24510

   i i 39000   ii 38600   iii 38600

   j i 43000   ii 42800   iii 42780

   k i 76000   ii 76100   iii 76060

   l i 39000   ii 39500   iii 39490

2 a i 5   ii 4.8   iii 4.85

   b i 3   ii 3.1   iii 3.11

   c i 8   ii 8.4   iii 8.40

   d i 8   ii 8.2   iii 8.24

   e i 25   ii 25.0   iii 24.97

   f i 23   ii 22.6   iii 22.62

   g i 3   ii 3.4   iii 3.42

   h i 8   ii 8.0   iii 8.05

   i i 3   ii 3.1   iii 3.14

   j i 3   ii 3.0   iii 3.01

   k i 2   ii 2.4   iii 2.43

   l i 5   ii 4.5   iii 4.55

---

3 a 305   b 596   c 1999   d 2630

   e 418   f 1323   g 398   h 2724

4 a 19.2   b 11.58   c 10.72   d 6.05

   e 4.82   f 3.5   g 3.91   h 5.82

5 a 189.4   b 596.77   c 799.11   d 1213.81

   e 678.32   f 326.82   g 592.86   h 2397.53

   i 1395.84   j 720.4   k 1282.28   l 2229.89

6 a 30   b 600   c 9   d 4

   e 380   f 17   g 0.67   h 49.7

   i 0.75   j 0.061   k 0.0482   l 0.32

7 a 0.4   b 0.6   c 0.07   d 0.02

   e 3.4   f 6.5   g 0.3   h 0.58

   i 0.854   j 0.073   k 6800   l 0.3

8 a 80.3   b 57.6   c 72.8

   d 403   e 98.7   f 58

   g 44.2   h $33\frac{1}{2}$ or 33 r1   i 58.75

   j 30.4   k 72   l 55.5

9 a 97.5   b 94.5   c 76.8   d 365.5

   e 358.8   f 381.9   g 288.8   h 677.6

10 a 69.92   b 112.86   c 254.38   d 450.12

   e 278.81   f 634.18   g 661.23   h 989.01

11 a 9.5   b 4.4   c 14.2   d 11

   e 14   f 16.5   g 18.5   h 16.2

12 a 4.6   b 6.2   c 4.3   d 4.0

   e 2.1   f 1.4   g 3.8   h 2.5

13 a 4   b 10   c 44

   d 2   e 162   f -7.2

14 a 2.18   b 21.55   c 252.81   d 38.8

   e 2.36   f 80   g 8   h 1

15 a 45 packs (with 3 hot cross buns left unpacked) so 46 packs

   b £2666.67

Summary   page 132

2 a 4410   b 2.5

Chapter 9

Check in   page 133

1 a (3, 1)   b (-3, -1)   c (-1, -2)

2 a D   b C   c B   d A

Exercise 9a   page 135

1 a   b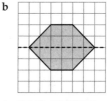

   c   d

2

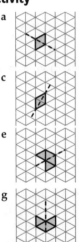 Square

3 a

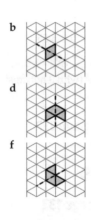

b A, B, E, F are rectangles

C is a scalene triangle

D is an isosceles trapezium

c A′ will have coordinates (9, 7), (10, 8), (8, 10), (7, 9) and (9, 7)

B′ will have coordinates (10, 8), (11, 7), (13, 9), (12, 10) and (10, 8)

C′ will have coordinates (11, 9), (10, 10), (9, 6) and (11, 9)

D′ will have coordinates (12, 3), (11, 5), (9, 5), (8, 3) and (12, 3)

E′ will have coordinates (8, 4), (10, 6), (9, 7), (7, 5) and (8, 4)

F′ will have coordinates (12, 4), (13, 5), (11, 7), (10,6) and (12,4)

Activity

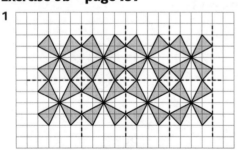

Exercise 9b   page 137

1

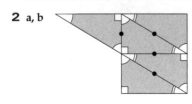

2 a, b

3 a, b

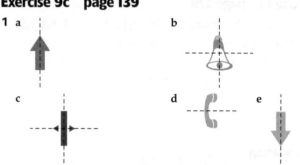

c translation 1 unit right and 4 units down

4 a, b

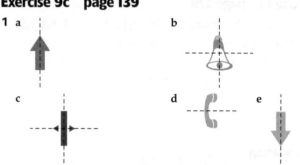

c rotation of 90° anticlockwise about the black dot

Activity

Check student's work

Exercise 9c   page 139

1 a   **b**

   c   **d**   **e**

The only figure with rotation symmetry is **c** which has order 2.

2 a 2 lines of symmetry

Rotation symmetry of order 2

b 4 lines of symmetry

Rotation symmetry of order 4

c 4 lines of symmetry

Rotation symmetry of order 2

d 1 line of symmetry

No rotation symmetry

3 100°, 80°, 100°, 80°

4 a   **b**

   c   **d**

   e   **f**   **g**

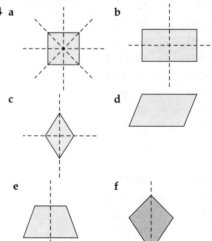

Order of rotation symmetry **a** 4 **b** 2 **c** 2 **d** 2 **e** none **f** none **g** none

Activity

b They all have at least one line of symmetry

Problem

Exercise 9d   page 141

1 a No   **b** Yes

    c Yes   Only **b** and **c** show enlargements

2 a 3   **b** 3   **c** 2

3 Check the enlargements are as follows:

    a An L-shape with short sides 2 and long sides 4

    b A 4×4 square

    c A trapezium with base 6 and vertical sides 3 and 6

    d A T-shape with top 6 and all other sides 2

    e A rectangle with base 3 and height 6

    f A downward arrow with top width 4, widest part 8 and depth 4

    g An isosceles triangle with vertical side 4 and width 4

    h A rhombus with diagonals 6

Activity

g The triangles are similar

    the scale factor is 2

Exercise 9e   page 143

1 a, b

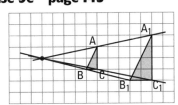

    c $BC = 1$, $B_1C_1 = 2$ $AC = 2$, $A_1C_1 = 4$ **d** 2

2 a   **b**

c   **d**

e   **f**

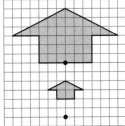

3 (6,0), (6,9)

Activity

Open ended

Exercise 9f   page 145

1

	Real life	Scale drawing
a	40 cm	2 cm
b	80 cm	4 cm
c	60 cm	3 cm
d	120 cm	6 cm
e	90 cm	4.5 cm

2 a 3.3 cm   **b** 3.3 m, 36°

3 a

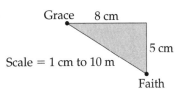

    b 94.3 m

Activity

Check heights are as follows:

HSBC Tower	4 cm
Chamberlain Clock	2 cm
Sutton Coldfield Moot	5 cm
Rotunda	1.6 cm

Consolidation   page 146

1

    a (1,-2), (2,-2), (2,-1), (3,-1), (3,-3), (2,-3)

    b (-2,-1), (-3,-1), (-3,-3), (-2,-3), (-1,-2), (-3,-2)

    c (-2,1), (-2,2), (-3,2), (-2,3), (-1,3), (-1,-1)

2

3 a, b

    c Rotation of 180° about the origin (i.e. the intersection of M_1 and M_2

4 a   **b**   **c**

d **e**

Orders of rotation symmetry are

a 2 **b** 4 **c** 1
d 2 **e** 2

5 An equilateral triangle

6 a 2

b

	Object	Image
forehead	1 cm	2 cm
nose	2.2 cm	4.4 cm
top of head	3.2 cm	6.4 cm
neck	3 cm	6 cm
mouth	1 cm	2 cm

c Check face has vertical depth 8 cm, i. e. diagram is extended 3 squares down from that given

7 a, b

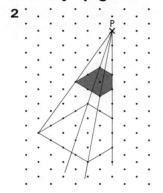

c (2, 1), (4, 1), (4, 2), (2, 2)   Object
(4, 2), (8, 2), (8, 4), (4, 4)   Image

d The Image coordinates are double the Object coordinates

8 a 2 m **b** 0.75 m **c** 1.5 m²

Case Study   page 148

1 All except **a**, the 'triangular' knot, which uses two strings.

2 No reflective symmetry
Order of rotational symmetry

a 3 **b** 4 **c** 4
d 6 **e** 4 **f** 4

3 Students' own sketches.
Reflective symmetry is not possible.

Summary   page 150

2

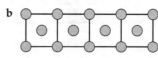

Chapter 10

Check in   page 151

1 $2^2 = 4, 3^2 = 9, 10^2 = 100$

2 a 7 11 15 19
b This is still the same sequence with 19 at the end
c 23

3 a 2 **b** 10

Exericse 10a   page 153

1 a 3, 6, 9, 12
The term-to-term rule is *Add* 3
15, 18, 21
b 4, 8, 12, 16
The term-to-term rule is *Add* 4
20, 24, 28
c 3, 5, 7
The term-to-term rule is *Add* 2
9, 11, 13
d 6, 10, 14
The term-to-term rule is *Add* 4
18, 22, 26
e 4, 7, 10
The term-to-term rule is *Add* 3
13, 16, 19
f 8, 14, 20
The term-to-term rule is *Add* 6
26, 32, 38

2 a add 3 14, 17, 20
b add 5 24, 29, 34
c add 4 24, 28, 32
d subtract 4 34, 30, 26
e subtract 3 18, 15, 12
f add 11 64, 75, 86
g × by 2 32, 64, 128
h double and add 1 47, 95, 191
i double and add 1 63, 127, 255
j × 3 and subtract 1 203, 608, 1823

3 a 6, 10, 14, 18, 22, 26 **b** 8, 10, 12, 14, 16, 18
c 60, 55, 50, 45, 40, 35 **d** 5, 11, 23, 47, 95, 191
e 2, 8, 20, 44, 92, 188 **f** 100, 89, 78, 67, 56, 45
g 1, 4, 13, 40, 121, 364 **h** 0, 2, 8, 26, 80, 242

4 a 40 **b** 21 **c** 15 **d** 12
e 175 **f** 81 **g** 100 **h** 38

Challenge

204

No. Add 49 lots of 4 to 8, or multiply the position by 4 and add 4.

Exercise 10b   page 155

1 a 3

b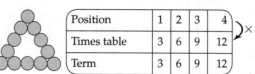

c

Position	1	2	3	4	
Times table	3	6	9	12	} ×3
Term	5	8	11	14	} add 2

d Multiply the position by 3 then add 2
e 14 **f** 62

2 a

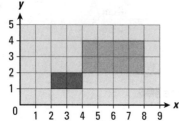

Position	1	2	3	4	
Times table	3	6	9	12	} ×3
Term	3	6	9	12	

Position to term rule is multiply the position by 3
50th term = 150

b

Position	1	2	3	4
Times table	2	4	6	8
Term	3	5	7	9

×2
add 1

Rule is multiply the position by 2 and add 1

50th term = 101

c

Position	1	2	3	4
Times table	4	8	12	16
Term	5	9	13	17

×4
add 1

Rule is multiply the position by 4 and add 1

50th term = 201

d

Position	1	2	3	4
Times table	2	4	6	8
Term	5	7	9	11

×2
add 3

Rule is multiply position by 2 then add 3

50th term = 103

3 a × by 3 and add 1, 19, 151

b × by 2 and subtract 1, 11, 99

c × by 4 and subtract 3, 21, 197

d × by 4 and add 2, 26, 202

e × by 3 and subtract 2, 16, 148

f × by 7 and add 1, 43, 351

g × by 8 and subtract 3, 45, 397

h add 3 to the position, 9, 53

Research

Open ended

Exercise 10c   page 157

1 a 9, 11

b

Position	1	2	3	4	5
Times table	2	4	6	8	10
Term	3	5	7	9	11

×2
add 1

Rule is × position by 2 and add 1

c 25

2 a Add 8   **b** 52, 60

c

Position	1	2	3	4
Times table	8	16	24	32
Term (£)	20	28	36	44

×8
add 12

Rule is × position by 8 and add 12

d £332

e £352 (as he already had £20 before he started work)

3 a 130, 150   **b** × by 20 and add 30

c 430   **d** 630

Investigation

a 3   **b** 6, 10, 15   **c** 1, 3, 6, 10, 15, 21, …

d Add 1 more than you added last time

---

e The differences between successive terms are not constant so the term-to-term rule does not work.

The position-to-term rule is not straightforward. It is $\frac{1}{2}n(n+1)$

Exercise 10d   page 159

1 a 4 cm   **b** 5 cm

2 a 4   **b** 7   **c** 9
  d 10   **e** 12

3 a 4 and 5   **b** 7 and 8   **c** 5 and 6
  d 9 and 10   **e** 10 and 11

4 a 8.9   **b** 7.4   **c** 6.2
  d 8.4   **e** 9.7   **f** 4.5
  g 3.2   **h** 10.2   **i** 11.8

5 122 cm

Challenge

When $h = 10$, $x = 11.07$ km (2 dp)

Exercise 10d²   page 159²

1 a 64, 4   **b** 8, 2   **c** 125, 5
  d 27, 3   **e** 216, 6   **f** 1000, 10

2 a i 64 cm³   **ii** 216 cm³   **b i** 2 cm   **ii** 3 cm

3 a 3.68   **b** 3.91   **c** 3.98   **d** 4.02
  e 4.99   **f** 5.01   **g** 2.96   **h** 4.64

4 a 3.36   **b** 4.31   **c** 4.06   **d** 4.12
  e 4.56   **f** 5.08   **g** 2.15   **h** 4.72
  i 5.19

5 3 cm

6 2.1 m

Challenge

a 0 and 1 (and -1)

b 3.16 (also -3.16 and 0 of course)

Consolidation   page 160

1 a

b 6, 11, 16, 21   **c** add 5

d 26, 31, 36

2 a add 4.   18, 22, 26   **b** add 5.   23, 28, 33

c double the previous term and add 1.
  47, 95, 191

d subtract 3.   18, 15, 12

3 a 5, 9, 13, 17, 21, 25   **b** 12, 18, 24, 30, 36, 42

c 40, 34, 28, 22, 16, 10   **d** 2, 5, 11, 23, 47, 95

e 3, 4, 6, 10, 18, 34   **f** 100, 79, 58, 37, 16, -5

4 a 3

b

c

Position	1	2	3	4
Times table	3	6	9	12
Term	1	4	7	10

×3
− 2

d × position by 3 and subtract 2

e 28

5 a × position by 3 and add 1
   19, 22
  b i × position by 2 and subtract 1, 11, 39
   ii × position by 3, 18, 60
   iii × position by 5 and add 2, 32, 102
   iv Take position from 20, 14, 0

6 a add 2, 11, 13
  b

Position	1	2	3	4	5	
Times table	2	4	6	8	10	×2
Term	3	5	7	9	11	+1

   × position by 2 and add 1
  c 41   **d** 201

7 a 6   **b** 8   **c** 12

8 a 5 and 6   **b** 4 and 5   **c** 8 and 9

9 a 8.4   **b** 7.4   **c** 6.2

10 a 1000, 10   **b** 8, 2   **c** 64, 4

11 a 3.4   **b** 4.6   **c** 3.7

12 10 cm

Summary   page 162

2 a $4n + 2$ **b** $3n + 3$ **c** $2(5n - 3)$ or $10n - 6$

Chapter 11

Check in   page 163

1 a Number of books

0

1

2

3 or more

 = 2 books

  b

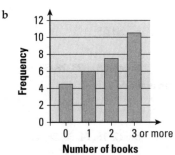

2

Number of dice	Frequency
1	10
2	8
3	12
4	14
5	9
6	7

3 a 5, 5   **b** 6, 7   **c** 5, 4

Exercise 11a   page 165

1 a secondary   **b** primary
  c primary   **d** secondary
  e primary

2 a secondary. Search the internet.
  b primary. Ask a sample of pupils in year 8
  c secondary. Search the internet.
  d secondary. Look at the scores across a sample of leagues on a random Saturday.
  e primary. Ask a sample of teenages how many hours TV they watch per evening

3 *e.g.* If you are looking for information about pupils in your school, you are unlikely to find it on the internet (e.g. are pupils in favour of school uniform in the sixth form)

4 Many organisations (e.g. the office of National Statistics) take great care in selecting samples that are representative of the whole population and have the resources available to them to ensure they take a cross-section of all areas of the country, social classes and ages.
  Kysia would not be able to collect her own data to get an accurate measure of infant mortality rates in the UK etc.

Discussion

She needs to define light and heavy.
She needs to weigh the bags.
She needs to take a random sample of pupils, not just her friends or her class.

Exercise 11b   page 167

1 a discrete   **b** continuous
  c continuous   **d** categorical

2

Occupants	Tally	Frequency
1	⊞⊞⊞⊞ I	21
2	⊞⊞ III	13
3	⊞ IIII	9
4	⊞ II	7

3

Height h metres	Tally	Frequency
$1.00 \leq h < 1.20$	I	1
$1.20 \leq h < 1.40$	⊞ I	6
$1.40 \leq h < 1.60$	⊞ ⊞ ⊞	15
$1.60 \leq h < 1.80$	⊞ II	7
$1.80 \leq h < 2.00$	I	1

4

		orange	green	blue
size	small	1	7	2
	large	4	1	3

colour →

Exercise 11c   page 169

1 a Sketch should have 3 vectors all roughly equal (120°)
  b Sketch should show approx. Japan 180°, Spain 90°, Cuba and USA 45°
  c Sketch should show approx. Fire, Police, Ambulance 110° and Coast Guard 30°

2 a Angles are Dogs 120°, Cats 110°, Birds 130°
   Check pie chart sectors are labelled, and check sector angles are correctly measured
  b Angles are Spain 84.4°, Cuba 44.4°, USA 48.9° and Japan 182.3°

Check pie chart sectors are labelled with countries and angles are correctly measured.

c Angles are Fire 117.6°, Police 102.9°, Ambulance 110.2° and Coast Guard 29.3°

Check pie chart sectors are labelled with Services and angles are correctly measured.

3 a Because the sectors are the wrong sizes. The sector for reference looks larger than that for History when it should be the other way around.

b Pie chart drawn should have these angles: Fiction 80°, History 120°, Reference 90°, Biography 70°

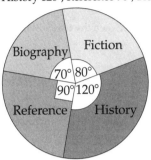

Discussion

Student's own answers.

Exercise 11d   page 171

1

Colour	Tally	Frequency
Green	IIII	4
Grey	IIII IIII I	11
Yellow	II	2
Red	IIII	5
Black	III	3
Blue	III	3
White	II	2
		30

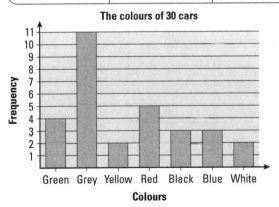

The colours of 30 cars

2

Size	Tally	Frequency
3	I	1
4	III	3
5	IIII	5
6	IIII III	8
7	IIII	5
8	IIII	5
9	II	2
10	I	1
		30

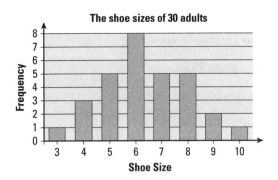

The shoe sizes of 30 adults

3

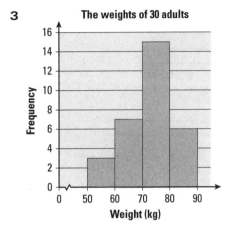

The weights of 30 adults

Discussion

The first diagram is far better for comparing the number of wildcats in each region as the bars are side by side. However, the 2nd diagram is better for seeing how the total number of wildcats has grown over the years.

Exercise 11e   page 173

1

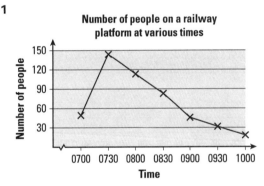

Number of people on a railway platform at various times

2

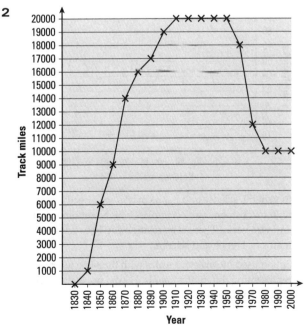

Discussion

Students discuss features of graph.

Exercise 11f   page 175

1

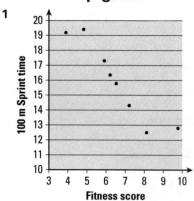

2 a

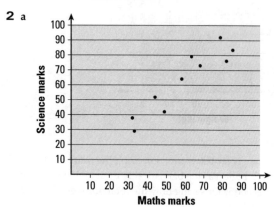

b

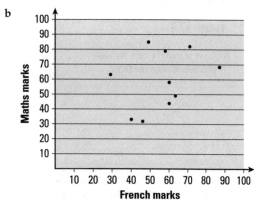

c

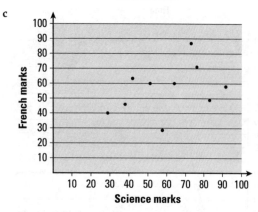

In general, science and maths marks are directly related; a high mark in one subject implies a high mark in the other and vice versa. There is no relationship between French and maths or French and science.

3 a Simon used inconsistent scales with the same internals representing different amounts. He did not even label his scales in numerical order.

b

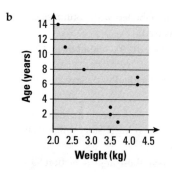

Discussion

No. There is a direct relationship between the two variables but it is not causal

Consolidation   page 176

1 Discuss students ideas

2

Number of books	Tally	Frequency
0	IIII	5
1	IIII I	6
2	IIII II	7
3	IIII III	8
4	IIII	4
	Total	30

3

		Colour		
		Red	Blue	Green
Size	Large	4	5	5
	Small	8	3	5

4 United: sector angles are Won 168° Drew 96° Lost 96°

Wanderers: sector angles are Won 120° Drew 72° Lost 168°

5

No. of calls	Tally	Frequency
0	II	2
1	II	2
2	II	2
3	II	2
4	IIII	5
5	IIII	5
6	I	1
7	I	1
		20

The number of calls received each day for 20 days

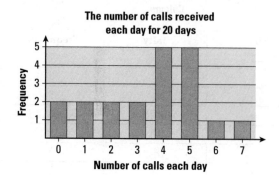

6 This answer will depend on the internals chosen by the students. A suitable answer is illustrated

Length of call (t) mins	Tally	Frequency
$0 \leq t < 1$	|||	3
$1 \leq t < 2$	||||	4
$2 \leq t < 3$	||||	5
$3 \leq t < 4$	|||| ||	7
$4 \leq t < 5$			1
		20

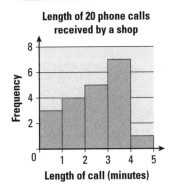

Length of 20 phone calls received by a shop

7

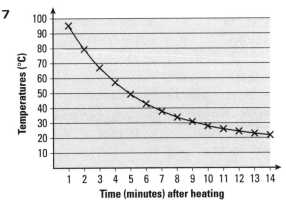

8

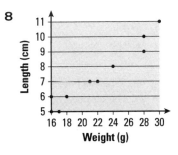

Summary   page 178

2 a $60 \div 5 = 12°$ per student, $96° \div 12° = 8$ students

 b $360° \div 24 = 15°$ per student, $15° \times 9 = 135°$

Chapter 12

Check in   page 179

1 a 12   **b** £12   **c** 72 g

2

Fraction	Decimal	Percentage
$\frac{17}{20}$	0.85	85%
$\frac{39}{50}$	0.78	78%
$\frac{24}{25}$	0.96	96%

Exercise 12a   page 181

1 a 2:7   **b** 2:3   **c** 3:5   **d** 2:3

 e 8:5   **f** 5:8   **g** 8:5   **h** 8:7

2 a 2:5   **b** 9:20   **c** 5:8   **d** 3:10

 e 8:5   **f** 3:4   **g** 9:2   **h** 4:9

3 a 3:2   **b** 5:22   **c** 1:25

4 a 39   **b** 480 g   **c i** 30 m   **ii** 20 cm

 d 1.5 m

Investigation

a 3:1   **b** 11:7   **c** The ratio gets closer to 1

Exercise 12b   page 183

1 a £20 : £30   **b** 25 cm : 35 cm   **c** 32 MB : 40 MB

 d 15p : 75p   **e** 45 : 75 seconds   **f** £150 : £90

2 a

 b 9

3 a

£5		£5	
£5	:	£5	£5
£5		£5	£5

 b £25

4 a 36   **b** £24

5 a Correct. If the ratio is 2 : 3 from 25 then each part of the ratio represents 5 children and so $2 \times 5 = 10 : 3 \times 5 = 15$ is the correct number of boys and girls.

 b Incorrect. $\frac{65}{4+9} = 5$ $5 \times 4 = 20$ $9 \times 5 = 45$
 Jack receives £20 and Oprah receives £45.

Investigation

a Something like this.

Blue	Green	Blue	Green	Blue

 b Similar to the rectangle above but with slightly different dimensions.

 c If their lengths are easily divisible by 8.

Exercise 12c   page 185

1 a £5.52   **b** £2.76   **c** 23p

 d £1.15   **e** £11.50   **f** £1.38

2 D-Mobile and codaphone

3 a £2.76   **b** 21 g   **c** 100 cals

 d i 875 g   **ii** 1625 g

4 a i £3.33   **ii** £2.96   **b** 24.5 litres   **c** 84p

 d 350 Chinese yuan   **e** £43.60   **f** 525 g

Challenge

kg	lb
1	2.2
1.8	4
5	11
10	22
23	50.6
50	110

2.2 lbs = 1 kg
0.45 kg = 1 lb

Exercise 12d page 187

1 a i $4:1$

ii $\frac{4}{5}$

iii red section = 4 × yellow section

yellow section = $\frac{1}{4}$ × red section

b i $3:1$

ii $\frac{3}{4}$

iii red section = 3 × yellow section

yellow section = $\frac{1}{3}$ × red section

c i $5:1$

ii $\frac{5}{6}$

iii red section = 5 × yellow section

yellow section = $\frac{1}{5}$ × red section

2 a $12:20$ **b** $\frac{20}{32}$

3 a $20:25$ **b** $\frac{20}{45}$

4 a 24 kg : 16 kg **b** £20 : £100

c 160° : 200° **d** 75 : 105 cats

e £1.50 : £2.50 **f** 3.5 m : 2.5 m

5 a 16 **b** i 42 **ii** 12

6 a $2:3$ **b** £48

Challenge

margarine 210 g

sugar 350 g

Exercise 12e page 189

1 a 8 **b** £12 **c** 9 g **d** £45

e 128 N **f** 40 mm **g** £1540 **h** 57.20%

2 a £6.50 **b** £28.14 **c** 43 kg **d** $3.84

e £16.65 **f** 14 km **g** £149.50 **h** 18.24 kg

3 a £57.50 **b** £42.50 **c** 200 m **d** 324°

e 63 kg **f** £1479 **g** £348 **h** 153.6 J

4 a 62.930 **b** 197.6 KB **c** 644 ml

Investigation

a TV £15.80

CD Player £67.15

Rug £3.95

Computer £71.10

Chair £2.37

b multiply by 0.79

c Discuss student's ideas as a class.

Exercise 12f page 191

1 a 70% **b** 46% **c** 56% **d** 125%

e 42.5% **f** 41.7% **g** 94.3% **h** 71.4%

i 80% **j** 65%

2 a i $\frac{2}{5}$ **ii** 40% **b** i $\frac{2}{3}$ **ii** 66.7%

c i $\frac{9}{16}$ **ii** 56.3%

3 a i $\frac{7}{10}$ **ii** 70% **b** i $\frac{4}{11}$ **ii** 36.4%

4 a Hilary scores 65.7% of the time, whereas Jodie scores 69.0% of the time, so Jodie is the better goal scorer

b Tina got 6.7% whereas Harriet got 7.8%. So Harriet got the better rate

Problem solving

Marks in percentage are:

Name	History	Geography	RS	Best subject
Zak	40%	42.9%	41.3%	Geography
Wilson	33.3%	20%	20%	History
Yvonne	75%	71.4%	65%	History
Ulf	91.7%	91.4%	91.3%	History
Veronica	50%	54.3%	41.3%	Geography

Consolidation page 192

1 a $1:3$ **b** $1:3$ **c** $2:3$ **d** $2:5$

e $5:4$ **f** $4:7$ **g** $7:10$ **h** $7:6$

i $6:25$ **j** $3:13$ **k** $1:2$ **l** $4:11$

2 a $1:15$ **b** $1:8$ **c** $13:16$ **d** $7:20$

e $10:1$ **f** $19:40$ **g** $1:1$ **h** $7:17$

3 a 24 **b** 702 **c** i 300 m **ii** 50 cm

4 a 30 km : 35 km **b** £120 : £105

c 96 MB : 160 MB **d** 2000 N : 2500 N

e 80 secs : 100 secs **f** 70p : £1.30

5 a 36 **b** £1400 **c** 120 g

6 a Yes, the cost increases by 15p with every extra 100 g.

b No, the cost changes at a different rate than the weight.

c Yes, the cost increases at the same rate as the weight.

7 a £2.46 **b** 80 cals **c** i 490 g **ii** 210 g

d i £2.03 **ii** £4.93

8 a $5:3$ **b** $8:1$ **c** £110

9 a £33 **b** €665 **c** 9.6 miles **d** 153°

e 350 kg **f** £97 000 **g** 325 rabbits

h 1950 kJ **i** 76.3 g **j** £1.17

10 a 82 156 **b** 455 g **c** £29.70

11 a Maths

b Megan won 71.4% of her matches, whereas Jane won 70% of hers. This suggests Megan is the better player (although her result are based on a small sample and it depends on the quality of the opposition in each case)

Case Study page 194

Book answers

1 white : wholemeal = 612 : 314 ≈ 2 : 1

2 Students' own description. For every 3 loaves that are sold, 2 are white (so $\frac{2}{3}$), and 1 is wholemeal (so $\frac{1}{3}$).

3 £599.76

4 £376.80

5 a Roughly 40% of the income comes from wholemeal loaves.

b Students' own description. Should say something about wholemeal loaves being more expensive than white loaves.

Ready reckoner answers

6

No. of loaves	flour (g)	yeast (tsp)	salt (tsp)	sugar (tsp)	butter (g)	water (ml)
1	750	1.5	2	1.5	30	480
2	1500	3	4	3	60	960
5	3750	7.5	10	7.5	150	2400
10	7500	15	20	15	300	4800
15	11 250	22.5	30	22.5	450	7200
20	15 000	30	40	30	600	9600
30	22 500	45	60	45	900	14 400

Notepaper answers

7 Students' answers, dependent on assumptions made.
Only sell fresh bread
186 loaves require 7 batches, plus 66 loaves require 3 batches.
(or 252, 9 batches).
White loaves same baking times as wholemeal loaves.
Mix + knead + prove + bake takes 2 hour 5 minutes.
Start at 07:25 am for first batch at 09:30
Repeat every 30 minutes (no time wastage).
If one oven, last (tenth) batch ready at 15:00

Clipboard answers

8 a white: 20 bags (28 kg spare)
b wholemeal: 10 bags (6 kg spare)
9 Students' answers.

Summary   page 196

2 Black area = $9 \times 9 + 3 \times 3 = 90\,\text{cm}^2$
Grey area = $9 \times 3 + 9 \times 3 = 54\,\text{cm}^2$
Black area : Grey area = $90 : 54 = 10 : 6 = 5 : 3$

Chapter 13

Check in   page 197

1 a 3   **b** 4
2 a €20   **b** €50
3 10

Exercise 13a   page 199

1 a $4x$ **b** $3y$ **c** $5z$
   d $10y$ **e** $6n$ **f** $12z^2$
2 a $8x + 5y$ **b** $2m + 6n$ **c** $8p + 3q$
   d $5y + 8z$ **e** $10s - 3t$ **f** $3x - 2y$
   g $9c + 5$ **h** $3s - 2$ **i** $6x - 2y - 2$
   j $3x - 5y - 1$
3 a $6x + 9$ **b** $12n - 8$ **c** $6a - 3$ **d** $7x + 14$
   e $6p - 12$ **f** $15 + 10q$
4 a $19x + 16$ **b** $8x + 2$ **c** $10n + 3$ **d** $9m + 1$
   e $9y - 2$ **f** $14z - 2$
5 a $2x + 3$ **b** $x + 4$
   c $2x + 3$ **d** $x - 2$
   e $x - 3$ **f** $2x - 1 + 2 - x = x + 1$

Challenge

	$7x + 8$	
$5x + 3$		$2x + 5$
$2x + 4$	$3x - 1$	$6 - x$

	$12x$	
$8x - 2$		$4x + 2$
$3x + 2$	$5x - 4$	$6 - x$

Exercise 13b   page 201

1 a 150   **b** 650   **c** 290
2 a 6 mins   **b** 13 mins   **c** 10 mins
3 $P = 3x + 10$ cm, 28 cm
4 a $P = 4n + 4$ cm   **b** $A = n(n + 2)$ cm^2
   c $P = 24$ cm, $A = 35$ cm^2 **d** $n = 4$ cm
5 a $A = 3h + 4t$ m^2
   b $A = 5(3h + 4t) = 15h + 20t$ m^2
   c $A = n(3h + 4t)$ m^2 **d** 40   m^2

Challenge

$N = 2x + 6$
If pond is 2 m wide, then $N = 2x + 8$

Exercise 13c   page 203

1 a 4   **b** 6   **c** 2   **d** 3
   e 2   **f** 3   **g** 7   **h** $3\frac{1}{2}$
   i 3   **j** 6   **k** $2\frac{1}{2}$ **l** 2
   m 10   **n** 12   **o** 12   **p** 15
   q $3\frac{1}{2}$ **r** 20   **s** $10\frac{1}{2}$ **t** 15
2 a 4   **b** 7   **c** 3   **d** 5
   e 5   **f** 6   **g** 3   **h** 0
   i $8\frac{1}{2}$ **j** $3\frac{1}{2}$ **k** $-2\frac{1}{2}$ **l** $-1\frac{1}{3}$
   m 3   **n** 2   **o** 4
3 a -1   **b** -3   **c** -2   **d** $2\frac{1}{2}$
   e $3\frac{1}{2}$ **f** $2\frac{1}{4}$ **g** -2   **h** -2
   i -4   **j** $2\frac{1}{2}$ **k** $3\frac{1}{4}$ **l** $-1\frac{1}{4}$
4 8
5 10

Challenge

a $18 - x$ and $16 - x$ **b** $18 - x + 16 - x = 26$
     or $34 - 2x = 26$
     so $x = 4$

Exercise 13d   page 205

1 a 8   **b** 2   **c** 2
   d 5   **e** 6   **f** 5
   g 6   **h** $\frac{1}{2}$ **i** $1\frac{1}{2}$
2 a 9   **b** 4   **c** 3   **d** 17
   e $2\frac{1}{2}$ **f** 0   **g** 11   **h** 22
   i $3\frac{1}{2}$
3 a 2   **b** 3   **c** 1   **d** 2
   e 5   **f** 0   **g** 3   **h** $\frac{1}{2}$
4 a 2   **b** 1   **c** 3   **d** 2
   e 0   **f** 1   **g** 2   **h** 0
5 a 3   **b** 2   **c** 4   **d** 3
   e 2   **f** 2   **g** 2   **h** 3
   i 5
6 9
7 11

Challenge

A $8 - x$ **B** $3 - x$
C $x - (3 - x) = 14 - (8 - x)$
         $x = 9$

Exercise 13e page 207

1 a

x	0	1	2	3	4
y	1	2	3	4	5

Graph is a straight line through $(0,1)$ and $(4,5)$

b

x	0	1	2	3	4
y	-1	0	1	2	3

Graph is a straight line through $(0,-1)$ and $(4,3)$

c

x	0	1	2	3	4
y	1	3	5	7	9

Graph is a straight line through $(0,1)$ and $(4,9)$

d

x	0	1	2	3	4
y	-2	0	2	4	6

Graph is a straight line through $(0,-2)$ and $(4,6)$

e

x	0	1	2	3	4
y	8	7	6	5	4

Graph is a straight line through $(0,8)$ and $(4,4)$

f

x	0	1	2	3	4
y	4	3	2	1	0

Graph is a straight line through $(0,4)$ and $(4,0)$

2 a yes **b** yes **c** yes
d no **e** no **f** yes

3

t mins	0	10	20	30	40	50
$y°C$	20	15	10	5	0	-5

Graph is a straight line through $(0,20)$ and $(40,0)$
a 30 mins **b** 40 mins **c** 20°C

Challenge

Graph is a straight line through $(0,9)$ and $(3,15)$

Exercise 13e² page 207²

1 a

x	0	1	2	3
y	3	5	7	9

b

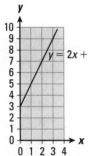

2 a

x	0	1	2	3
y	-2	1	4	7

b

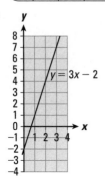

3

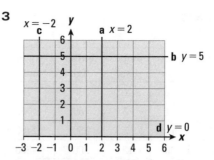

a $x = 2$ **b** $y = 5$ **c** $x = -2$ **d** $y = 0$

4 a vertical straight line
b horizontal straight line
c sloping straight line
d not a straight line
5 A $y = 2x - 3$
  B $y = x + 1$
  C $x = -1$
  D $y = 1$

Exercise 13f page 209

1 a £5 **b** £4 **c** £2.50 **d** $6
  e $7 **f** $5 **g** $\frac{1}{2}$

2

Kilometres	8	16	24	40	44	48
Miles	5	10	15	25	27.5	30

Graph will be a straight line through $(0,0)$
and $(40,25)$

a almost 19 miles $\left(18\frac{3}{4}\right)$ **b** 32 km

3 a 200 metres
  b 5 minutes
  c 10 minutes
  d 2 minutes
  e To eat when the food is as hot as possible

Research

Discuss students' results

Exercise 13g page 211

1 a The baby loses weight slightly over the first
    4 weeks and then starts to gain, reaching her original birth
    weight after 8 weeks.
  b Her weight increases gradually over the first 13 to 14 years,
    with a sudden increase through puberty until the late teens.

2 Jim started off quickly and then gradually slowed over the next
  2 hours. He then had an hour's rest, before riding for another
  2 hours, but this time he slowed down much more.

3 a P **b** Q
    a is P because it blinks on and off at regular time intervals.

4

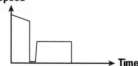

Challenge

Compare and discuss students' answers.

Exercise 13h page 213

1 AB 3, CD 1, EF -2
  GH -1, IJ $\frac{1}{2}$
2 KL 2, MN 1
  PQ -3, RS $-\frac{1}{2}$
3 a $(3,4)$ **b** $(3,2)$ **c** $(5,\frac{1}{2})$ **d** $(6\frac{1}{2}, 3)$

4 AB $(4, 9)$ PQ $(4, 4)$

5 a $(4, 4)$ **b** $(5, 3)$ **c** $(4, 3)$ **d** $(4, 2\frac{1}{2})$

e $(15, 30)$ **f** $(5, 3)$ **g** $(2, 3)$ **h** $(4, 2\frac{1}{2})$

Challenge

$(6, -6)$

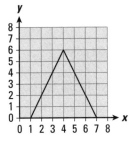

Isosceles triangle

Consolidation   page 214

1 a $4y$ **b** y^4 **c** $15y$ **d** $15y^2$

2 a $6x + 8y$ **b** $8p - q$ **c** $4z - 2$

3 a $10x + 20$ **b** $11y + 1$ **c** $5 + 5y$ **d** $-x + 18$

4 a $R = 20, S = 12, T = 14, U = 100$

5 a $A = x(x - 3)$ cm^2 $= x^2 - 3x$ cm^2

b $P = 2(2x - 3)$ cm $= 4x - 6$ cm

when $x = 5$, $A = 10$ cm^2 and $P = 14$ cm.

6 3

7 a 4 **b** 3 **c** 3 **d** 3

8 a $1\frac{1}{2}$ **b** $2\frac{1}{4}$ **c** 10

9 a 3 **b** 1 **c** 3 **d** 4

e $3\frac{1}{2}$ **f** 3

10 a 1 **b** 2 **c** 2 **d** 5

11 -4

12 a

x	0	1	2	3	4
y	1	2	3	4	5

Graph is a straight line through $(0, 1)$ and $(4, 5)$

b

x	0	1	2	3	4
y	-1	0	1	2	3

Graph is a straight line through $(1, 0)$ and $(4, 3)$

c

x	0	1	2	3	4
y	-1	1	3	5	7

Graph is a straight line through $(1, 1)$ and $(4, 7)$

d

x	0	1	2	3	4
y	8	7	6	5	4

Graph is a straight line through $(0, 8)$ and $(4, 4)$

13 a

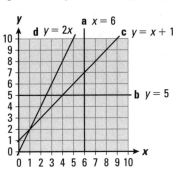

14 a

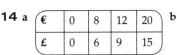

€	0	8	12	20
£	0	6	9	15

b

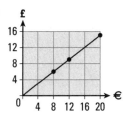

b i £12 **ii** €4

15 a

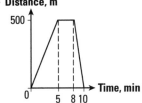

b 100 m per minute.

16

Temp

Time (minutes)

17 a 2 **b** 3 **c** $\frac{3}{4}$ **d** $-2\frac{1}{2}$

e -1 **f** $\frac{1}{2}$ **g** 3 **h** 0

Summary   page 216

2 I was walking at a steady speed.

Chapter 14

Check in   page 217

1 a $45°$ **b** $135°$ **c** $225°$ **d** $315°$

2 a 18 cm **b** $4\frac{1}{2}$ m **c** 15 cm

Exercise 14a   page 219

1 Check accuracy of constructions

a isosceles **b** right angled

c equilateral

2 Check constructions. Missing lengths are

a 8 cm **b** 9.1 cm **c** 4.2 cm

3 a Third angle = $45°$. Check accuracy of construction using ASA with $45°$, 6 cm, $40°$

b $110°$. Check construction uses ASA with $43°$, 5.5 cm, $110°$

c $57°$. Check construction uses ASA with $33°$, 6.5 cm, $57°$

4 a Check construction of triangle using ASA with $40°$, 5.5 cm, $105°$

b 209.2 m

Activity

b 9.4 cm and 3.9 cm

Exercise 14b   page 221

1 Check accuracy of constructions

Angles are **a** $60°, 60°, 60°$

b $34°, 34°, 112°$

c $44°, 57°, 79°$

2 Check constructions. Quadrilaterals are:

kite,   rhombus,   parallelogram

3 The triangles can all be different sizes as no side was given. However, they are all similar

4 a Check construction

   b It is right angled

Activity

a tetrahedron (triangular-based pyramid)

b tetrahedron

c square-based pyramid

Exercise 14c page 223

1 Check constructions

2 Check constructions

3 a, b Check constructions

   c The 2 lines of symmetry coincide with the perpendicular bisectors

4 Check constructions

5 a *OACB* is a rhombus (but not that it could be a kite—this would still give the angle bisector.)

     As constructed, *OAC* and *OBC* are isosceles triangles

   b Because the diagonals of a rhombus are lines of symmetry and therefore bisect the angles

Activity

Check constructions

Exercise 14c² page 223²

1 Check construction. Angle $P \times B = 90°$

2 Check construction

3 a, b Check construction

   c   2.7 units

4 Check construction

   a a kite

   b $A = 90°, B = 53°, C = D = 108.5°$

Activity

a, b Check constructions

c The lines of symmetry are the three perpendiculars constructed in **b**

Exercise 14d page 225

1 a a straight line

   b a circle

   c a straight line

   d a series of straight lines and curves

   e as shown

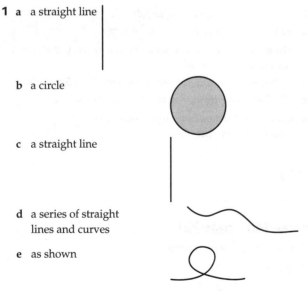

2 a, b Check locus is the angle bisector of *AOB* i.e.

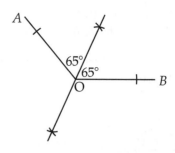

3 Check that the locus is a circle, centre the fixed point, radius 3 cm

4 Check that the locus constructed is the perpendicular bisector of *AB*

5 a Check scale drawing – moat will be 3 cm wide

   b

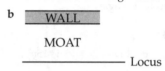

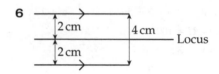

6

Activity

Discuss students' results. Should get a curve like this (a circle radius half the length of the ladder)

Exercise 14e page 227

1 a College   **b** Windmill   **c** Westwick farm

   d Moss Farm   **e** Well   **f** Church

   g Camping barn **h** Bus station   **i** Beech Hill

   j Youth Hostel

2

Direction	N	NE	E	SE	S	SW	W	NW
Bearing	0°	45°	90°	135°	180°	225°	270°	315°

3 Check students' constructions

Challenge

a Check scale drawing

b 225°

c The shaded angle shown is 45° (alternate angles). Measuring from North gives the required bearing as 180° + 45°

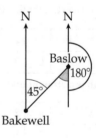

Exercise 14f page 229

1 a 16 cm²   **b** 22 cm²   **c** 32 cm²

2 a 40 cm²   **b** 48 cm²   **c** 30 cm²   **d** 236 cm²

3 a 72 cm²   **b** 250 mm²   **c** 58 m²

4 a 150 mm²   **b** 486 m²   **c** 1350 cm²

5 a 10 cm   **b** 8 cm   **c** 1.5 cm

Challenge

8 cm × 5 cm × 3 cm

Exercise 14g page 231

1 a cm³   **b** mm³   **c** cm³

   d m³   **e** cm³

2 a 36 m³　**b** 240 cm³　**c** 30 m³

3 15.625 cm³

4 a 2 cm　**b** 6 cm　**c** 0.5 m

5 a 6000 cm³　**b** 4500 cm³　**c** 12 cm

Challenge

Possible cuboids	Surface area
20 × 1 × 1	82
10 × 2 × 1	64
5 × 4 × 1	58
5 × 2 × 2	48←smallest

Consolidation　page 232

1 a 5.5 cm, 7.1 cm　20.1 cm
　b 12.3 cm　26.8 cm
　c 4.0 cm, 6.8 cm　16.3 cm

2 a Check construction　　Scalene
　b Check construction　　Isosceles
　c Check construction　　Right-angled

3 Check construction

4 Check that a perpendicular
to the base line has been
constructed from the dot i.e.

5 a 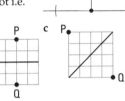　**b**　**c**

6 a 055°　**b** 260°　**c** 115°　**d** 315°
　e 045°　**f** 235°

7 a 126 cm²　**b** 162 cm²　**c** 304 cm²

8 a 7 cm　**b** 16 cm　**c** 6.26 cm

9

	Length	Width	Height	Volume
a	9 m	8 m	8 m	576 m³
b	15 mm	10 mm	20 mm	3000 mm³
c	2.5 cm	2 cm	8 cm	40 cm³
d	5 m	3.5 m	3.2 m	56 m³
e	6 cm	2 cm	3.5 cm	42 cm³
f	4 m	4 m	3 m	48 m³
g	7 cm	2 cm	20 cm	280 cm³
h	9 cm	1 cm	32 cm	288 cm³

Summary　page 234

2 a 60 cm³　**b** $x = 6$ cm

Chapter 15

Check in　page 235

1 a Pete –　The range is 9
　　　　　　There are two modes – 5 and 9
　　Maria –　The range is 3
　　　　　　The mode is 7

　b Maria has the better scores because her mode is high and
there is a small range in her data set

2 8b2 did best because they gained the highest score in three out
of four weeks. Although they did much worse in week 3, in
week 4 they compensated by doing well

Exercise 15a　page 237

1 a Amy – 7　　　　　　　Jen – 6
　　Selma – 12　　　　　　Jo – 6
　b Amy – 5　　　　　　　Jen – 9
　　Selma – 6　　　　　　Jo – no mode

2 a James　　**b** Kalid – 5　Dan – $6\frac{1}{2}$
　　James – 8　　Emily – 4　Steph – 6

3 a Sandra – 4,　Abbie – 5,　Lizzie – 4,
　　Jill – 6,　Maya – 4
　b Sandra – 5,　Abbie – 8,　Lizzie – 4.5
　　Jill – 9.5,　Maya – 5

4 a 8　　**b** 6 students　　**c** 9　　**d** 4.2

Discussion

Compare students' answers in group discussion (multiple answers
possible)

Exercise 15b　page 239

1 a 31　**b** 1　**c** 3
2 a 32　**b** 4　**c** 4
3 a 30　**b, c** Total 66　**d** 2.2

Discussion

a $150 \leq h < 160$

b You can't find mean, median or range easily because you don't
know exact values, only ranges of data

Exercise 15c　page 241

1 A 25%　　B 30°　　C 25%　　D 20%
2 a 1　**b** 50　**c** 70　**d** 1.4
3 a 81　**b** 2 h 15 min– 2 h 20 min
　c 69　**d** 21

Discussion

Check pie chart has segments:

0	86°
1	130°
2	72°
3	58°
4	14°

It would be harder to answer **b, c** and **d** as the pie chart only
illustrates proportions not actual numbers

Exercise 15d　page 243

1 a Low tide is 0.8 m at just after 14:30. The depth of water then
increases steadily to high tide of 7.8 m just before 20:30. The
depth of water then decreases steadily to 1 m (low tide) at
03:00

　b 16:15 ; 00:25

2 Prices rose fairly steadily are the first 70 years of the 20th Century.
Prices were doubling roughly over each 20 year period during
that time, apart from a period of stability between 1920 and
1940. Since 1970, inflation has been rampant with prices nearly
quadrupling between 1970 and 1980, almost doubling between
1980 and 1990, before inflation slowed down again between 1990
and 2000.

Discussion

Class discussion.

Exercise 15e   page 245

1 a

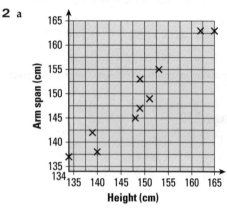

b Maths and science scores show positive correlation; a student who gets a good score in one tends to get a good score in the other

2 a

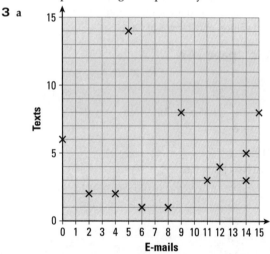

b Arm span and height are positively correlated

3 a

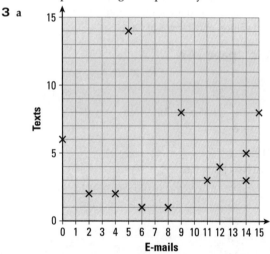

b There is no correlation between e-mails and texts

Discussion

She would have needed to measure the heights of her original ten people so that she could plot the same person's height against their weight

Exercise 15f   page 247

1 a 225   b 250 – 259   c 56

2 a

Test 1		Test 2
	50	000
88642	40	113489
9765542210	30	025688
7441	20	45569
8	10	

Key | 20 | 4 | = 24

b The scores in Test 2 tended to be higher with a median of 38 compared with a median of 34.5 for Test 1. The scores in Test 1 are more spread out then Test 2 with a range of 30 compared to 26

Discussion

With a stem and leaf diagram you can see the actual data (rather than a grouping)
However the bar chart or pie chart is clearer at showing the relative sizes of the various categories, because it is pictorial and drawn to scale

Consolidation   page 248

1

	Mean	Median	Range
a	14.4	16	21
b	75.25	79.5	74
c	6.48	7.3	5.7
d	83.44	88.9	47.6

2 a 4   b 4   c 4.2   d 6

3 Generally more houses are sold in Region 2 and the number is fairly consistent. Region 1 appears to be gaining popularity and by 2010, has overtaken Region 2

House sales in Region 1 are doubling every year, but in Region 2, despite a slight rise in 2009, they have remained at about 50

4

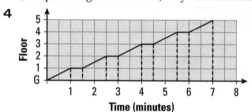

5 a   b

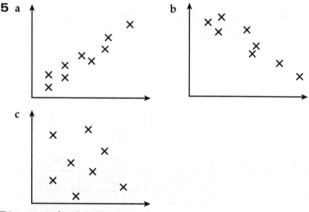

c

Discuss student's answers.

6

Ages of customers in years	
60	01
50	35
40	1234459
30	2389
20	269
10	57

Key | 40 | 3 | = 43

7

Office 1		Office 2
	30	24666666666666
555554321	20	479
9753310	10	16
9876	0	8

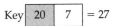

Key 20 7 = 27

Office 2 hires out far more cars per day with a modal group of 30 – 39 compared with 20 – 29 for Office 1. The range for Office 2 is higher at 28 compared with 19 for Office 1 indicating greater variation in the number of hires

Case Study   page 250

Electricity usage answers

1 a Lighting uses the most electricity

b Students' answers: many bulbs that are used a lot

c $\frac{1}{4}$

2 a Microwave uses the least electricity

b Students' answers: only used occasionally

3 Students' answers: fridge and freezer permanently on

4 Students' answers: turn off lights, don't use 'stand by' *etc.*

Switch off answers

5 and 7

Item	power used per day (Wh)	Power used no standby
Television	1422.5	1300
Satellite TV	397	150
DVD player	175.5	18
Main light	600	600
Microwave oven	259	140
Desktop computer	800	500
Laptop computer	156	116

6 a Television uses the most electricity

b Lap top uses the least electricity

8 a Television uses the most electricity

b DVD player uses the least electricity

Fridges answers

9 A £15.29

  B £18.04

  C £25.74

10 Students' answers

11 Indicates how cold the freezer compartment is (* -6 °C, ** -12 °C, *** -18 °C, **** -18 °C or less) and consequently how long food can be safely stored. Only **** freezers should be used to freeze fresh food.

12 Students' answers: capacity and freezer power are key factors

13 a Students' answers: efficiency compares power consumption/storage capacity

b See http://en.wikipedia.org/wiki/European_Union_energy_label

Summary   page 252

2 a 15th February

b It was colder for longer or it became warmer later in the year.

Chapter 16

Check in   page 253

1 a 21.1   **b** 6.47   **c** 4.81   **d** 3.36

2 a 561.39   **b** 233.07

3 a 6.3   **b** 21   **c** 16.8

  d 22 r10 or $22\frac{2}{3}$ **e** 149.1

4 a 33.57   **b** 20.04   **c** 39.1   **d** 43.2

5 a 7.9   **b** 4.4   **c** 28   **d** 45

Exercise 16a   page 255

1 a 36.16 km

b Further from Preston to Accrington (26.14 km as opposed to 22.71 km)

c 252.5 km (only travels Mon – Fri)

2 a Stephanie 1.7 m   **b** 0.4 m   **c** 0.05 m

3 a £7.91   **b** £6.45

c He doesn't have enough money – he is £0.83 short

d Valerie spends £13.84. Natasha spends £14.40 so Natasha spends most by £0.56

4 a 0.7 seconds   **b** Barry and Cris

Exercise 16b   page 257

1 Risotto   1621.936 g

  Spiced Rice 757.527 g

2 a 9.35 m   **b** 157 m

Puzzle

a Ayton → Isay → Hodear → Gewizzle → Eckerslike 47.453 km

b Not necessarily. Depends on traffic conditions, roadworks etc

c Multiple answers – check students' working

d e.g. Drat – Crikey → Biheck → Ayton → Isay → Hodear → Flipeneck → Gewizzle → Eckerslike → Drat

Exercise 16c   page 259

1 When she cycles (1764 beats as opposed to 1748 beats)

2 36 pages

3 a Yes, as cost will be less than £9 × 5 + £4 × 4 = £61

b No. To the nearest penny, cost will be £12 × 12 + £11 × 7 = £221. But this is so close, we need to do the actual calculation which gives £216.73

c Yes, as cost will be less than £2 × 9 + £3 × 4 + £5 × 12 = £90

4 a $6\frac{2}{3}$ **b** 4.2 litres, 5 cartons

Exercise 16d   page 261

1 a 14 p   **b** 54 p   **c** £28.08

2 a £7.20

b 14.8 miles (assume that Javed walks the same distance each day)

3 a £7.20 per hour

b £8.20 per hour

c £13.20 per hour

Exercise 16e   page 263

1 Packet A : each biscuit costs 5.5 p

      B : each biscuit costs 5 p

      C : each biscuit costs 5.2 p

  So packet B is the best value

2 i 8.4 miles   **ii** £4088.36

3 Jayne's driving distance is 110 km per day. So in 46 weeks, she drives 25 300 km (5 day week). This requires 2056.9 litres of petrol at a cost of £2342.82

Therefore, her cost of driving to work is

Petrol	2342.82
Insurance	280.45
Road tax	180.00
Servicing	390.00
MOT	50.63
Total	£3243.90

Her daily travelling time is *1 hour 50 mins*.

By train Jayne requires 11 monthly season tickets (excluding August) at a cost of £3138.19

Her daily travelling time is *2 hours 20 mins*.

Jayne could save money by selling her car and travelling by train. However, her journey times would be longer and she would not have the convenience of the car for her holidays, shopping etc.

Summary   page 264

2 $50 \times 31 = 1550$, $20 \times 22 = 440$, $10 \times 41 = 410$, $5 \times 59 = 295$, $155 + 440 + 410 + 295 = 2695$, $2695 \div 55 = 49$ cans of drink

Chapter 17

Exercise 17a   page 266

1 a Accommodation: £147·50   Coach: £519·00
Activities: £425·00   Ferry Births: £175   Insurance: £64·50

2 a £1331·00
b £27·00 Ms. Perry has rounded the money to the nearest £1·00 or, £30·00 rounded to the nearest £10·00

3 a £190 to the nearest £10
b £190·00 − answer to question 2b
c £16·30 per week or more sensibly, £16·00 per week

4 a Coach: 37%   Ferry Births: 9%   Accommodation: 16%
Food: 15%   Insurance: 5%   Activities 18%
b Should be 100%

5 a Those properly converted are: Coach
b The two confused companies are: The Active Company and the Camp Company

6 a €3·00 = £2·50   €2·40 = £2·00   €45·00 = £37·50
€119·99 = £99·99 (£100·00)

Exercise 17b   page 268

1 3·3 m²   5·25 m²   2 m

2 Tent A: John   Tent B: Carl   Tent C: Magnus
Tent D: Cherry   Tent E: Kadeja

3 a (0·5, 2·7): Pool   (2·2, 3·7): Sports Hall
(-2·5, 2·8): Pitch   (-0·4, 2·1): Shop
(-1·5, 2·7): Play Area
b (1·4, 0·6): Tent N   (3·0, 1·8): Tent H
(-2·3, 0·4): Tent A   (-1·6, 0·5): Tent C
c Office

4 (-1·8, 2·8)

5 a (-2·6, 2·0)   **b** (-1·8, 1·7)   **c** (0·4, 0·7)
d (1·1, 0·4)   **e** (2·7, 2·8)   **w** (2·1, 1·0)
x (0·4, 1·0)   **y** (1·0, 1·5)   **z** (3·1, 1·5)

6 a **a** (2,1)   **b** (1,2)   **c** (-3,2)
b **a** (2·6,1·7)   **b** (1·6,2·3)   **c** (-2.2,2·3)

Exercise 17c   page 270

2 a i Tennis
ii $\frac{1}{2}$ of the students played football
b i Archery
ii Approximately 12 people
c Football

2 a Total of goals in the competition = 37
b The modal score per game is 2 goals
c

		Games won	Games drawn	Games lost	Goals for	Goals against	Points
All Stars	4	0	2	2	7	10	2
Champions	4	2	2	0	7	4	8
High 5	4	1	2	1	7	8	5
Superstars	4	2	2	0	9	5	8
Cheetahs	4	1	0	3	8	10	6

d 3·7 goals
e The team who scored most goals wins

3 a i Red → 8   **ii** Blue → 2
b The mean is 20

4 a i 16·4s   **ii** 14·4s   **iii** 17·3s
iv 15·1s   **v** 14·6s
b 14·4   14·6   15·1   16·4   17·3

5 a Carl, Darren, Hussain, Reece, Hamed
b Carl, Darren, Hussain and (with the same score) Reece and Hamed

Exercise 17d   page 272

1 a Delicia: 9 kg   Lau It: 7 kg   Ahmed: 14 kg
Eddy: 8 kg   Dan: 12·2 kg   Maggie: 10 kg
b 21 kg
c Maggie

2 a i B to C, 500 m North East
ii C to D, 250 m South East
iii D to E, 500 m North
iv E to F, 250 m South West
v F to G, 100 m South
b Anticlockwise
c Clockwise 90°

3 a 100 years
b 2010 years plus current year (e.g. if 2009, then 2010 + 2009 = 4019 years)

4 a i 120°   **ii** 45°   **iii** 50°   **iv** 95°
b i 155°   **ii** 90°   **iii** 105°
c i 250°   **ii** 235°   **iii** 255°

5 a i 5 m   **ii** 3 m   **iii** 5·5 m   **iv** 4.5 m
b 12 m

Exercise 17e   page 274

1 a 26 m²   **b** 16 m²

2 a 5 litres
b i 3 litres   **ii** 4·5 litres

3 a Teacher to visually check
b Teacher to visually check

4 9 kg + 13 kg + 9 kg

5 5·5 hours

Index

Book 8B contains roughly 75% of the material needed for Year 2 of a condensed Key Stage 3, focusing mainly on level 6.

The remaining 25% can be found in book 9B, which consolidates level 6 and stretches to level 7.

If an exhaustive coverage of level 7 is required, this can all be found in 9B.

Table 1 shows how you can cover the condensed Key Stage 3 using Mathslinks in two different ways, depending on class ability.

	Y7	Y8	Y9
Fast-track route 1 **Book B**	Levels 4-5 Fast-track Year 1	Levels 5-6 Fast-track Year 1/2	Levels 5-7 Fast-track Year 2
Fast-track route 2 **Book C**	Levels 5-6 Fast-track Year 1	Levels 6-7 Fast-track Year 2	Levels 6-8 Fast-track Year 2

Table 2 shows how you can plan Year 2 for fast-track route 1, using books 8B and 9B.

Medium-term plan unit	Recommended no. of hours	8C lessons	Content from 9C
Autumn term			
A6	6	10a, 10b, 10c, 13e, 13f	distance-time graphs
N5	9	8h, 8i, 9d, 9e, 12a, 12b, 12c, 12d, 12e,	
S5	9	6a, 6b, 6c, 6d, $6d^2$, 14a+b, 14c, $14c^2$, 14d	
D4	6	11a, 11b, 11c, 11d, 11e, 11f	
A7	6	5b, 5c, 5f, 7a, 7b	trial-and-improvement to solve non-linear equations
Spring term			
S6	6	2a+b, 2e, 14f, 14g	circumference and area of a circle (2 lessons)
N6	8	4c, 4d, 8a, 8b+c, 8d, 8e, 8f, g	
A8	9	1d, 7e, 7f, 10d, $10d^2$, 13g, 13h	simple index laws; implicit functions
D5	4	3a, 3b, 3c, 3d	
S7	6	6e, 6f+g, 9b, 9c, 9f	maps
Summer term			
A9	7	5a, 5d, 5e, 13c, 13d, $13e^2$	further factorising
D6	6	15a, 15b, 15c, 15d, 15e, 15f	
SP2	11	17a, 17b, 17c, 17d, 17e	year 9 functional maths chapter (4 lessons); year 9 general problems (2 lessons)
S8	6		regular polygons; properties of a circle; enlargement (negative scale factor - could stretch to fractional); Pythagoras (2 lessons); could cover further loci here
D7	6		uncertainty and prediction; mutually exclusive outcomes; two events; tree diagrams; experimental probability; comparing experiment with theory
Total hours	**105 hours**	**79 hours**	**26 hours needed from 9B**

Further level 7 topics that you may wish to cover (mainly algebra):
compound measures (speed, density)
accuracy of measurement
estimation of statistics for grouped data
significant figures
quadratic expressions
quadratic sequences
simultaneous equations
linear inequalities
curved graphs